THE HANDBOOK OF CREATIVE DATA ANALYSIS

Edited by
Helen Kara, Dawn Mannay, and Alastair Roy

First published in Great Britain in 2024 by

Policy Press, an imprint of
Bristol University Press
University of Bristol
1–9 Old Park Hill
Bristol
BS2 8BB
UK
t: +44 (0)117 374 6645
e: bup-info@bristol.ac.uk

Details of international sales and distribution partners are available at
policy.bristoluniversitypress.co.uk

British Library Cataloguing in Publication Data
A catalogue record for this book is available from the British Library

ISBN 978-1-4473-6956-1 hardcover
ISBN 978-1-4473-6958-5 ePub
ISBN 978-1-4473-6959-2 ePdf

Cover design: Robin Hawes
Front cover image: iStock/xxmmxx
Bristol University Press and Policy Press use environmentally responsible print partners.
Printed and bound in Great Britain by CPI Group (UK) Ltd, Croydon, CR0 4YY

FSC
www.fsc.org
MIX
Paper | Supporting responsible forestry
FSC® C013604

We dedicate this book to Mary Roy 1938–2023.

Contents

List of figures and tables

Figures

Tables

Notes on contributors

Charlotte Barratt has been described as a collector of degrees. She holds a BA (Hons) in art history from University of Wales, Aberystwyth, an MA in museum studies, an MRes in social sciences research methods, and PhD exploring museum visitor experiences in super-diverse Leicester from the University of Leicester. She currently works in student and education equality, diversity, and inclusion and attempts to get colleagues to be more creative with their evaluation methods.

Katherine Broomfield (PhD, MSc, BMedSci, MRCSLT), University of Gloucestershire/Manchester Metropolitan University, Bristol Speech and Language Therapy Research Unit, North Bristol NHS Trust, is a UK-based clinical academic: a speech and language therapist and researcher working with health-care providers and universities. She approaches research with a clinical perspective and casts a critical scientific lens over clinical practice with people who have complex communication disability. Through applying a creative lens to her research practice, she is challenging traditional boundaries that exist between people who have communication disabilities and professionals, the paradigms of art and science, and the roles of clinician and academic.

Nicole Brown is Director of Social Research & Practice and Education Ltd and Associate Professor at University College London, working on the cusp of research/practice/teaching. Her creative and research work relate to representations of experiences, the generation of knowledge, and research methods and approaches to explore the same. Her methodological publications include her books: *Making the Most of Your Research Journal*, *Embodied Inquiry: Research Methods*, and *Photovoice Reimagined*. She shares her work at https://www.nicole-brown.co.uk and her social media handles are @ncjbrown and @AbleismAcademia.

Nathalie Busschaert, originally born in Antwerp, Belgium, completed a BSc in chemistry at KU Leuven (Katholieke Universiteit Leuven), Belgium, and continued studying for an MSc in chemistry at the same university. She then moved to the University to Southampton (UK) in 2010 to undertake a PhD under the supervision of Professor Philip A. Gale. In 2015 she joined the group of Andrew Hamilton at the University of Oxford to work as a post-doctoral researcher. In January 2016 she followed Prof Andrew Hamilton to New York University. Since 2017, she has been working as Assistant Professor at Tulane University (New Orleans). Her research interests are lipid binding and transmembrane ion transport and its applications.

Anuja Cabraal is an independent researcher and trainer in qualitative research, including software tools, methods, data analysis, reflexivity, and journalling. She provides consultation and mentoring services and works across a wide variety of disciplines. Organisations and bodies she has worked with include royal commissions, universities, hospitals, banks, government departments, and not-for-profits. She enjoys discovering innovative ways to enhance data analysis, including using emoji as part of the analytical toolkit.

Claudia Caltagirone is Associate Professor in Inorganic Chemistry at the University of Cagliari (Italy). Her research interests are in the field of supramolecular chemistry. She is co-author of more than 120 papers, and has been Co-Chair of the Women in Supramolecular Chemistry (WISC) network since its inception in 2019.

Kate Carruthers Thomas is Senior Research Fellow at Birmingham City University; an academic, artist, and author. She researches gender, contemporary higher education (HE), and equalities, as well as creative research methodologies. Her own creative academic practice includes graphic and visual essays, research poetry, performance, and the graphic novella *Five 'Survive' Lockdown* (2021) and the podcast 'A Feminist Family History in Eight Lives' (2024).

Adam Carter is Leverhulme Early Career Fellow in the Department of Sociological Studies at the University of Sheffield. He is interested in the connections between affect/emotion and power relations in society, particularly with regard to the social life of laughter, humour, and comedy. His current research utilises video methods to explore the ways families facing challenging times employ humour in everyday life.

Naomi Clarke is currently an Economic and Social Research Council (ESRC)-funded PhD candidate at the University of Bristol. Crafting and sewing are her main areas of focus personally, professionally, and within academia. As such, sewing is an integral part of everyday life for her, whether it be doing it, reading about it, or a combination of them both! Her work is often shared through her Instagram account @naomialicec.

Claire Coleman is currently the lead writer for the Arts Curriculum with the Ministry of Education, Aotearoa New Zealand. She is passionate about embodied and creative pedagogies for transformative education. She is a performer, creator, and educator dedicated to inclusive practices enhancing opportunities for the *tamariki* (children) of Aotearoa NZ. Her areas of research interest include process drama, social justice, quality learning environments, creative pedagogies, and research methods. She has co-edited a special issue of *Teachers and Curriculum: The Arts Will Find a Way: Breaking Through and Moving Forward*, dedicated to creating vital arts education experiences for all *ākonga*. She is happy to hear from anyone interested in critical pedagogy, the arts, and/or creative pedagogies: claire@playfulpedagogy.nz.

Kathryn Coleman is Associate Professor in Visual Arts and Design Education and co-lead of Speculative Wanderings in Space and Place (SWISP) Lab with Dr Sarah Healy. She is the lead investigator on 'The Learning with the Land' Social Sciences and Humanities Research Council (SSHRC) global a/r/tographic project project in the Faculty of Education. Her pluridisciplinary research and teaching are positioned in the intersection of art, design, digital, practice, culture, and data. Her research into practice includes digital practices and immersive data sites, creativities, speculative enquiry, and data creation with young people living in the midst of climate collapse.

Anne Collis is the neurodivergent founder of PinkGold Ltd. She is also Visiting Research Fellow at University of Southampton and Wrexham University and co-founder of NeuDICE Community Interest Company (CIC). In 2021, she received her PhD in organisational management from Bangor University. Her academic work is transdisciplinary and collaborative, bringing together people from the worlds of academia, lived experience, policy making, and civic society organisations. She has previously co-authored academic publications with colleagues from Barod CIC. She has also written books for The Fostering Network.

Peter J. Cook is Head of School and Dean, Education, and Associate Professor Arts Education at University of Southern Queensland, Australia, and substantively focuses on dance and creativity in teacher education. His interest and expertise in dance education developed through his 20-year dance teaching career in secondary and primary schools, including state-wide dance/arts positions. His research centres on understanding learning with and through dance, education, creativity, and their associated practices. His work extends into the interdisciplinarity of creativity in education and, more broadly, transitions, initial teacher education, and creative administration. His creative practice explores interactive digital choreographed works and their pedagogical potential.

Louise Couceiro is a postdoctoral researcher at the University of Oxford's Department of Education. She is currently undertaking ethnographic research as part of an ESRC-funded project investigating the relationships between equity, digital technologies, and teaching and learning. Prior to this, she completed her PhD at the University of Glasgow. Utilising online and creative methods, her research explored how a group of children responded to and engaged with contemporary biographies about women. She is an experienced educator, having taught in the UK, China, and Australia, and has worked on numerous research projects with children in Canada and the UK.

Andrew James Davies is Professor of Education at Swansea University, and teaches in the fields of educational leadership, research methods, and professional enquiry. Among his research interests are recruitment and retention in education, the professional roles of teachers, headteachers, and other educational leaders, and post-compulsory bilingual education.

Katherine Davies is Senior Lecturer in Sociology at the University of Sheffield. Her research focuses on relationships and personal life and she has a particular interest in creative qualitative methodological approaches that are able to capture everyday lives and connections. Her recent projects include a monograph – *Siblings and Sociology* – which explores the sociological significance of sibling relationships, published in 2023 by Manchester University Press, and two projects (funded by the British Academy and ESRC's Governance After Brexit scheme) that qualitatively investigate how Brexit has been experienced in everyday lives and relationships in the UK.

Jacqueline Dodding is Lecturer in Education at the University of Central Lancashire and holds a doctorate in education. She works with cross-disciplinary research teams. Her research interests are the climate emergency, barriers and enablers to education, and well-being and empowerment in young people. She encourages participants in her research to be co-researchers and uses creative data collection methods and analysis to empower her co-researchers to share their experiences.

Emily R. Draper is Senior Lecturer in Chemistry at the University of Glasgow. Her research interests include the self-assembly of organic materials. She works with small angle neutron scattering and rheology. She has two children and is passionate in making academia a more welcoming and accommodating place for parents.

David Duncan is an experienced primary teacher and has worked as a theatre practitioner for over ten years. He completed his doctorate of education (EdD) with the University of Strathclyde. His main area of interest is in children's experiences of death and dying, particularly in schools. He also has a keen interest in creative methods, with a specific focus on facilitating theatre in participatory research. His work aims to promote children's voice around sensitive issues in creative ways.

Chloe East taught drama in secondary schools for 20 years. In search of a new challenge, she completed an MSc in psychology at Kingston University before embarking on a PhD at the University of Southampton. This PhD was generously funded by the South Coast Doctoral Training Partnership (SCDTP). She is part of the Autism Community Research Network at Southampton (ACoRNS) and was expertly supervised in her PhD by Dr Hanna Kovshoff and Professor Sarah Parsons. She now works for a large local authority supporting schools to be inclusive of all pupils.

Mar Estupiñán Fdez. de Mesa is a public health specialist and pharmacist with over 20 years' experience working with local, national, and international organisations. She holds a PhD in health sciences from the University of Surrey, UK.

Mar's research focuses on using transformative mixed methods, intersectionality, and co-production to understand the mechanisms driving health inequities, and knowledge translation to reduce the research-policy gap. Committed to inclusion and diversity, Mar works with and for communities to find actionable solutions to improve health outcomes and transform services so these are sustainable and equitable for all.

Nick Fuller (RN, PhD) combines experience as a registered nurse, counsellor, qualitative researcher, and educator with his training as a shamanic practitioner and holistic healer. During his early career he gained a wide range of clinical nursing experience, before specialising in additions. This led to PhD research into the spiritual dimensions of addiction recovery, through which he developed approaches to applying shamanic journeying techniques within qualitative research processes. He is currently employed as a lecturer at the University of the West of Scotland (UWS), combining teaching with researching spirituality, addictions, labyrinths, and shamanism.

Louise Gascoine is Associate Professor in the School of Education at Durham University. As a former secondary school teacher who has psychotherapy training, she is particularly interested in student and teacher views and experiences, and her work has considered how these are captured in the process evaluations of large-scale educational interventions across a range of school settings. She has been engaged in a range of research focused on metacognition and is also interested in systematic reviewing.

Lauren Gawne is a linguistics researcher and Senior Lecturer at La Trobe University. She is interested in documenting and analysing how people speak and gesture, and her current research focus is the cross-cultural variation in gesture use. She co-hosts the podcast *Lingthusiasm* with Gretchen McCulloch and runs the generalist linguistics website *Superlinguo*.

Makeda Gerressu holds a PhD in public health and social studies from University College London. Makeda has extensive experience of mixed-methods research in the field of sexual health with a particular interest in young people, health service provision, and ethnic health inequalities.

Karen Gray is a senior research associate at the University of Bristol. Her research interests focus at the intersection between arts and culture and health and well-being. She has particular expertise and interest in methodologies used for evaluation and research in this area and involving older people and people living with dementia.

Karen Hammond is Lecturer in Criminology and Criminal Justice within the School of Education and Social Sciences at UWS, teaching critical criminology and victimology and research methods. Her doctoral research, completed in 2021, was a narrative ethnography of a hidden population of women who use drugs, exploring the creative ways that women resist and reinforce processes of stigmatization and marginalisation through storytelling. Growing up in a deprived area of Glasgow in the 1980s, she now lives with her older children and animals in Lanarkshire, where she spends most of her time quite peacefully and contently.

Cally J.E. Haynes is Lecturer in Organic Chemistry and Chemical Biology at University College London. Her research is in the field of supramolecular chemistry for biological and materials chemistry applications. She is a co-founder and Vice Chair (Organisation) of the WISC network.

Sarah Healy is co-lead of SWISP Lab with Associate Professor Kathryn Coleman. She is an inaugural Melbourne Postdoctoral Research Fellow in the Faculty of Education, the University of Melbourne. Best known for her contributions to the fields of critical affect studies, digital methods, and the post-humanities, her interdisciplinary programme of research involves research collaborations with academics, artists, practitioners, and educators from around the world.

India Henry is passionate about improving the health and well-being of individuals from marginalised communities. Driven by her lived experience as a Black Caribbean woman and strong passion for community engagement, she strives to make sexual health information available and attainable for all, improve community engagement, and create informative content.

Karen Henwood (School of Social Sciences, Cardiff University) has research interests spanning the social science of risk and identity studies. As a qualitative research specialist, her work draws on in-depth, in-situ, longitudinal methods, using a range of lenses to create more dynamic understandings of how people live with sociocultural and environmental change.

Steve Higgins is Professor of Education at Durham University. As a former primary school teacher he has a particular interest in how children's thinking and reasoning develops and how teachers can be supported to promote this. His research focuses on the use of evidence from research to improve learning and he is responsible for the development of the Education Endowment Foundation's Teaching and Learning Toolkit.

Jennifer R. Hiscock is Professor of Supramolecular Chemistry at the University of Kent, UK Research and Innovation (UKRI) Future Leaders Fellow, and Chair of the Royal Society of Chemistry (RSC) Macrocyclic and Supramolecular Chemistry Interest Group (MASC). Her research focuses on the translation of supramolecular innovation towards use in the real world.

Amie Hodges is Senior Lecturer in the School of Healthcare Sciences, Cardiff University. She is a sociologist with a clinical background in health care. She has a long-standing interest in improving health and well-being outcomes. Her research focuses on the lived experiences and social worlds of children, young people, and families in relation to chronic illness trajectories. She uses participatory, visual, and creative methods within her work. In her research she has used dramaturgy to explore the family-centred experiences of siblings living in the context of cystic fibrosis.

Gwenda Hughes is Professor in Epidemiology and Public Health at the London School of Hygiene and Tropical Medicine and Head of Research with the UK-Public Health Rapid Support Team. She leads a programme of multidisciplinary research in outbreak response in low-and middle-income countries.

Kristin M. Hutchins is Associate Professor in the Department of Chemistry at the University of Missouri and is affiliated with the MU Materials Science & Engineering Institute. Her research focuses on controlling the structure and properties of organic materials. She is a member of the WISC advisory board.

Helen Kara (FAcSS) has been an independent researcher since 1999 and an independent scholar since 2011. She writes about research methods and research ethics and teaches doctoral students and staff at higher education institutions worldwide. Her books include *Creative Research Methods: A Practical Guide* and *Research Ethics in the Real World: Euro-Western and Indigenous Perspectives*, and she edits the Creative Research Methods in Practice series, all for Policy Press. In 2021, at the age of 56, she was diagnosed autistic. Her neurodivergence explains her lifelong special interest in, and ability to focus on, words, language, and writing.

Neil Kenny is Assistant Professor at the School of Inclusive and Special Education at Dublin City University where he is the Programme Chair of the Master of Education in Autism. His research interests are in the areas of autism, participatory research, and inclusive education.

Marion Kieffer works as a scientist in the pharmaceutical industry, where she specialises in all aspects of chemistry, manufacturing, and controls for drug development. She holds a PhD in chemistry and cultivates ties with academia through her role as Vice Chair (First Generation cluster) of the WISC network.

Sarah Koops is a cultural scientist with a master's degree in migration and diversity. She worked as a research assistant at WISC until August 2022. She currently works as consultant for women's and social policy at the German Rural Women's Association in Berlin.

Larissa K.S. von Krbek is Emmy Noether Group Leader at the University of Bonn, Germany, and WISC Vice Chair (Communications). After obtaining a

PhD at Freie Universität Berlin in 2016 and a postdoctoral stay at the University of Cambridge, she started her independent career in 2020 working on stimuli-responsive supramolecular systems.

Rachel Kurtz entered academia following a career teaching and performing dance and aerial, including extensive work in schools. Drawing on phenomenological experience, her research combines multiple creative and therapeutic dance methodologies to bring attention to the body in academic contexts, where it is often unsupported and overlooked – nurturing the instinct to move!

Emma Lazenby is a social entrepreneur, artist, and BAFTA-winning filmmaker who has worked in the animation industry for over 20 years. She is the Director of ForMed Films CIC, which makes animated films and illustrations that help people better understand medical and health issues, make informed choices, and engage in treatment.

Jennifer S. Leigh is an interdisciplinary academic with a particular interest in using embodied, reflective, and creative practices for social justice. Her recent work includes addressing and highlighting experiences of marginalisation in academia, science, and society due to intersectional factors including disability, gender, race, and caring responsibilities.

Anna J. McConnell is Junior Professor of Organic Chemistry at the University of Siegen in Germany, where her research focuses on stimuli-responsive metal-organic cages and the dynamic covalent chemistry of amidoboronates. She is a co-founder and, since 2023, the Chair of WISC.

Abbey MacDonald is Senior Lecturer in Arts Education at the University of Tasmania. Her interdisciplinary arts-based research is used to inform arts, cultural, and education policy development, and museum and gallery education resources. She partners with environmental conservation and social change organisations to tackle some of the most pressing challenges educators face today.

Dawn Mannay is Professor of Creative Research Methodologies in the School of Social Sciences at Cardiff University, and her interests include education, inequalities, and children and young people. She authored the book *Visual, Narrative and Creative Research Methods* (2016), co-authored *Creative Research Methods in Education* (2021) with Helen Kara, Narelle Lemon, and Megan McPherson, and co-edited *The SAGE Handbook of Visual Research Methods* (2020) with Luc Pauwels. She is committed to working creatively with communities to produce data and disseminate the messages from research findings in innovative ways to increase the potential for social, educational, and policy change and support informed practice.

Jess Mannion (PhD) is a health and social care lecturer and a researcher in the Learning Disability and Autism Research Cluster, in the School of Social Care and Social Work at Manchester Metropolitan University. She specialises in collaborative research with people with intellectual disabilities on sexual citizenship. She works with a team of researchers with intellectual disabilities and neurodivergence in Ireland, called the Relationships and Sexuality Research Team. The research team use visual and creative research methods to collect, analyse, and disseminate data. These methods include art/drawing, participant poetry, rap, creative performance and practice, third objects, joint sandboxing, and LEGO® SERIOUS PLAY®.

Laura Mazzoli-Smith is Associate Professor in the School of Education at Durham University. Her work primarily focuses on educational identities and progression across the life course, with an anti-reductionist stance to educational questions that foregrounds inclusion, health, and well-being. She is co-lead of the Narrative Practices Lab, within the Discovery Research Platform for Medical Humanities and she contributes to Digital Storytelling Durham.

Catherine H. Mercer is Professor of Sexual Health Science at University College London, and co-leads Britain's National Surveys of Sexual Attitudes and Lifestyles (Natsal). A statistician and demographer by training, she develops and employs robust methods to advance the scientific study of one of the most socially sensitive disciplines: sexual health.

Emmajane Milton is Professor in Educational Practice at Cardiff University, and has held a wide range of senior leadership roles within academia, policy development, and the statutory school sector. An Advance-HE National Teaching Fellow, she is passionate about learner experience and how to best support and enable learning both in schools and in HE contexts. Her research interests are focused around professional learning, educational practice, teacher enquiry, adopting a learner-centred orientation, educative mentoring, and the recruitment and retention of education professionals.

Alexandra Morgan is Lecturer in Education at Cardiff University where she co-ordinates the Professional Doctorate Programme. She has over 20 years of experience of teaching in HE, particularly in the areas of research methods, professional learning, teacher enquiry, and educational practice. Her current research interests include educational leadership, teacher enquiry, professional learning, and developing innovative approaches to data analysis.

Stuart Neilson lectures and writes about the autism spectrum as a health statistician and from his personal perspective of an autism diagnosis in 2009, at the age of 45. He was a founder member of the team that developed the innovative Diploma in Autism Studies at University College Cork, Ireland. He has a degree in computer science and a doctorate in mathematical modelling of inherent susceptibility to fatal disease. His images explore autistic portrayals and reactions to sensory stimulus and motion, particularly in public spaces, and complement his written work.

Hazel Partington is Senior Research Fellow in Ethics and Health, working within the Centre for Professional Ethics at the University of Central Lancashire. A qualitative researcher with a doctorate in education, she is interested in breaking down power dynamics to further the democratisation of research and believes that creative research and data collection methods can contribute to this.

Nick Pidgeon (PhD, MBE, FBA), is Director of the Understanding Risk Research Group at Cardiff University and Professor of Environmental Risk. His research looks at public engagement, communication of, and decision making for environmental and technological risks.

Jacquie Ridge is Senior Lecturer within the Field of Advanced Nursing at the University of Northampton. She is also a founder member of the Non-Traditional Research Methods (NTRM) Network, which provides a meaning space for cross-disciplinary research debate, encouraging creativity in research design, novel approaches, and unorthodox ideas about research by those using non-traditional methods to elicit new ways of thinking and knowing. Motivating her work is the drive for authenticity and legitimacy of the voice represented within research and its (re)production using creative methodologies, such as in her 2020 doctoral research, 'Authentic identity: a visual grounded theory of constructing and sustaining of professional identity in adult nursing'.

Erin Roberts is a sociocultural geographer based in the School of Social Sciences at Cardiff University, whose research interests span the intersections and interactions between environmental/climate policies and wider concerns within contemporary society. Through the use of qualitative and creative methodologies, her work has predominantly focused on how these entanglements are understood and imagined by decision makers and experienced by local communities, with clear implications for a just and fair transition to a more sustainable society.

Mel Roberts is an experienced creative practitioner who runs Mazin Mosaics. She is also a parent carer for her learning-disabled autistic son. She has been making mosaics for ten years and enjoys using different materials and textures. Her passion lies in making art for health and well-being, and she runs regular workshops for this purpose with a variety of groups, including other parent carers, people with mental health issues, and community groups. As a facilitator, she enjoys enabling people to express their thoughts, feelings, and ideas through different art mediums.

Alastair Roy is Professor of Social Research and Co-Director of the Centre for Children and Young People's Participation at the University of Central Lancashire. Before becoming an academic he was a youth worker and residential social worker. He is an inter-disciplinary scholar with a particular interest in research methodology and has worked on innovations in mobile, visual, narrative, and participatory methods. He co-edited *Knowledge on the Move: Studies on Mobile Social Education* (2022) with Mariusz Kwiatkowski and Sinan Gulhan and *Social Work and the Visual Imagination: Seeing with the Mind's Eye* (2019) with Lynn Froggett, Julian Manley, and Martin Smith. He has undertaken applied research in the fields of youth homelessness, substance misuse, and mental health, as well as in participatory and collaborative arts.

Chantel Sealey is Health Promotion Specialist and Equality, Diversity and Inclusion Champion at Central and North West London NHS Foundation Trust, with over 15 years of experience working within sexual health, including with both young people and adults (some with multiple complex needs) from the Black Caribbean community.

Anna G. Slater is a Royal Society University Research Fellow and Professor of Chemistry at the University of Liverpool. She is a chemist with interests in the design, discovery, and applications of organic materials, sustainability in research, and research culture and ecosystems.

Merryn Thomas uses interdisciplinary, mixed-methods research to draw on geography, psychology, and social science, and she has a long-standing interest in using creative and innovative methods to engage diverse publics. She is particularly interested in intergenerational research and the potential of creative methods to help people connect with nature and articulate intangible relationships with their environments. She is currently based in the Centre for Ageing and Dementia Research at Swansea University.

Kyla Tully is currently in the last stages of her PhD, an Applied Research Collaborative Studentship with Queen Margaret University, Edinburgh, and the University of Glasgow. Her doctoral research utilised ethnography and creative methods to explore the relationship between care and professionalisation within rural arts management practice. Prior to this, she obtained an MA in arts, festival, and cultural management from Queen Margaret University and a BA in dance studies from Knox College in Galesburg, Illinois.

Kate Wall is Professor of Education at the University of Strathclyde. A primary teacher by trade, her work is characterised by enquiry-based partnership with children and practitioners of all ages and stages. She is interested in methodologies for gathering learner perspectives on experience, curriculum, and metacognition, and is recognised for her work on how visual and creative approaches with pedagogic orientations can facilitate this kind of dialogue.

Davita Watkins is Associate Professor at The Ohio State University. Her research interests focus on using non-covalent interactions to construct nanostructured polymeric materials. In 2014, she began her independent academic career at the University of Mississippi. She joined the faculty at The Ohio State University in 2022.

Lauren White is Lecturer in Social Research Methods based in the Sheffield Methods Institute at the University of Sheffield. She is particularly interested in participatory, creative methods and co-production. Her work focuses on experiences of health, illness, and disability. She cares about meaningful participation and engagement in both research and teaching contexts.

Dawn Wink explores the dynamics and stories at the heart of language, culture, and place. Wink's transdisciplinary work and creative research explores language through the lenses of wildness, beauty, and imagination. She is passionate about multilingualism, linguistic human rights, and ecolinguistics. Wink is Academic Director of the Department of Teacher Education at Santa Fe Community College, New Mexico.

Melvina Woode Owusu is Honorary Senior Research Fellow, Patient and Public Involvement and Engagement (PPIE) lead on National Institute for Health and Care Research (NIHR)-funded programmes, and co-founder of a cross-study PPIE initiative at University College London's Institute for Global Health.

Shehr Bano Zaidi completed her PhD in sociolinguistics in 2020, focusing on how English and Urdu are played out in Pakistani English and Urdu newspapers. All PhD journeys are hard but some become more arduous when a researcher decides to go against the flow. Taking a creative plunge, she adapted the grounded theory method to suit the demands of her study and generated a theory and a model to explain the complex relationship between the two major languages.

Fizzacc were a research group that consisted of six autistic young people: Feydakeen, Iris, Zaynab, Zoe, Annie, and Carys; and Chloe, a postgraduate researcher. When they first formed, the young people were aged between 12 and 17. Over the course of 11 months, they met weekly online to explore their experiences in schools. They created body maps, wrote stories, shared celebrations, and talked about mutual interests including concerts, animals, and robotic arms! They also produced resources to be used in schools to help other autistic young people feel better supported and understood. Most importantly they worked collectively: 'All of us together sort of having an input has been really nice.' For more information about Fizzacc please refer to ACoRNS Southampton (acorns-soton.org.uk).

Acknowledgements

It has been a pleasure and a privilege to bring together this edited collection, *The Handbook of Creative Data Analysis*, and there are many people who should be thanked and acknowledged for their role in this process.

It is important to thank all of the individual chapter authors for engaging with the project and offering a set of varied and thoughtful accounts and reflections, drawing on their extensive knowledge and expertise. The authors have detailed their own experiences of creative data analysis, generously sharing their approach to their craft and the uncertainties, concerns, enjoyments, and questions that it entails. Many of the contributors are friends and colleagues that we met at conferences, events, or through social media. However, we feel that we have come to know all of the authors more closely through the editing process. We have a deep respect for all the authors in this collection, and we were pleased that they were able to contribute to *The Handbook of Creative Data Analysis*. Overall, the authors' enthusiasm for the collection and their carefully crafted responses have engendered an inspiring set of chapters, and, in reading them, we have gained a wealth of knowledge and developed more nuanced understandings of the ways in which we can work creatively with the data that we generate.

It is also important to thank all of the researchers, participants, practitioners, and students that we have worked with, who have inspired us to edit this book, offered ideas, and engaged with us in informal conversations that have been central in bringing this project to fruition. However, we need to offer a special mention to the attendees at the 'Creativity in Research' webinar facilitated by Policy Press in February 2021, who asked so many questions about our strategies of analysis. Those questions inspired us to take on this project. We hope that this book will contribute to answering many of these questions and be useful in supporting an understanding of different ways to analyse research data.

We are also grateful to the commissioning, editorial, production, and marketing team at Policy Press for their guidance throughout the process of bringing the book to completion. We could not have produced this book without the support of our ever-helpful editors at Policy Press, first Philippa Grand and then Paul Stevens, as well as Senior Editorial Assistant Isobel Green, and their colleagues and Polly Chester, Desk Editor at Bourchier, whose input has been essential in preparing the manuscript for publication.

Lastly, we would like to thank all our great colleagues, friends, and family, and in particular those that are most involved in our everyday lives for their loving presence and enduring support. On this basis, Helen would like to thank Zoë Clarke, Su Connan, Leigh Forbes, Radhika Holmström, Sharon Inglis, Su-ming Khoo, Hazel Larkin, Lucy Pickering, Jon Rainford, Janet

Salmons, and Katy Vigurs; Mark Miller, Rosalind Round, and Jamie Round; and her beloved partner, Nik Holmes. Dawn would like to thank David, Jamie, Jordon, Sherelle, Tahlia, Taya, Tilleah, Tim, Toyah, Travis, and Travis Jay. Ali would like to thank Kate, Joe, Freddie, Bobby, Steve, Carolyn, Rob, Cath, and Dad.

1

Introduction

Dawn Mannay, Alastair Roy, and Helen Kara

Moving beyond the mystery and magic of data analysis

All research projects require a research design, phases of data collection or data production, and careful processes of analysis that lead to the dissemination of key findings. The activities of analysis have been defined as a process of attention, reflective systematic inquiry, and exploration.

> To analyse is to examine or study something closely to understand it better or discover something about it. It means to break it down into its components to study its structure and find out what it is made up of by identifying its constituent parts and how they are put together. Data analysis involves all these. It involves the examination of data in detail in order to first understand it better and be able to draw conclusions from it. (Akinyoade, 2013, p 4)

Given its central importance, there could be an assumption that the nuts and bolts of analytical procedures are clear to researchers, and that they are well aware of the steps required to move beyond the mass of data accumulated in a study to the distillation and refinement of key points that attend to research aims and central questions. However, data analysis has been positioned as 'the most complex and mysterious of all of the phases' of research and 'one that receives the least thoughtful discussion in the literature' (Thorne, 2000, p 69). This impression of mystery and mystical journeys is sustained in the reporting of research studies that only offer a glimpse of the concrete steps of the analytical process.

> Some authors use language that accentuates this sense of mystery and magic. For example, they may claim that their conceptual categories 'emerged' from the data –almost as if they left the raw data out overnight and awoke to find that the data analysis fairies had organised the data into a coherent new structure that explained everything. (Thorne, 2000, p 69)

This mysterious phenomenon, and the lack of insight into the processes by which conceptual leaps that generate abstract theoretical ideas from empirical data, have also been noted by other scholars who call for 'greater openness and legitimacy for reflexive accounts' of the 'how' of data analysis (Klag and Langley, 2013, p 149).

This call for detail, transparency, and openness was made apparent to us when we presented at the Creativity in Research webinar facilitated by Policy Press in February 2021. The webinar attracted 205 attendees and we were prepared to share our combined experiences of creative forms of data production, working with participants in the modes of collage, sandboxing, film, walking, drawing, and other modes of engagement. We were also well practised in communicating our approaches to dissemination and impact, and in the value of moving beyond the academic article and adopting creative techniques to communicate our research findings to wide and diverse audiences. However, what the attendees really wanted to know about were our strategies of analysis, asking us to provide insights into the dark art of making meaning from the vast and often confounding realms of data that we had generated in our studies. Given the title of the webinar, Creativity in Research, these questions evolved around how to analyse creative data and how to analyse data in creative ways.

These questions should have been expected as many publications only have a few perfunctory paragraphs on methodology with some fragmentary sentences on what may have constituted a broad strategy of data analysis. However, in the confines of a time-restricted question-and-answer segment, in the online setting of a webinar, it was difficult to adequately respond to the desire for clarity on how to approach analysis in a creative way. Once the webinar came to its conclusion, we realised that we had not responded well to the questions posed, and that these queries could not have been addressed by our immediate and constricted responses within the online webinar format. However, these questions deserved answers, both for us and for the webinar attendees, and this led us to develop this *Handbook of Creative Data Analysis*.

The initial call for abstracts generated an enthusiastic response from potential authors, making it difficult to select a restricted number of chapters that could be fitted into one book. The journey of the handbook was a privileged one, which introduced us to a vast range of innovations and creative approaches to working with data, which were sometimes entirely novel and in other cases inventive adaptations of existing practices. Creative data analysis refers to both 'the analysis of data gathered using creative methods' and the creative analysis of 'data gathered using creative or conventional methods' (Kara et al, 2021, p 83); accordingly, the chapters submitted emerged from a broad range of disciplines and contexts and engaged with diverse topics, philosophies, and methodologies.

Contributors to the book are based in Australia, England, Germany, Ireland, Italy, New Zealand, North America, Pakistan, Scotland, and Wales. The resultant international collection of 30 chapters reflecting on creative data analysis offers an assortment of possibilities, which we are eager to share with readers and consider for our own future research practices. In Chapter 24, Kyla Tully refers to her search for the 'secret research handbook', which would set out a clear set of guidelines about the correct way to conduct analysis. There is no such authoritative code that all should follow, but each chapter provides insights from the authors' practices, offering sufficient detail for the reader to explore how they could work with their data in different ways.

Analysing data is not easy. In attempting to unpick the complexities of data, many researchers will have experienced 'the frustration at data that will not yield easy answers' (Loughran and Mannay, 2018, p 2). In part this is because analysing data often demands a level of creativity. We are not simply following the dictates of a research handbook, secret or not, but finding a route to understand data in nuanced ways and making it communicable to others. One directive we gave to all the authors was to 'outline a case example of a method in enough detail for another researcher to be able to use the method'. Consequently, each chapter in this handbook offers a clear route map of the journey taken to understanding, and the side roads, blockades, and moments of serendipity that contribute to the process of creative data analysis.

The book is presented across nine themed sections.

1. Creative Analysis of Quantitative Data
2. Creative Embodied Analysis
3. Creative Performative Analysis
4. Creative Visual Analysis
5. Creative Written Analysis
6. Creative Arts-Based Analysis
7. Existing Methods Adapted In Creative Ways
8. Analysis with Participants
9. Pushing the Boundaries

The thematic categories are not mutually exclusive as there is considerable overlap between these section divisions. Analysis is complex, often engaging with multiple techniques, disciplines, and theoretical frames, and many of the chapters could have been comfortably positioned in two or more sections. For example, David Duncan's focus on theatre-based analysis in Chapter 11 aligned with the section 'Creative Performative Analysis', but it could easily have been placed in 'Analysis with Participants' to reflect the role of children in exploring and interpreting data. These themes then are inevitably interrelated and overlapping, reflecting the intricacies that researchers encounter in their fields, but they are designed to enable readers to find their way towards chapters that emphasise particular elements of creative methods of analysis.

Across the chapters, there are reflections on where analysis takes place in the research process. The neatness of the linear steps of data generation, analysis, and reporting findings is complicated by accounts where analysis is not a separate stage but something that underlies and informs various phases of the research journey. This is illustrated in Chapter 2, where Kate Carruthers Thomas returns to data after the completion of statistical analysis and the presentation of key findings in graphs and tables, developing a creative approach to understand the data differently. In Chapter 18, Naomi Clarke also challenges the idea of analysis as a discrete stage, explaining how, as a researcher and crafter, she embedded stitching throughout every stage of her doctoral work as a tool of reflexivity to reconsider her data.

Rather than a fixed or rigid undertaking, the chapters also demonstrate how analysis can be a playful process, of experimentation and learning. For example, in Chapter 19, Karen Gray and Emma Lazenby share how their analogue journey method, involving the storyboarding approach with an 'evidence wall', provided dynamic collaborative engagements with research data. Elements of co-production are also central to creative assemblages featuring a teacup and a toy unicorn in the work of Mel Roberts and Anne Collis, documented in Chapter 12. These approaches to analysis are creative and playful, but at the same time rigorous and carefully considered to support a more nuanced understanding of the data generated in research studies. Furthermore, creative approaches to analysis also help researchers to examine their positionality and subjectivity, and to be more reflexive in questioning initial interpretations and developing novel perspectives.

Accordingly, the book invites the reader to consider a range of creative approaches to analysis and offers insights that will be useful for an international readership of both new scholars and those with significant experience of working with data. As documented in the following sections of this chapter, each contribution is written by authors with a genuine commitment to furthering understandings of methods of creative data analysis, and developing an analytical toolkit from which researchers and practitioners can refine and develop their skills, knowledge, and craft.

Section One: Creative analysis of quantitative data

Our first section, 'Creative Analysis of Quantitative Data', is an important beginning to the edited collection as its positioning is deliberate in foregrounding the relationship between creativity and quantitative research. The popularisation of creative methodologies has been mainly focused on qualitative forms of research involving visual and art-based approaches (Mannay, 2015; Kara, 2020; Kara et al, 2021; von Benzon et al, 2021). This easy marriage between the qualitative and the creative can be divisive as it often overshadows the imagination necessary to work effectively with quantitative forms of research data. In response to this troublesome boundary work, this section showcases the creativity involved in making sense of data generated using large-scale surveys, examining frequency and word vector techniques in 'big data' and exploring the content of multiple public documents – approaches that would be categorised as fundamentally quantitative.

Chapter 2, the opening chapter of this section, is Kate Carruthers Thomas' '*Five "Survive" Lockdown*: revisualising survey data as a graphic novella', which describes the creative analytic process of transforming quantitative online survey data into a graphic novella. Kate generated responses from an online survey (n = 543) relating to the gendered experiences of academic and professional services staff during the COVID-19 pandemic. Expert statistical skills within the team established patterns of significance within the quantitative data and the findings were reported in a combination of textual explanations, pie charts, graphs, and tables. However, Kate wanted to extend their immersion in the minutiae of this quantitative data. The chapter demonstrates how creating storylines, characters, and illustrations for a graphic novella offered both new analytical insights and a vehicle to communicate the key findings to wider audiences.

Stuart Neilson's and Neil Kenny's work featured in Chapter 3, 'Visually mapping the interplay between pandemic interest groups and "the vulnerable" in newspaper accounts, 2018–2022', similarly focuses on the COVID-19 pandemic. The authors worked with a data set of 29 million words in 34,249 news items. Despite the challenge of this scale of information they took an approach that was exploratory and playful, with the intent of fostering imaginative readings of the data. Word frequencies were examined to investigate variability across times and location, and word embeddings highlighted dominant themes. However, as in Kate's work, Stuart and Neil also recognised the potential of moving beyond textual and numerical representations, drawing on video-photographic composite artwork to expose the inner structures of word models and consider which messages were made mainstream and who remained silenced.

In Chapter 4, Alexandra Morgan, Andrew James Davies, and Emmajane Milton offer reflections on 'Using discourse analysis to inform content analysis: a pragmatic, mixed methods approach exploring how the headteacher role is articulated in job descriptions'. The data set consisted of 67 headteacher job descriptions in Wales, which contained a total of 2,970 individual statements of essential and desirable competences. Adopting a creative approach that challenged strict epistemological boundaries and prescriptive ideas about how data should be analysed, the authors worked innovatively to evoke the affordances of lenses of both discourse analysis and content analysis. In combining the strengths of both these well-established analytical methods, the authors argue that they were able to both zoom out and zoom in throughout the process of analysis, to attend to previously unanswerable research questions.

Chapter 5, 'Word clouds as creative data analysis – what can they tell us about student views of learning something new?', considers the analysis of over 10,000 words generated in Pupil Views Templates (n = 374). Louise Gascoine, Kate Wall, and Steve Higgins discuss how this data was generated to explore developmental trends in metacognition. The research design included a statistical analysis of the codes across different age groups, exploring developmental trends using a series of one-way analyses of variance (ANOVA) with post hoc testing to explore relationships between age groups and the individual dependent variables from the deductive coding. However, word clouds were also included as a creative method of analysis, not simply to provide frequency statistics or to engage at surface level with the data visualisations that they generated, but to provide a structural and strategic means of eliciting and comparing the stories within the data. This offered an opportunity for the authors to think about the data in a different way.

Section Two: Creative embodied analysis

In embodied research, the 'primary objects of investigation are the possibilities and potentials of bodies, individually or together' (Spatz, 2017, p 5), and following the embodied turn there has been a growing interest in the body and its active role in knowledge creation (Thanem and Knights, 2019). 'Stories are told not just about bodies but also through bodies' (Chadwick, 2017, p 55), and researchers have been concerned that the process of transcription and interpretation can negate the centrality of the body. In response there has been a focus on developing strategies of embodied analysis, and the chapters in this section contribute to these developments.

Chapter 6, 'Analysing creative multi-modal data for a scientific audience', is authored by Jennifer Leigh, Jennifer Hiscock, Sarah Koops, Anna McConnell, Cally Haynes, Claudia Caltagirone, Marion Kieffer, Emily Draper, Anna Slater, Kristin Hutchins, Davita Watkins, Nathalie Busschaert, and Larissa von Krbek. The chapter discusses a qualitative project undertaken by the International Women in Supramolecular Chemistry Network. The authors drew on reflective methods as part of an Embodied Inquiry that sought to raise awareness and effect change in relation to the underrepresentation of women and other marginalised groups in science. Data was generated through collaborative autoethnography and reflective research group meetings, facilitated with art-based activities. Fictional vignettes were created to represent the lived and embodied realities of women working within a discipline in which they are marginalised, and the authors argue that their approach to analysis enabled an implementation of transdisciplinary perspectives, multilayered accounts, and a space to process lived experiences.

Chapter 7, titled 'Object-work as a creative approach to data analysis in Embodied Inquiry', is authored by Nicole Brown. As in the previous chapter, the methodological and theoretical framing for Nicole's work is Embodied Inquiry, which is concerned with lived experience, the researcher's body in the field, the body as communicator, and the body in interaction. Nicole discusses why Embodied Inquiry requires engagement with data that goes beyond the conventional coding and theme-finding, and offers a range of concrete examples to illustrate how using objects as a creative form of analysis constitutes a systematic analytical process within Embodied Inquiry.

The work of Rachel Kurtz and Laura Mazzoli-Smith is featured in Chapter 8, 'A composite approach to movement analysis for embodied methodology'. The chapter shares the composite method of analysis for embodied movement data, developed in Rachel's doctoral project, which was concerned with children's embodied experiences within schools. The project used elements of creative dance practice to develop a non-verbal means of interpreting the intangible experience of the learner, and the children were observed and filmed. The data was then analysed using a composite approach that drew on the analytical lenses of 5Rhythms®, Ways of Seeing, Open Floor, and Laban Movement analysis, which are outlined within this chapter.

Section Three: Creative performative analysis

This section has an emphasis on creative performative analysis. Theatre, as a genre of professional and community performance, is a communitive mode to share all aspects of social life including 'the description and analysis of social ills' (Becker, 2007, p 8). It is unsurprising then that theatre-based performance has been adopted by researchers who want their work to make a difference in the everyday world by connecting with policy makers, practitioners, and communities (Stuttaford et al, 2006; Byrne et al, 2016; O'Neill et al, 2018). For Richardson (2015, p 621), theatre is an important mechanism for shifting dissemination 'beyond the confines of the "Ivory Tower", outside the bounded walls of words in books and texts – and through spoken word and performance'. Theatre, then, can be particularly useful for engagement and social action, 'bringing research findings to life and connecting with audiences who would not necessarily read a related text-based academic output' (Mannay, 2020, p 662). However, as the chapters in this section demonstrate, elements of theatre such as the drama-based techniques of script writing and characterisation can provide creative ways to analyse data.

In Chapter 9, 'Rewriting in role: inviting readers through imagination', Claire Coleman discusses how she employed drama strategies to build meaningful fictional characters with specific viewpoints, knowledges, and questions. As noted earlier in Section One, Chapter 2 documents how Kate Carruthers Thomas created characters and storylines to explore and represent elements of large-scale survey data. Claire also created imaginary characters but rather than sitting within their own storyline they interacted with each other in a fictional dialogue scripted by Claire. This rewriting in role was a reconsideration of the initial data generated in her doctoral thesis, which explored critical pedagogy. Claire dedicated time to a familiarisation with and interrogation of the initial data, and then drew on rewriting in role to move beyond her habitual responses and engage with the data from different perspectives.

Chapter 10, 'A mosaic of siblings of cystic fibrosis: a creative dramaturgical analysis', draws on conceptual frames introduced in the work of Erving Goffman. Amie Hodges used dramaturgy as a creative analytical framework to make sense of her research with children that drew on interviews, observations, and a range of arts-based methods. Amie worked with her data from the perspective that life is a theatrical play in which actors take the stage to give a group or solo performance, examining their own positionality and interactions within the research sites as front stage, backstage, and role performances. In conducting her analysis Amie explored her data via theatre stage settings and this dramaturgical lens enabled a focus on presentation of character, performance within the family, and the risks of going out of character.

David Duncan's perspective on the analytical insights gained in an attention to performance feature in Chapter 11, 'Theatrical research-based performance: an analytic method using theatre in an educational context'. Reflecting on a participatory arts-based research project with bereaved children aged 10–11, David discusses how theatre can be used to analyse data with children. David initially drew on thematic analysis and, when themes were developed, worked with children to explore these themes through improvisation, voice and movement, mime, characterisation, physical theatre, and monologues. For David, drama-informed methods provided space and tools to delve deeply into such data and create rigorous findings, as well as involving children, who would usually be excluded from the process of analysis. The involvement of participants in creative data analysis is also a key feature of the chapters in Section Eight.

Section Four: Creative visual analysis

This section discusses a range of creative approaches to visual analysis. There are longstanding semiotic frameworks that support the analysis of the visual in relation to photographic images (Barthes, 1972; Floch, 2001) as well as more contemporary directions in visual semiotics (Aiello, 2020). The analysis of still images and film has also been supported with visual content analysis (Parry, 2020), visual rhetorical analysis (Gries, 2020), and iconology (Bohnsack, 2020). However, rather than an analysis of existing visual images, the chapters in this section discuss the ways in which the creation of visual and multisensory materials can provide an opportunity to better understand photographs and text-based data generated in research studies. The chapters explore the affordances and limitations of assemblage, drawing, and emoji to provide an alternative lens through which to analyse and understand data.

In Chapter 12, Mel Roberts and Anne Collis reflect on the 'Co-creation of a sensory assemblage as data analysis'. Drawing on data generated in a doctoral study that explored public involvement in policy making, in this collaborative process Anne contributed their lived experience of generating research data, while Melanie, as a creative practitioner, provided the expertise to make this knowledge and lived experience visible and accessible. Anne and Melanie co-created a sensory assemblage with materials including a teacup, CDs, cotton wool, wire mesh, and a toy unicorn. The developing relationship between Melanie and Anne enabled the production of a sensory assemblage, which was both a method of sense making with the data and an opportunity to communicate key aspects of the doctoral findings and journey to wider audiences.

Charlotte Barratt also considers her doctoral research in Chapter 13, 'Creating artworks from data'. Charlotte was interested in the experiences of visitors in museum spaces and their ethnographic research involved documenting interactions between visitors and artefacts through photographs, film, and field notes. In analysing these notes and images, Charlotte began to draw some of the photographs. The process of drawing enabled Charlotte to spend time concentrating on the captured moments represented in the photographs, opening up new ways of thinking about these images, deepening the process of interpretation of both the image and its context, and offering a more nuanced analysis of the data generated in her doctorate.

Anuja Cabraal and Lauren Gawne authored Chapter 14, 'Using emoji as a creative tool for data analysis'. The secondary data in the project consisted of interview data concerned with the relationship between linguistic education and subsequent employment. The visual emoji set available in Google Docs were assigned descriptors and drawn on in the coding process, with each emoji choice made with the focus and attention on how the researcher was understanding and analysing the data. The chapter suggests that emoji work provides distinct perspectives on data by requiring deeper thinking than conventional coding about the meanings in the data, their context, and the stories being told.

Section Five: Creative written analysis

While the previous section illustrated how data can be analysed with visual approaches, this section focuses on the ways in which text can be employed creatively in data analysis. Poetry and other forms of creative writing have been positioned as 'exclusive and technical' and not conducive to the rigour required in data analysis (Cahnmann-Taylor, 2009, p 14). However, there has been a growing recognition and appreciation of the utility of creative writing as an effective tool of data analysis (Kara et al, 2021), and scholars have drawn on techniques including poetic analysis (McDonough, 2018), script writing (Saldaña, 2011), and ethnodramatic writing (Cannon and Blair, 2014) to interrogate their data. The chapters in this section build on these approaches to analysis and the authors contemplate their experiences of working with I poems, character development, and Lilyology.

Chapter 15, 'I poems and polyvocality: experiences of using a combined qualitative creative analysis technique to strengthen the voices of research participants and aid reflexivity' opens the section on creative written analysis. Jacqueline Dodding and Hazel Partington utilised thematic analysis as well as drawing from Carol Gilligan's Listening Guide (1993) to construct I poems with their doctoral data. Introducing a project that they later collaborated on to explore the transition to working from home during the COVID-19 pandemic, and examples from their earlier doctoral studies, the authors offer an insight into the construction of I poems. They argue that constructing I poems engenders opportunities for researcher reflexivity, which is a central aspect of effective qualitative analysis.

In Chapter 16, 'Composite narratives, developing characters: a method of creative data analysis in developing public engagement artefacts', Lauren White, Adam Carter, and Katherine Davies discuss the process of creating composite narratives and fictional participants. The authors outline two projects where these analytical approaches were adopted: one researched how Brexit was experienced in the everyday lives of UK families, and the other worked with people living with the health condition irritable bowel syndrome. Composite narratives combined the data from multiple interviews and then presented it as a story from a single fictional participant. The authors demonstrate how this approach extended the analytical process through the assembly of pertinent themes, in considering difference and representation, and by their attempts to capture and distil the 'spirit' of their research projects and participants.

Chapter 17, authored by Dawn Wink, is 'Beyond the brick wall: transdisciplinary and creative research through scholarly personal narrative and Lilyology'. The chapter draws on her personal journals as a source of data. Embracing the theoretical framework of Lilyology (Blair, 2015), which reclaims and repositions Australian Indigenous Knowings, Dawn shares her research journey, communicating the reading and writing research dedicated to her inquiry, and examining the strengths and limitations of personal narrative scholarship.

Section Six: Creative arts-based analysis

This section considers creative ways to analyse data using arts-based techniques. While the previous sections focused on visual and written approaches to analysis, this section is concerned with the multimodal process of working with arts-based materials. This form of hands-on working to make sense of data is illustrated in the work of Marsh (2022), who created large-scale models reflecting interview accounts around experiences of rheumatoid arthritis, which expressed their material and form. Similarly, the chapters in this section reflect on art-based and crafting approaches of data analysis, including stitching, the analogue journey method, and mapping, which demonstrate how materials carry the potential to engage differently with data.

In Chapter 18, 'Slow-stitch: reflexive creative analysis and meaning making', Naomi Clarke focuses on analysing data creatively through the reflexive technique of stitching. As a crafter, as well as a researcher, Naomi embedded stitching throughout every stage of her doctoral work. The Crafting during Coronavirus research project generated data from 313 makers including textual diaries, images, videos of making, and tangible crafted items. The slow-stitch analysis approach was informed by reflexive thematic analysis and trajectory thematic analysis. The process featured the use of grouping and tables to develop an aerial view of the data, but, accompanying this material, hexagons were hand-made and stitched to represent the data generated by each participant. For Naomi, the time, thought, and energy in making each hexagon facilitated a slow process of thinking, reflexivity, and craft, which helped to discover meanings in the data.

Chapter 19, 'The analogue journey method' reflects on a collaboration between researcher Karen Gray and filmmaker Emma Lazenby. They worked together to create a graphic narrative to communicate Karen's research findings evaluating the impact of arts and creative engagement for people living with dementia. Influenced by film storyboarding and the 'evidence wall', the analogue journey method is a novel creative analysis tool that can be used with almost any kind of qualitative or mixed methods research data, making sense of the data by visualising them in the shape of a journey, and with the end goal of communicating this sense to others. Through creating the graphic narrative using the analogue journey method, Emma and Karen were able to engage and re-engage reflexively with the data together and ask some important ethical and theoretical questions of this data.

In Chapter 20, 'Using creative mapping methods to analyse multimodal data', Erin Roberts, Merryn Thomas, Karen Henwood, and Nick Pidgeon discuss a project that aimed to better understand the well-being benefits of saltmarsh habitats. Participants were invited to provide photographs representing their experiences of the coast, guide the research team in local walks, and take part in interviews. The multimodal data set became increasingly difficult to manage, and to develop a more nuanced understanding the authors drew on mapping techniques. The chapter illustrates how visually organising, or mapping out, the thematic data, each mode, and each kind of data (narratives, photographs, annotated maps, route plots), helped to build a picture of how a particular aspect of the marsh was important to the participants' well-being.

Section Seven: Existing methods adapted in creative ways

Researchers can be creative at all stages of the research process, and creative research can include engagement with the arts but is also a broader way of thinking about and conducting research 'where imagination, questioning and uncertainty are welcomed as part of the process' (Mannay and Turney, 2020, p 233). This creativity can be helpful in disrupting and challenging assumptions of conventional research methods, how research should be done, and what counts as knowledge (Kara, 2020). Creativity then can lead to variation, revision, and adjustment and Section Seven is concerned with how existing methods can be adapted in creative ways.

In Chapter 21, Louise Couceiro shares her journey of 'Embracing creativity in familiarisation'. Familiarisation is an established stage in analysis, with the standard steps of coming to know the data through review, reading, and listening (Lacey and Luff, 2001). However, Louise adopted a creative familiarisation approach and rather than relying on reading and rereading, she also experimented with writing analytic memos and poems. Louise's study explored how children responded to and engaged with collective biographies about women published between 2016 and 2020. Louise found that re-presenting participants' responses to the biographies, in the form of analytical memos and poems that rewrote these biographies, provided a vantage point from which to view the participants' most salient responses, and to begin familiarising herself with the contradictions and complexities within the data set.

Shehr Bano Zaidi contributes Chapter 22, 'How to be creative when creativity is policed'. The chapter draws from Zaidi's doctoral study that explored the relationship between Urdu and English in the context of Pakistan. It vividly demonstrates the difficulties some researchers have in using creative methods in the geo-political South, particularly younger female researchers. Against considerable opposition, Zaidi was able to creatively adapt the grounded theory method to study the relationship between Urdu and English in the Pakistani mainstream media.

Chapter 23 focuses on 'Visual grounded theory: a new way of seeing, knowing, and constructing theories grounded in data'. Jacquie Ridge investigated professional identity in adult nursing in a series of photo-elicitation interviews. In the chapter, Jacquie explains that employing visual imagery as the preferred tool for data production among grounded theory researchers has been slow to develop, and accordingly she worked with grounded theory in a novel way to ensure that both textual and photographic data were considered. In setting out her approach to analysis, Jacquie illustrates the ways in which data production and analysis of imagery can be used to substantiate a visual grounded theory, with participants as image makers creating a subjective knowing of self in a visual format.

Chapter 24, 'But is it professional? Pairing creative practice and thematic analysis to illustrate organisational culture', documents Kyla Tully's experiences of designing and developing an analytical framework during a doctoral study within a rural arts organisation. Data was generated through participant observation, interviews, and drawings as a substitute for documentational photography, and the field notes were in a mixture of conventional annotations and poetic forms. Kyla describes how processes of generating and analysing data are entangled together, and illustrates the benefits of taking a thematic approach while also including the creative practices of compiling and pairing select poems and drawings. This served to both draw out specificities and enable the identification of the broader patterns across the data set.

Section Eight: Analysis with participants

Participant-centred techniques are often only partially collaborative or partially participatory, as the perspectives of participants are often missing in the transfer from research activity to analysis, and then later in the dissemination of the findings to different publics and contributions to policy change (Mannay, 2015; Roy et al, 2021). However, this section is concerned with studies that actively involve participants in the analysis of data. The chapters in this section align with earlier studies that have positioned participants as experts by experience and involved them in multiple stages of the research process (Staples et al, 2019) or involved key stakeholders in validating and questioning data sets (Roberts et al, 2021). The chapters illustrate how analysis can be shifted from an activity that is restricted to researchers to one where wider groups and communities are central to making sense of data.

Opening this section, Chapter 25 is a contribution from Mar Estupiñán Fdez. de Mesa, Melvina Woode Owusu, Makeda Gerressu, India Henry, Chantel Sealey, Gwenda Hughes, and Catherine H. Mercer titled 'Using patient and public involvement and engagement and co-production to enrich the analysis, interpretation, and utilisation of sexual health research'. Reflecting on their research in the area of sexual health, the authors outline how they worked with participants in knowledge translation and co-production during the interpretation phase. The initial research analysis and key priority areas were presented to stakeholders, representatives from Black Caribbean communities and health professionals, who then critically reflected on the data produced, discussed alignments and omissions, and contributed to the development of creative film-based outputs to foster effective knowledge exchange.

Chapter 26, 'Creative collaborative data analysis: co-constructing and co-analysing the data together', documents how Jess Mannion devised a creative approach to analysis in collaboration with a group of co-inquirers with intellectual disabilities – the R&S (Relationships and Sexuality) Research Team. For Jess, it was important that the co-inquirers had the opportunity to be meaningfully and authentically engaged in all aspects of the study. Accordingly, the co-inquirers selected research topics to explore and chose a range of techniques of data production, including drawing, sandboxing, performance, poetry, and rap. In making sense of the data and developing the key findings, the co-analysis involved multiple stages that were designed to maximise accessibility and centred around visual interpretation, member checking, and content, art-based, and embodied forms of analysis. The chapter illustrates how the co-inquirers took ownership of the data, and Jess contends that without their inclusion, they would have omitted this important data in the analysis process.

The work of Chloe East, Feydakeen Smith, Iris, Zaynab Charafi, Zoe Fordham, Annie Champion, and Carys Jones is featured in Chapter 27, 'Perspective Narrative Analysis: analysing data creatively within a participatory research group'. The chapter discusses how Chloe, a postgraduate researcher and previous secondary school teacher, worked with a participatory research group named Fizzacc, comprised of six autistic girls between the ages of 13 and 18. The group was created in relation to a project that explored the transition from primary to secondary school for and with autistic young people. The data was generated using body mapping and the accompanying descriptions, and the analysis of this data incorporated Perspective Narrative Analysis. First-person narratives were developed from the perspective of an imagined autistic contributor. These drew on a composite of key themes generated in the study, alongside third-person narratives. This offered an imagined outside perspective that provided contrast in order to provoke reflection. The reading and rereading of these narratives led to a series of collaborative questions and talking points, which helped to further interrogate the data.

Section Nine: Pushing the boundaries

This final section focuses on creative methods of analysis that have elements aligned with pushing the boundaries. It is important that research continues to evolve, and this evolution very much depends on the 'continued effort to cross disciplinary boundaries and engage in a constructive dialogue with different schools of thought' (Margolis and Pauwels 2011, p xxi). This discourse has been evident across all of the sections, but here there is an emphasis on the adoption of techniques of analysis not widely discussed in the academic literature such as the creative use of dialogical analysis, currere, and shamanic journeying.

This section opens with Chapter 28, 'Analysing the unspoken: finding the richness created in dialogue with people who cannot speak'. Katherine Broomfield's work focuses on understanding the experiences of people who rely on augmentative and alternative communication, who are frequently excluded from or misrepresented in research. Incorporating her embodied experience as a speech and language therapist and researcher, Katherine worked with a dialogic conceptualisation of communication that encompassed more than speech and language. The analysis of the interview and observational data was informed by Interpretative Phenomenological Analysis. However, a dialogical analytic process helped to explore the accounts of the participants, the influence of those supporting the interviews, Katherine's embodied responses, and the reactions within and between the multimodal data.

Chapter 29, 'Speculative a/r/tography' provides an insight into the work of Kathryn Coleman, Sarah Healy, Abbey MacDonald, and Peter J. Cook. The authors are educational researchers working in Australian teacher education contexts, and position themselves as artist-researcher-teacher or a/r/tographic inquirers. The data generated is within collaborations and between art and theory, and includes sonic, moving, filmic, digital, tracing, and patterning, and prioritises currere. Currere is personal, responsive, active, and always contextual, and in both generating and analysing data, currere is central in this approach, as are other techniques, such as 'the hundreds' (Berlant and Stewart, 2019) and a range of sensitising concepts that are introduced with concrete examples within the chapter.

In Chapter 30, 'Between two worlds: a (shared) shamanic journey of creative qualitative data analysis and researcher identity', Karen Hammond and Nick Fuller offer a conversational autoethnography that reflects on their doctoral projects. They discuss how spiritual and alternative healing practices outside of the academy shaped their approaches to data analysis, with a focus on Nick's unique use of the shamanic journeying method. The chapter explores the challenges of developing creative approaches to qualitative data analysis in the context of mainstream research spaces. It also considers the ethical issues and implications arising from the shamanic approach, including concerns about cultural appropriation, researcher identity, and positionality in the analytical process.

Summary

This chapter has offered an introduction to the aims of this edited collection and provided an overview of the contributions to follow. Each chapter contains reflections on creative analysis. In editing the book, we have been inspired and learnt new things about how to manage and work with data creatively, and within this editorial process we have experienced moments of learning, understanding, and serendipity. We invite you now to read from start to finish, or to move across the entries following the threads that resonate with your interests and questions. At the end of the book the concluding chapter draws together key strands from across the collection and considers the potential for further developing *creative data analysis*.

References

Aiello, G. (2020) 'Visual semiotics: key concepts and new directions', in L. Pauwels and D. Mannay (eds), *The SAGE Handbook of Visual Research Methods* (2nd edn), London: SAGE Publications, pp 367–80.

Akinyoade, D. (2013) 'Approaches to data analysis in social research', Paper presented at the Research Methods Seminar of the College of Social and Management Sciences, Afe Babalola University, Ado-Ekiti, 19–20 March.

Barthes, R. (1972) *Mythologies*, New York: The Noonday Press.

Becker, H. (2007) *Telling About Society*, Chicago: University of Chicago Press.

Berlant, L. and Stewart, K. (2019) *The Hundreds*, Durham, NC: Duke University Press.

Blair, N. (2015) *Privileging Australian Indigenous Knowledge: Sweet Potatoes, Spiders, Waterlilys, and Brick Walls*, Champaign, IL: Common Ground.

Bohnsack, R. (2020) 'Iconology and documentary method in the interpretation of divergent types of visual materials', in L. Pauwels and D. Mannay (eds), *The SAGE Handbook of Visual Research Methods* (2nd edn), London: SAGE Publications, pp 397–412.

Byrne, E., Elliott, E., and Williams, G. (2016) 'Performing the micro-social: using theatre to debate research findings on everyday life, health and wellbeing', *The Sociological Review*, 64: 715–33.

Cahnmann-Taylor, M. (2009) 'The craft, practice, and possibility of poetry in educational research', in M. Prendergast, C. Leggo, and P. Sameshima (eds), *Poetic Inquiry: Vibrant Voices in the Social Sciences*, Rotterdam: Sense Publishing, pp 13–29.

Cannon, A. and Blair, A. (2014) 'Dramatic data: the possibilities of ethnodramatic writing for understanding the experiences of English learners in community college settings', *Critical Inquiry in Language Studies*, 11(4): 307–33.

Chadwick, R. (2017) 'Embodied methodologies: challenges, reflections and strategies', *Qualitative Research*, 17(1): 54–74.

Floch, J.-M. (2001 [1990]) *Semiotics, Marketing and Communication: Beneath the Signs, the Strategies*, New York: Palgrave Macmillan.

Gilligan, C. (1993) *In a Different Voice: Psychological Theory and Women's Development*, Cambridge, MA: Harvard University Press.

Gries, L. (2020) 'Advances in visual rhetorical analysis', in L. Pauwels and D. Mannay (eds), *The SAGE Handbook of Visual Research Methods* (2nd edn), London: SAGE Publications, pp 381–96.

Kara, H. (2020) *Creative Research Methods: A Practical Guide* (2nd edn), Bristol: Policy Press.

Kara, H., Lemon, N., Mannay, D., and McPherson, M. (2021) *Creative Research Methods in Education: Principles and Practices*, Bristol: Policy Press.

Klag, M. and Langley, A. (2013) 'Approaching the conceptual leap in qualitative research', *International Journal of Management Reviews*, 15(2): 149–66.

Lacey, A. and Luff, D. (2001) *Qualitative Research Analysis*, Trent: Trent Focus Group.

Loughran, T. and Mannay, D. (2018) 'Introduction: why emotion matters', in T. Loughran and D. Mannay (eds), *Emotion and the Researcher: Sites, Subjectivities, and Relationships*. Studies in Qualitative Methodology, Vol. 16, Bingley: Emerald, pp 1–18.

McDonough, S.L. (2018) 'Inside the mentors' experience: using poetic representation to examine the tensions of mentoring pre-service teachers', *Australian Journal of Teacher Education*, 43(10): 98–115.

Mannay, D. (2015) *Visual, Narrative and Creative Research Methods: Application, Reflection and Ethics*, Abingdon: Routledge.

Mannay, D. (2020) 'Revisualizing data: engagement, impact and multimodal dissemination', in L. Pauwels and D. Mannay (eds), *The SAGE Handbook of Visual Research Methods* (2nd edn), London: SAGE Publications, pp 659–69.

Mannay, D. and Turney, C. (2020) 'Sandboxing: a creative approach to qualitative research in education', in M.R.M. Ward and S. Delamont (eds), *Handbook of Qualitative Research in Education* (2nd edn), Cheltenham: Edward Elgar, pp 233–44.

Margolis, E. and Pauwels, L. (eds) (2011) *The SAGE Handbook of Visual Research Methods* (1st edn), London: SAGE.

Marsh, P. (2022) 'How models can translate the lived experience of rheumatoid arthritis into material and form', PhD thesis, University of Brighton.

O'Neill, M., Erel, U., Kaptani, E. and Reynolds, T. (2018) *Participatory Theatre and Walking as Social Research Methods: A Toolkit*. Available from: http://eprints.ncrm.ac.uk/4120/1/2.%20Toolkit%20PASAR%20Final%2030%20Jan%2018.pdf

Parry, K. (2020) 'Visual semiotics: key concepts and new directions', in L. Pauwels and D. Mannay (eds), *The SAGE Handbook of Visual Research Methods* (2nd edn), London: SAGE Publications, pp 353–66.

Richardson, M.J. (2015) 'Theatre as safe space? Performing intergenerational narratives with men of Irish descent', *Social & Cultural Geography*, 16(6): 615–33.

Roberts, L., Rees, A., Mannay, D., Bayfield, H., Corliss, C., Diaz, C., and Vaughan, R. (2021) 'Corporate parenting in a pandemic: considering the delivery and receipt of support to care leavers in Wales during COVID-19', *Children and Youth Services Review*, 128, article number: 106155.

Roy, A., Kennelly, J., Rowley, H., and Larkins, C. (2021) 'A critical discussion of the use of film in participatory research projects with homeless young people: an analysis based on case examples from England and Canada', *Qualitative Research* 21(6): 957–74.

Saldaña, J. (2016 [2011]) *Ethnotheatre: Research from Page to Stage*, Abingdon: Routledge.

Spatz, B. (2017) 'Embodied research: a methodology', *Liminalities: A Journal of Performance Studies*, 13(2). Available from: http://liminalities.net/13-2/embodied.pdf

Staples, E., Roberts, L., Lyttleton-Smith, J., and Hallett, S. (2019) 'Enabling care-experienced young people's participation in research: CASCADE Voices', in D. Mannay, A. Rees, and L. Roberts (eds), *Children and Young People 'Looked After'? Education, Intervention and the Everyday Culture of Care in Wales*, Cardiff: University of Wales Press, pp 196–209.

Stuttaford, M., Bryanston, C., Hundt, G.L., Connor, M., Thorogood, M., and Tollman, S. (2006) 'Use of applied theatre in health research dissemination and data validation: a pilot study from South Africa', *Health*, 10(1): 31–45.

Thanem, T. and Knights, D. (2019) *Embodied Research Methods*, London: SAGE.

Thorne, S. (2000) 'Data analysis in qualitative research', *Evidence Based Nursing*, 5: 68–70.

von Benzon, N., Holton, M., Wilkinson, C., and Wilkinson, S. (2021) *Creative Methods for Human Geographers*, London: SAGE.

Section I

Creative analysis of quantitative data

2

Five 'Survive' Lockdown: revisualising survey data as a graphic novella

Kate Carruthers Thomas

Introduction

This chapter describes the creative analytic process of transforming quantitative data gathered via an online survey into a graphic novella. The survey, Living and Working in Lockdown: What's Gender Got To Do With It?, focused on the experiences of UK university staff living and working through the first COVID-19 pandemic lockdown. The graphic novella: *Five 'Survive' Lockdown* is an intertwining of words and images in sequential form (Baetens and Surdiacourt, 2020) and narrates the stories of five fictional female academics during the period of March to September 2020. This revisualisation of data is located between two contrasting philosophical approaches to knowledge, meaning, and the social world: a positivist emphasis on statistical significance and proof derived from quantitative data and an interpretivist approach deploying subjective and situated knowledge to understand social phenomena.

A common refrain I hear when talking to other academics about using graphic or visual elements within research dissemination is 'But *I* can't draw so this isn't something I could consider.' It seems important to clarify that while I have developed a practice of graphic social science, that is, using graphics and/or visuals to communicate social research (Carrigan, 2017), I am not a trained – or even a particularly competent – graphic artist, nor am I an expert in visual theory or semiotics. I am an interdisciplinary academic, bringing sociological, geographical, and educational theories and concepts to bear on primarily qualitative research into gendered issues in contemporary higher education (HE). Ontologically I am an interpretivist, viewing social phenomena as subjective, plural, and socially constructed. Epistemologically I align with a feminist perspective, which rejects a positivist stance of objective and universal knowledge and argues that how and what we know is shaped by our social location and positionality.

This chapter aims to walk the reader through a specific example of the revisualisation of an online survey data set into a graphic novella. It first contextualises the original survey research (Carruthers Thomas, 2020) and the nature of my own graphic social science practice. It then outlines my motivations for revisualising the survey data in this way, referring to the work of Mannay and McLure, among others. I detail the methods and challenges of the project, from plotting narrative and characterisation, the laborious progression from pencil sketches, to pre-set panels, to inked A3 sheets, and thereafter to publication. Finally, I offer suggestions and resources for researchers at any level who are interested in applying and adapting similar methods in their own contexts.

Contexts

The survey: Living and Working in Lockdown: What's Gender Got to Do with It?

In June 2020, the UK was still in lockdown against the COVID-19 pandemic and large sections of the workforce were working from home. Multiple accounts and analyses of these experiences were flooding mainstream and social media channels as well as workplace and personal conversations. A rapidly emerging academic literature pointed to the exacerbation of existing social inequalities and the differential impact of the pandemic on men and women (as well as different ethnic groups) in all spheres of society, including the HE sector. To better understand the gendered nature of these experiences of academic and professional services staff in my own urban, modern university, I designed and ran an online survey open to all academic and professional services staff at the university. The online survey is a method I would normally use as a supplementary data collection tool in small-scale research, in combination with qualitative data collection methods such as individual interviews. In these circumstances, the choice of online survey method was a pragmatic one, offering a fast, accessible, and low-cost method to gather data from geographically distributed participants. Survey questions covered household and working environments; care and homeschooling responsibilities; voluntary and mutual aid activity; changes in workload and challenges, feasibility and impact on a range of working tasks; work/life balance; professional development and well-being; and experiences of work-related tasks such as student support, professional development, and research. Forty-five of the 52 questions were multiple choice, three were Likert scale, and four were free text.

The survey was live for one month and attracted 543 responses (20 per cent of the university workforce). Of these, 67 per cent of respondents identified as female, 31 per cent as male (compared to the university gender profile of 55 per cent female, 45 per cent male); 46 per cent of respondents were in academic roles, 41 per cent in professional services, 9 per cent in management, and 4 per cent 'other'; 67 per cent of academic respondents identified as female and female respondents also predominated in all age groups; and White staff were overrepresented in the respondent sample at 84 per cent (compared to the university workforce profile of 70 per cent). While pleased with the scale of the response, I was aware that the volume and complexity of largely quantitative data presented a challenge to my skillset as a qualitative researcher. I was capable of running standard data analysis reports through the survey software, but more sophisticated statistical analyses would be required to establish whether lockdown experiences varied with gender or other demographic characteristics. At this point, a generous colleague, an expert statistician, stepped in to establish the statistical significance of the findings.

Our analyses aligned with the wider literature in finding that the impacts of living and working in the COVID-19 lockdown were asymmetric and gendered. Our data showed that the university's female staff were disadvantaged in relation to access to dedicated working space at home; the burden of care including homeschooling and household management; and disruption of work–home boundaries. Moreover, female academic staff were more likely to report a negative impact on their capacity to conduct research and write for publication. This was particularly acute for those with school-age children and those with caring responsibilities for the over-70s during the pandemic. I packaged these findings neatly in a report, using text, pie charts, graphs, and tables. Excerpts of respondents' free text responses were included to illustrate the statistics. The report concluded with a set of recommendations for mitigation and action and was duly circulated among the university's senior management committees. This was 'a conventional approach to dissemination underpinned by technical-rational models that are based on positivist understandings of social reality … the end-point of a linear research design … a relatively complete, apolitical, formal act or process' (Barnes et al, 2003, pp 148 and 160).

Job done.

But the job did not feel 'done' to me.

On reflection, I can identify two reasons for this. Firstly, the use of automated reports to obtain preliminary analyses, then depending on my colleague's expert statistical skills to establish 'significance', distanced me from the findings and the emotional labour I had performed as a researcher in designing and conducting the research (Hochschild, 1983). Also sidelined was the emotionally sensed knowledge I habitually employed as a researcher interpreting qualitative research data (Hubbard et al, 2001). Secondly, the report highlighted gendered experiences and disadvantages, but the complex, affective stories behind the statistics played a distant second to 'statistical significance'. As an interpretivist, I considered these to be equally if differently 'significant' in understanding lived experiences of the pandemic. My graphic social science practice offered a way to resolve these disparities through creative data analysis.

A graphic social science practice

As an academic my research practice involves collecting, analysing, distilling, and presenting data. My research is a form of inquiry seeking enhanced intelligence and evidence to advocate organisational, structural, and cultural change. As a graphic social scientist, I follow a similar process to create a visual piece. More or less consciously, I collect data: ideas, questions, emotions, sense phenomena, then manipulate form and line to distil the data into visual form. I have been developing this practice since using visual and participatory mapping as a research tool in my doctoral study, when I had been struck by 'the capacity of participant-generated materials to help participants distance themselves from what they are usually immersed in and explore the taken-for-granted things in their lives' (Rose, 2014, p 27). As a postdoctoral researcher, I continued to combine visual mapping and interview methods in my research, further recognising the potential for visual, oral, and text-based data to interact in a mutually enriching way.

I had not picked up a pencil with intent since school art classes but began to experiment with simple graphic representations such as cartoons, caricatures, and visual metaphors to illustrate research findings and conference presentations. I found I enjoyed not only the act of drawing itself, but also the opportunity to approach data from different, creative angles: 'comics as a way of thinking' (Camden and Zullo, 2023). When deciding how to disseminate the findings of a research investigation into gender and career trajectory in HE, these experiments culminated in a 'graphic essay': a non-fictional, visual piece structured in the same way as an academic article (introduction, context, methods, etc). Entitled *My Brilliant Career? An Investigation*, this large-scale comic strip on four A3 panels was exhibited at conferences and seminars and laid the foundations for the graphic novella *Five 'Survive' Lockdown* (see Figure 2.1).

Why (not) revisualise data?

Academic convention dictates that text is the dominant medium of research dissemination. Academics work in a neoliberal, metricised milieu in which particular categories of publication (notably the peer-reviewed academic journal article) are essential career currency. Depending on the discipline, some techniques of visualising data – for example, charts, graphs, and infographics – commonly play an illustrative or supporting role in academic publications. However, the same metricised milieu also prioritises and ranks the importance of research impact beyond the academic sphere. Mannay (2019, p 659) argues that the impact agenda can be a driver for a range of creative outputs which 'promote engagement, wider impact and a potentiality for change beyond the academic article'. Revisualising data through multimodal formats including the graphic narrative has the potential to extend the reach of research to audiences who are less likely to access standard academic texts. A key driver behind Tyler and Bailey's graphic essay Beyond Stigma was 'a concern with making rich historical lines of Black critical thought about stigma visible … to reach new readers' (Tyler and Bailey, 2018, n.p.).

If extending the reach of the survey findings beyond the committee tables within my own institution was an important factor in my decision to transform the graphs, charts, and bullet points of the survey report into an extended graphic narrative, so was my desire as a researcher to reconnect with the emotionally sensed knowledge I brought to the data as a female academic living and working through lockdown. This was underpinned by my interest in practicing 'immersion in and entanglement with the minutiae of the data … an experimentation or

Figure 2.1: *My Brilliant Career? An Investigation* (2018, Panel 1)

crafting … a very different kind of engagement with data from the distanced contemplation of the table (McLure, 2013, pp 174–5). Engaging with the data through the creation of a hand-drawn visual narrative subsequently invites audiences to engage or entangle affectively with research and scholarship. In conversation with Ernesto Priego (2016) about the research-informed comic *Higher Fees, Higher Debts*, Katy Vigurs argues the format has the potential: 'to bring social research findings alive … help a wider range of readers to empathise with research participants' experiences … prompt emotions that might be less likely or less obvious when reading the same information in a written report' (Priego, 2016, n.p.).

Vigurs highlights the point that revisualisation in itself is not sufficient to bring research findings to a wider readership. Publication, in the widest sense of the word, is also required. I mulled over different publication possibilities for the graphic novella, including approaching an academic publisher with a proposal or self-publishing in hard copy. Despite generous discussions with a couple of friendly academic publishers, the non-standard format did not fit academic publishing models or readerships. I resolved somehow to self-publish, drawn to the sense of agency this gave me to create as a sole creator, while keenly conscious of my own artistic limitations. The graphic novella in its final printed form was the product of two cross-disciplinary collaborations: with a graphic designer skilled in Adobe Illustrator and with a local printer with experience of small print runs and clients with limited financial resources. Both involved 'a respectful exchange of intellectual ideas and technical skills and aims' (Back and Puwar, 2012, p 11), which resulted in the publication of *Five 'Survive' Lockdown*.

(The many) processes of creative data analysis

This section of the chapter attempts to summarise the processes involved in creating *Five 'Survive' Lockdown*. It should be read with the proviso that 'methods' reporting tends to present a smooth, linear account of what is often a messy, emergent, and serendipitous practice.

Starting with the sketchbook

I begin every graphic social science project with several sketchbooks filled with very rough pencil sketches and scribbled notes. *Five 'Survive' Lockdown* was no different; my sketchbooks were crammed with lists of possible characters and events, thumbnail sketches, and diagrammatic representations of potential storylines (see Figures 2.2 and 2.3).

Survey data provided the starting points for this process, for example, 54 per cent of respondents had agreed with the statement 'It has been very challenging to balance homeschooling and other tasks in my home with my workload during lockdown.' The challenge of visualising that statistic through character, settings, and storyline involved creative data analysis.

Figure 2.2: Page from *Five 'Survive' Lockdown* sketchbook

Figure 2.3: Planning plotlines from *Five 'Survive' Lockdown* sketchbook

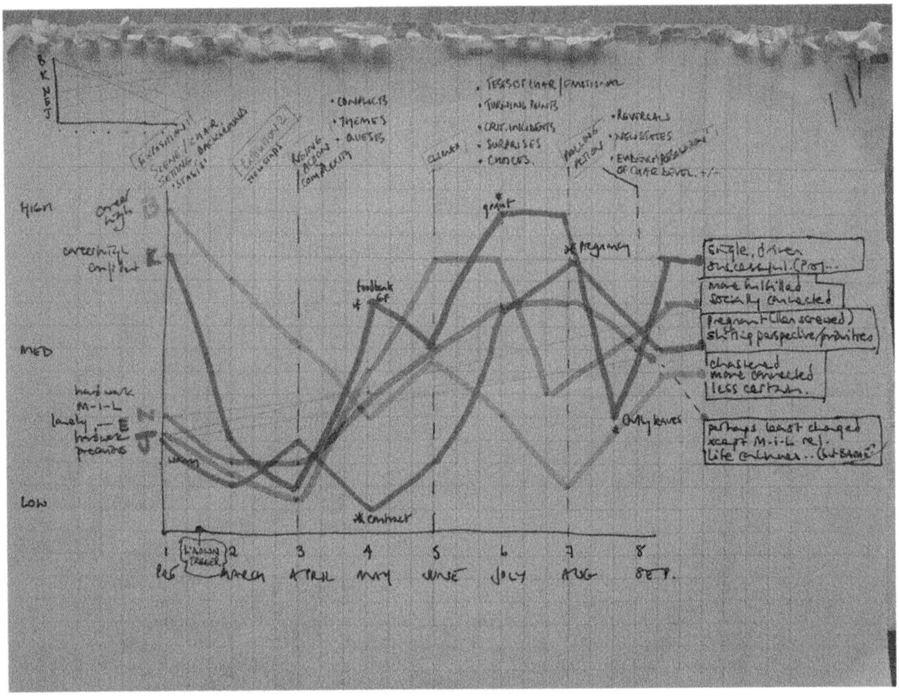

Creating narrative focus

Survey data ranged across multiple professional and personal aspects of living and working through the first lockdown. A total of 67 per cent of survey respondents and 67 per cent of respondents in academic roles had identified as female. In addition to the disproportionate disadvantages experienced by female university staff in comparison to their male colleagues, female academic staff had experienced specific challenges in terms of research productivity. This included completing research projects, writing and submitting bids for research funding, and writing for publication, all of which are critical to academic career progression. Throughout 2020 and into 2021, these challenges were surfacing regularly in the academic literature and social media and questions were beginning to be asked about longer-term implications for gender equality in the HE sector. These factors were influential in my decision to selectively focus the novella on the lives of female academics during the lockdown. I also reasoned that a more restricted narrative focus would make this ambitious project more manageable.

Characterisation

I remained ethically accountable to preserve respondents' anonymity in any revisualisation of the data. I created five composite female characters at different

Table 2.1: Characters

Character name	Career stage	Ethnicity	Age	Marital status
Brigid McNiff	Professor	White Irish	59	single
Kim Campbell-Bailey	Reader	Black British	43	partnered
Neela Gupta	Senior Lecturer	British Asian	41	married with 3 children
Jessica Wells	Lecturer	White British	32	married with 1 child
Emily Yang	PhD student	Taiwanese	23	single

stages of the academic career path: Professor, Reader, Senior Lecturer, Lecturer, and PhD student. All were based at the fictional Hightown University. Each character also embodied specific demographic characteristics and personal circumstances as set out in Table 2.1.

The characters reflected the survey sample's age ranges and varied marital and parental statuses but deliberately represented a wider range of ethnic identities than the survey sample and included an international student (students had not been eligible to complete the survey). This allowed me to explore a wider set of issues and experiences within the narrative than had explicitly emerged in the survey data; for example, living in extended family households (Neela), family separation and the student experience (Emily). Once characters were beginning to take shape in my imagination, my sketchbook pages became full of experimentations with their visual appearance, using the simplification and exaggeration of physical features typical of cartoon and comic styles. In due course this was repeated with secondary characters: partners, spouses, children, siblings, and neighbours.

Storylines

By the time I started working on the project in January 2021, it was in the knowledge that following the March–June 2020 lockdown, there had been a tentative return to university campuses in September 2020. I set the graphic novella's timeline from March–September 2020 and structured it into eight chapters, each between five and seven pages. The first chapter, entitled 'Before', introduced each character immediately prior to the announcement of the first UK COVID-19 lockdown. Storylines evolved through seven subsequent chapters entitled 'March', 'April', 'May', 'June', 'July', 'August', and 'September'. I plotted the five protagonists' narratives across this seven-month period, using four guiding influences. Firstly, each character was positioned in particular ways to the major survey themes of adapting to working or studying from home during lockdown, homeschooling, caring responsibilities, and well-being. Secondly, individual storylines were created in alignment with the overall timeline of the pandemic. I bookmarked a timeline file on my laptop and constantly referred to it, not least to accurately reflect changes to COVID-19 restrictions from June 2020. Thirdly, storylines reflected academic-specific practices relevant to each character's job role/career stage; for example, online teaching and doctoral supervision, Zoom meetings, funding bids, book launches, and the expiry of a fixed-term contract. Fourthly, the storylines incorporated aspects of the pandemic 'zeitgeist', which had been reflected to a limited extent in the survey data but by January 2021 had become established as part of the fabric of the pandemic experience; for example, the hour of exercise, panic shopping, mutual aid, and Clapping for Carers.

The development of these cross-cutting narrative arcs was linear to an extent, but also involved an affective, imaginative engagement as I got to 'know' each character over a period of months. I began to incorporate themes and issues not directly based on survey data but in the public domain during the time I was working on the novella. For example, the challenges of living in lockdown with family members or partners for an extended period (Neela; Kim), loneliness (Emily), elderly and vulnerable people in care homes (Brigid).

Visual design

Graphic narratives, like comic books, use sequential art and their visual structure and design are crucial to driving the narrative arc forward, creating and resolving narrative conflicts and developing protagonists with whom the reader can identify. Each page and panel is integral to this trajectory, assisting the reader to engage in a 'complex encounter of various kinds of literacy: verbal literacy, visual literacy, media literacy' (Baetens and Surdiacourt, 2020, p 516). I used a set of widely available Microsoft manga templates to structure the pages, each containing different numbers and arrangements of panels (see Figure 2.4).

I discovered that the variation in panel arrangements on each page suited different purposes. For example, the page template containing six symmetric rectangular panels was well suited to introducing each character in work and household settings and for simple narrative development. Asymmetric panel arrangements allowed for varying perspectives and close-ups, useful for moments of conflict, drama, and emotion. The final version also included two full-page illustrations. As a bimedial form in which 'the visual dialogue between words and images on the page … underscore the importance of the visual form of the words' (Baetens and Surdiacourt, 2020, p 516), the style and placing of text content need to clearly demonstrate its function: as titling, dialogue, and narration. Once a draft structure and content of individual chapters, pages, and panels were decided, I created a full version of the novella, in pencil on A4 copy paper. This was laboriously revised and repeated. Once I was satisfied with the draft, I created a new version on A3 Bristol Board, inking in the 40 sheets with Mars Matic technical pens and a black brush pen. Each sheet was scanned and sent to the graphic designer as an image file. They worked on each page in Adobe Illustrator, retaining the character (and flaws) of my illustrations but creating a smooth and consistent finish to the images I could never have achieved myself (see Figure 2.5).

Figure 2.4: Working with varied panel templates

Each page was returned to me as an individual pdf file for corrections and finally, a full pdf of the whole graphic novella was created for printing.

Publication – and beyond

I self-published 100 copies of *Five 'Survive' Lockdown* in September 2021, launching it via social media, my personal website, and personal and professional mailing lists. Copies sold so quickly I ordered a reprint! I mailed each copy out by hand. At the time of writing I have sold over 150 copies, to academics, students, university libraries, but also complete (non-academic) strangers who have 'found' the novella on social media and my website. Beyond this, I can trace ways in which *Five 'Survive' Lockdown* has 'promoted engagement, wider impact and a potentiality for change beyond the academic article' (Mannay, 2019, p 659). I have made several public presentations about the novella and the gendered experiences of female academics it focuses on. The experience of creating *Five*

Figure 2.5: Print-ready page (Chapter 1 'Before' – Brigid)

'Survive' Lockdown formed a bridge into my next funded research project, which uses diary and interview methods to explore the lived experiences of female academics during and beyond pandemic lockdowns, as well as the longer-term implications and challenges for gender equality in HE. Finally, I have now set up a COVID-19 and Higher Education Network, which will continue to amplify research into the lived experiences of the pandemic for staff and students as well as the longer-term impacts of the pandemic era.

Reflections and suggestions

Five 'Survive' Lockdown is my most ambitious work of graphic social science to date. Experience had taught me that the process of creative data analysis through the technique of hand-drawn images could be time and labour-intensive, but had I known quite *how* much time, labour, and learning would be involved I may never have started! That I did owes much to the volume and nature of the survey data, which inspired me to commit to other forms of data analysis, recording, and dissemination of this extraordinary, contemporary phenomenon. The continuation of the pandemic beyond 2020 meant that *Five 'Survive' Lockdown* became a historical document even as I worked on it. It is a fictional version of that period, grounded in research data, which amplifies the complexities and minutiae of the everyday that individuals lived out behind their front doors. It makes visible with frankness and some humour, elements of a collective but deeply unequal experience, which seem since to have been too easily forgotten.

Self-publishing a graphic novella is not for everyone. However, visual and graphic approaches to creative data analysis offer multiple possibilities for engaging with data in new ways and opportunities to engage intellectually, affectively, and effectively with wider audiences. I hope the following practical suggestions inspired by prompts from the chapter might inspire you to experiment, just as I did.

- Tyler states '*One of things which has guided our collaboration is a concern with making rich historical lines of Black critical thought about stigma visible.*' (Tyler and Bailey, 2018)

  What do you want to make visible in your research? What methods can you use to do this? Photography? Comics? Collage?
- *The complex, affective stories behind the statistics played a distant second to 'statistical significance'. As an interpretivist, I considered these to be equally if differently 'significant' in understanding lived experiences of the pandemic.*

  If you are presenting or reporting a large amount of quantitative data, think about ways of highlighting individual stories 'beneath' the statistics which will engage audience's empathy and emotion. This might include simple case studies, photographs, or images. Or comic strips, photo essays, and so on.

- (Research comics) '*bring social research findings alive … help a wider range of readers to empathise with research participants' experiences … prompt emotions that might be less likely or less obvious when reading the same information in a written report*' (Vigurs in Priego 2016, n.p.).

  Can you, or your students, create a research comic in response to a project or paper? Ask audiences to create their own images in response to your data.

It may be that creative, critical work in the academy can only fully gain legitimacy through text-based analyses of it and publication of those analyses in conventional academic formats – such as this chapter. However, in revisualising and disseminating original research in this way, the relationships of statistics, words, and images are disrupted and rebalanced. In so doing, the potential of the visual is extended from an illustrative tool to 'a sociology with the image in which the visual is understood as a source of knowledge construction in itself, as well as a means of sharing analyses with others' (Traue et al, 2019).

References

Back, L. and Puwar, N. (2012) 'A manifesto for live methods: provocations and capabilities'. *Sociological Review*, Special issue: *Sociological Review Monograph Series: Live Methods*, ed L. Back and N. Puwar, 60: 6–17.

Baetens, J. and Surdiacourt, S. (2020) 'How to "read" images with texts: the graphic novel case', in L. Pauwels and D. Mannay (eds), *The SAGE Handbook of Visual Research Methods* (2nd edn), London: SAGE Publications, pp 514–22.

Barnes, V., Clouder, L., and Hughes, C. (2003) 'Deconstructing dissemination: dissemination as qualitative research', *Qualitative Research*, 3(2): 147–64.

Camden, V.J. and Zullo, V.L. (2023) '"I'm reminding us of where we came from": an interview with Nick Sousanis', *Journal of Graphic Novels and Comics*, p 1. Available from: https://doi.org/10.1080/21504857.2023.2202871

Carrigan, M. (2017) 'What is graphic social science?' *The Sociological Review*. Available at: https://www.thesociologicalreview.com/what-is-graphic-social-science/

Carruthers Thomas, K. (2020) 'Living and working in lockdown: what's gender got to do with it?' Internal research report. Birmingham City University.

Hochschild, A. (1983) *The Managed Heart: Commercialization of Human Feeling*, Berkeley: University of California Press.

Hubbard, G., Backett-Milburn, K., and Kemmer, D. (2001) 'Working with emotion: issues for the researcher in fieldwork and teamwork', *International Journal of Social Research Methodology*, 4(2): 119–37.

McLure, M. (2013) 'Classification or wonder? Coding as an analytic practice in qualitative research', in R. Coleman and J. Ringrose (eds), *Deleuze and Research Methodologies*, Edinburgh: Edinburgh University Press, pp 164–83.

Mannay, D (2019) 'Revisualizing data: engagement, impact and multimodal dissemination', in L. Pauwels and D. Mannay (eds), *The SAGE Handbook of Visual Research Methods* (2nd edn), London: SAGE Publications, pp 659–69.

Priego, E. (2016) 'Comics as research, comics for impact: the case of higher fees, higher debts', *The Comics Grid: Journal of Comics Scholarship*, 6(1): 16. Available from: https://www.comicsgrid.com/article/id/3542/

Traue, B., Blanc, M., and Cambre, C. (2019) 'Visibilities and visual discourses: rethinking the social with the image', *Qualitative Inquiry*, 25(4): 327–37.

Tyler, I. and Bailey, C. (2018) *From Stigma Power to Black Power: A Graphic Essay*. Available from: https://thesociologicalreview.org/collections/guest-essays/from-stigma-power-to-black-power/

3

Visually mapping the interplay between pandemic interest groups and 'the vulnerable' in newspaper accounts, 2018–2022

Stuart Neilson and Neil Kenny

Introduction

The COVID-19 pandemic had an unprecedented impact on society, presenting a valuable potential case study of how media represents a societally important topic. We use a visual mapping approach to illuminate and discuss the patterns that emerge within media narratives and coverage relevant to the COVID-19 pandemic. Our approach involves the creative analysis of 'big data' extracted from media reportage in order to create visually accessible depictions of emerging patterns within the analysis. We analysed the contrasting roles played by 'vulnerable' groups, 'mental health', and people with 'underlying conditions' on the one hand, and social and business interest groups on the other hand, within this manufactured 'collective imaginary' (Bouchard, 2017, p 19). We visually exposed their often-conflicting impacts on shared media or policy narratives regarding public health, risk, and policy responses to the COVID-19 pandemic. The complex interplay of these themes creates an excellent case study for the visualisation of 'post-truth' media messaging (Fuller, 2020), expanding on techniques we have previously explored. The creative analysis, and resulting visually engaging presentation of the salient factors, effectively represent the complex interplay of narrative persuasion and policy response across the COVID-19 pandemic.

This study applied the term frequency and word vector techniques that are common in the 'big data' tools that underlie web searches and recommendation algorithms and guide many other aspects of 21st-century life. To make these underlying quantitative data analyses visible and intelligible to a non-mathematical viewer, this chapter uses a range of visual approaches to data representation to portray patterns and findings in an accessible manner. Our objective is to explore the potential within big data analysis to make observable the metadata underlying media narratives and messaging. Creativity is fostered by tools that are easy to use, converting statistical analysis into play and effort into image making. There is a tension between creating imagery with a high impact, in which the message is simplified and amplified into an aesthetic call for action, and the systematic application of statistical techniques. Any resolution of the tension must not misrepresent findings, and at the same time it must recognise the distorting influence of aesthetic choices.

Theoretical frame and study rationale

We adopt critical realism as a theoretical lens through which to explore the interplay of patterns that emerge within the data and their implications. Critical realism, as a philosophical framework for social scientific research, uses components of both the positivist and constructivist paradigms (Brown et al, 2002). Importantly, the framework of critical realism allows researchers to search for causal factors, which help explain social events but also suggest practical policy recommendations to address social problems (Yucel, 2018). The social world is comprised of layers of phenomena, from the 'real', to the social and the cultural (Fletcher, 2017). In essence, there is space for features of positivist 'causation' alongside socially constructed shared understandings within cultural spaces (Botha, 2021).

The domains of science and health have, across much of the 20th and 21st centuries, been characterised by an emphasis on 'evidence' and 'facts'. However, more recently, with the 'post-truth' era of social media and information saturation, there has been an increasing narrative regarding 'alternative facts' and divergent opinion (Gravino, 2022). The dissemination and interpretation of scientific or medical knowledge is being increasingly challenged by those with vested interests in distorting and denying the evidence-based claims of scientists (Lockie, 2016; Lubchenco, 2017). There has simultaneously been an explosion in conflicting and emotively toned public health messaging attempting to manage public impressions or understandings regarding the pandemic. Evidence from within the European Union (EU) has shown that lobbying by business and professional organisations increased during the pandemic (Junk et al, 2021), presumably in an effort to influence short-term policy responses to COVID-19. This is most pronounced among groups or industries which self-identify as being highly impacted by the pandemic, such as the hospitality, sports, and leisure industries, among many other examples. This lobbying has *visually identifiable* impacts within the intersection of economic, public health, political, and media messaging, which illustrate how media narratives are constructed or influenced. The volume of media content focused on the pandemic over a relatively short timeframe is unprecedented in scope and accessibility. Therefore, it is useful to provide a visually clear and accessible analysis of the features of media narratives relevant to COVID-19, shared public understandings, and how vulnerable citizens – who are the stated subjects of concern – have been represented.

While the pandemic has proved both complex and long-lasting, it has also illustrated a previously undetected lack of public health education among citizens in Ireland and internationally (Jarreau et al, 2021). Nearly half of all Europeans

(Kickbusch et al, 2013) and nine out of 10 adults in the US have inadequate health literacy skills and struggle to follow personal or public health information (CDC, 2021). This, unfortunately, has led to many consuming information regarding COVID-19 that is not accurate or trustworthy (Caballero et al, 2020). In other words, the audience for public health messaging during the pandemic are particularly vulnerable to consuming selective data, processed by well-resourced organisations and individuals. These sources may present analyses characterised by a narrow demographic representation, which have the potential to both encode and amplify specific messages under the guise of producing neutral data sets or universal 'facts'. Such analyses hide impactful decisions in what data to present to the public and how to present it, while hiding such decisions behind a veneer of scientific objectivity (Bryan et al, 2019). In the context of quantitative analysis of media narratives, such degrees-of-freedom decisions dictate which narratives are given media focus, or which slices of research findings are referenced within government messaging.

Newspaper text data

Our data set comprises every newspaper article in five large-circulation national newspapers mentioning key phrases identifying vulnerable cohorts. For contrast, we include the two years preceding and the two and a half years following the first public health measures in Ireland in March 2020. We also contrasted news reports in Ireland with news reports in Australia and New Zealand, where the pandemic response was very different. Following preliminary searches and discussions to expand and refine our keywords, our text corpus consists of the full text of every article mentioning terms related to 'mental health' (defined by a search for mental with health, illness, disorder(s), or well-being), 'underlying conditions' (underlying or pre-existing condition(s), illness(es), or disease(s)), and 'vulnerability' (vulnerable or vulnerability related to health).

We obtained the full text of all articles and wrote short Python and R programs to examine and clean the data, ensuring consistent (e.g. UTF-8 Unicode) character encoding, punctuation codes, date formats, and article delimiters. The cleaning phase provided an opportunity to skim the news, absorbing themes, significant terms, and time trends. The end result of the cleaning process was five plain text files, one per newspaper title, containing the title, date, and article text ready for analysis.

We examined these articles across both time and geography. We categorised time in three epochs: a pre-pandemic phase from January 2018 to March 2020, a pandemic crisis phase from April 2020 to December 2021, and a post-crisis phase from January 2022. We categorised geography into Ireland, represented by the *Irish Examiner*, *Irish Independent*, and *Irish Times*, and Oceania, represented by the *New Zealand Herald* and *The Age* (Australia). In total, our corpus contained 29 million words in 34,249 news items.

Procedures: quantitative models of collective understandings

The wide availability of and easy access to large volumes of text has stimulated the development of increasingly sophisticated and complex text analysis tools claiming to capture collective opinions, sentiments, and understandings. We chose to use code in Python and the statistical language R due to our familiarity with these, and their wide selections of well-documented open-source text analysis packages. Implementations of the same procedures are available in many computer languages and packages, and would reproduce the same outcomes. Quantitative word embeddings and their associated algorithms are usually black boxes with mysterious workings. Their outputs are taken on trust, providing the responses to online searches, selecting the posts at the top of social media streams, suggesting the next streaming TV show, locating housing, reviewing employee performance, and recommending prison sentences. 'Making visible' the word frequencies and word vectors (the frequency of every word that co-occurs with a word) in accessible form is a window into a large volume of text, and the knowledge or opinions a machine-learning system would claim to unearth.

Despite early warnings, the myth persists that analyses based on large data sets will transform science, bracket human biases or limitations in intelligence or knowledge, and generate insights with 'the aura of truth, objectivity, and accuracy' (boyd and Crawford, 2012 p 663). Indeed, analyses based on big data findings continue to be regarded as an oracle of the public zeitgeist. In both publicity puff pieces and academic papers, complex analyses based on machine-learning research are regularly presented as 'value-neutral' and possessing powerful potential for knowledge (Birhane et al, 2022). The reality, however, is that these analyses are based on social or information systems *designed by* humans. As such, the big data is likely also to reflect existing societal barriers, use the language of existing social categories or metrics, and perpetuate existing understandings or biases within their predictions or findings (Weidinger et al, 2021).

The current chapter harnesses these *weaknesses or challenges* related to the use of big data, reversing our analysis to use big data to *reveal* existing understandings, power structures, and influences within media reportage and ecology. Media text can be taken as a selective snapshot of society's current opinion that is collective but not representative, due to the influence of dominant and powerful voices. This unrepresentative collective both reflects *and* forms opinion, with the potential to codify and magnify bias, exclusion, and harmful expressions.

Our approach to interrogating this data set is, firstly, exploratory and playful, with the intent of fostering a wide variety of imaginative readings of the data, and an unrestrained dialogue in imagery. We winnow the large volume of quantitative and visual findings into those we feel are objectively substantiated expressions of our opinions. We are very conscious that we have formalised subjective storytelling within an envelope of collective truths. This *truths envelope* contains diverse, contrary, mutually incompatible, and opposing views. The objective envelope expressed in media narratives is a collective envelope that shrinks to narrower ranges of 'objective truth' with greater knowledge (when some opinions are widely recognised as invalid), and also as a consequence of compromise, compliance with more authoritative voices, and submission to (or silencing by) dominant voices.

The aim of our analysis is to expose the inner structures of word models, thus producing relatively easily understood visualisations. Within the images that are the outcome of our analyses we hope to explore the potential over-representation of powerful groups that could function to exclude vulnerable or less empowered people in society. This is prescient given the pandemic has been shown to have had a profound effect on vulnerable, disabled, and neurodivergent groups who may be less likely to be represented in media text. An additional and complementary aspect may be to uncover aspects indicating how these latter groups are rarely the creators or controllers of media narratives. Identifying these lacunae is an opportunity to consult these missing voices and bring to the foreground the lived experience of psychiatric care, health care, and the needs of vulnerable people during and following the COVID-19 pandemic. In Ireland, examples of such voices might be mental health patients, people with underlying conditions, Travellers, migrants in Direct Provision, prisoners, ethnic minorities, and neurodivergent and other vulnerable people.

Data analysis: making visible

One artistic influence on our approach is Philipp Steinwebber and Andreas Koller's 2007 exhibition 'Similar Diversity' (Lima, 2011), a large-scale information graphic visualising commonalities and differences in the world's five largest religions, derived from the text of their holy books. The clarity and surface-level simplicity of their infographics are windows into a detailed analysis of word frequencies and word embeddings. *Word embeddings* – the vector of co-occurring words defining the contexts in which a word is used – are said to represent meanings and understandings (Jurafsky and Martin, 2000), presented without interpretation by the artists as a supposedly objective visualisation of commonality and difference. Steinwebber and Koller's results are reproducible, but, in 'making visible' rather than interpreting, they invite us not to trust the objectivity of the algorithm, but to perceive how it works (Arnall, 2014).

The frequency of words and phrases (including nouns and names) represent one measure of relative dominance and power in written media. The quantities tell us who writes, who is quoted, and whose interests are identified. Our aim is to visually represent these features of the patterns within the objective envelope of the media through data-driven representations using a range of techniques, which we discuss later in this chapter. We present a suite of case studies of how our data can be used creatively to present patterns in a manner more easily accessible or discernible to a general viewer. Each case study provides exemplars of how art can effectively use visual images to represent artefacts of meaning, which accessibly represent actual patterns within the data so that quantitative variables influencing socially constructed narratives become overtly clear to the reader.

Case 1: Waxing and waning concern, and extreme variability in word frequencies

The *word frequency* of our three search terms varies over time periods and between geographic regions, as displayed in Figure 3.1. Our immediate observation was that Ireland displayed a substantial rise in interest or concern in our chosen terms during the pandemic crisis in the *Irish Examiner* (publishing 463 per cent of the pre-pandemic baseline number of items), *Irish Independent* (168 per cent), and *Irish Times* (157 per cent). The increased interest during the crisis was followed by a large fall in the post-crisis phase in the *Irish Examiner* (−70 per cent), *Irish Independent* (−36 per cent), and *Irish Times* (−49 per cent). In contrast, the rising and falling interest in Oceania was more restrained, with a pandemic crisis rise in the *New Zealand Herald* (publishing 134 per cent of the pre-pandemic number of items) and *The Age* (158 per cent), followed by smaller post-crisis changes in the *New Zealand Herald* (a continuing 7 per cent rise in concern) and *The Age* (−25 per cent). It is clear that Ireland exhibited a significant rise in concern regarding 'mental health', people with 'underlying conditions', and 'vulnerable' people, and

Figure 3.1: Frequency of newspaper mentions of search terms identifying 'mental health' (MH), 'underlying conditions' (UC), and 'vulnerability' (V)

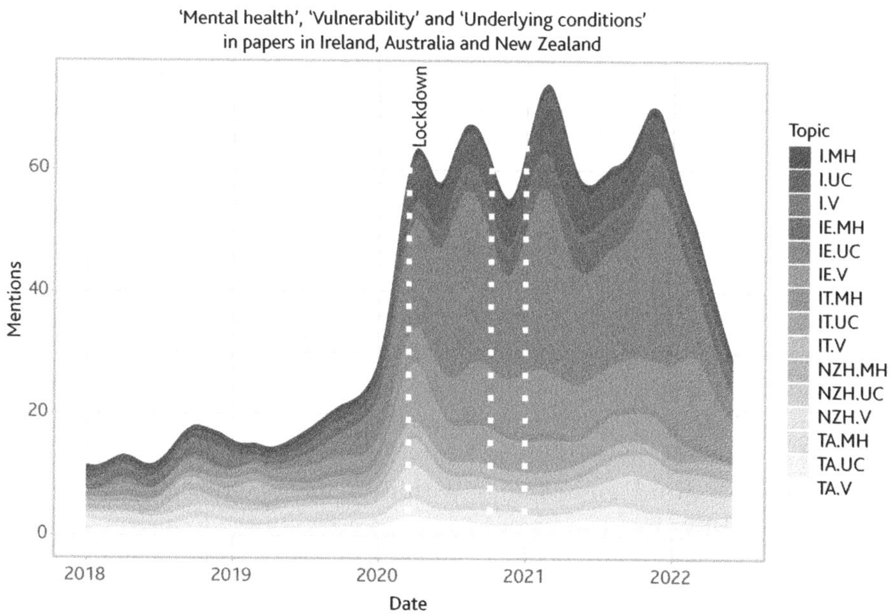

Note: Newspapers are the *Irish Independent* (I), *Irish Examiner* (IE), *Irish Times* (IT), *New Zealand Herald* (NZH), and *The Age* (TA).

that this concern declined with the declining sense of pandemic crisis. Australia (as represented by *The Age*) exhibited a more restrained change in interest, and New Zealand (as represented by the *New Zealand Herald*) exhibited the most moderate change in concern, allied with the strictest COVID-19 response in any English-speaking country.

It is also visually apparent that rising concern coincided with the imposition of COVID-19 restrictions in Ireland, particularly the three national lockdowns limiting movement, gatherings, and access to employment and leisure.

Case 2: Pandemic power structures

When we highlight specific *word embeddings*, we can highlight dominant themes (or interest groups) clustered around a theme of interest. Within our corpus of text we can identify who is permitted a platform to speak about, on behalf of, or from within our groups of interest. Figure 3.2 is a visualisation of the word vector associated with the word 'said', displaying co-occurring words and their frequencies. These related terms are typically the identities of the voices represented in print media when discussing these topics. It is striking that of the most frequently identified people who 'said' words quoted in print media, only one of the top 50 is female, and not one is a Traveller, a migrant, a pupil with special educational needs, a patient with an underlying condition, or a representative advocating on behalf of these groups of people. Identified individuals, with their pandemic-era roles, include former Teachta Dála (TD) Jim Daly, Minister for Health Stephen Donnelly, Deputy Chief Medical Officer Dr Ronan Glynn, Minister for Health Simon Harris, Chief Medical Officer Tony Holohan, Boris Johnson PM, Alan Kelly TD, Tánaiste Micheál Martin, Paul Murphy TD, chair of the Irish Epidemiological Modelling Advisory Group Philip Nolan, Chief Operations Officer of the Health Service Executive (HSE) Anne O'Connor, Director-General of the HSE Paul Reid, neuroscientist Dr Tomás Ryan, Taoiseach Leo Varadkar, and a wide variety of politicians, officials, and spokespeople for related entities. The lived experience of vulnerable people is notable for its absence, falling far from the ideal representation of 'nothing about us without us' (Charlton, 1998).

Our visual analysis similarly exposed a notable change in the embedding of 'mental health' across time. Both pre- and post-crisis, mental *health* or *illness* co-occurred with psychiatric diagnoses, suicide, and support. During the pandemic crisis, mental *well-being* co-occurred with access to gymnasia, hospitality, and the imperative to return to the workplace.

We expose the word embeddings within weighted semantic networks, displaying the relative frequencies of each word node (in the font size) and the strength of the embedding (in the line thickness or colour density of connecting arcs). Word vectors visually represent the self-reinforcing frequency with which terms or individuals are associated with media discourses. It is easier to repeat a familiar concept than to synthesise a novel concept, and ease increases with

Figure 3.2: Most frequent 50 contexts (word vector) of the word 'said'

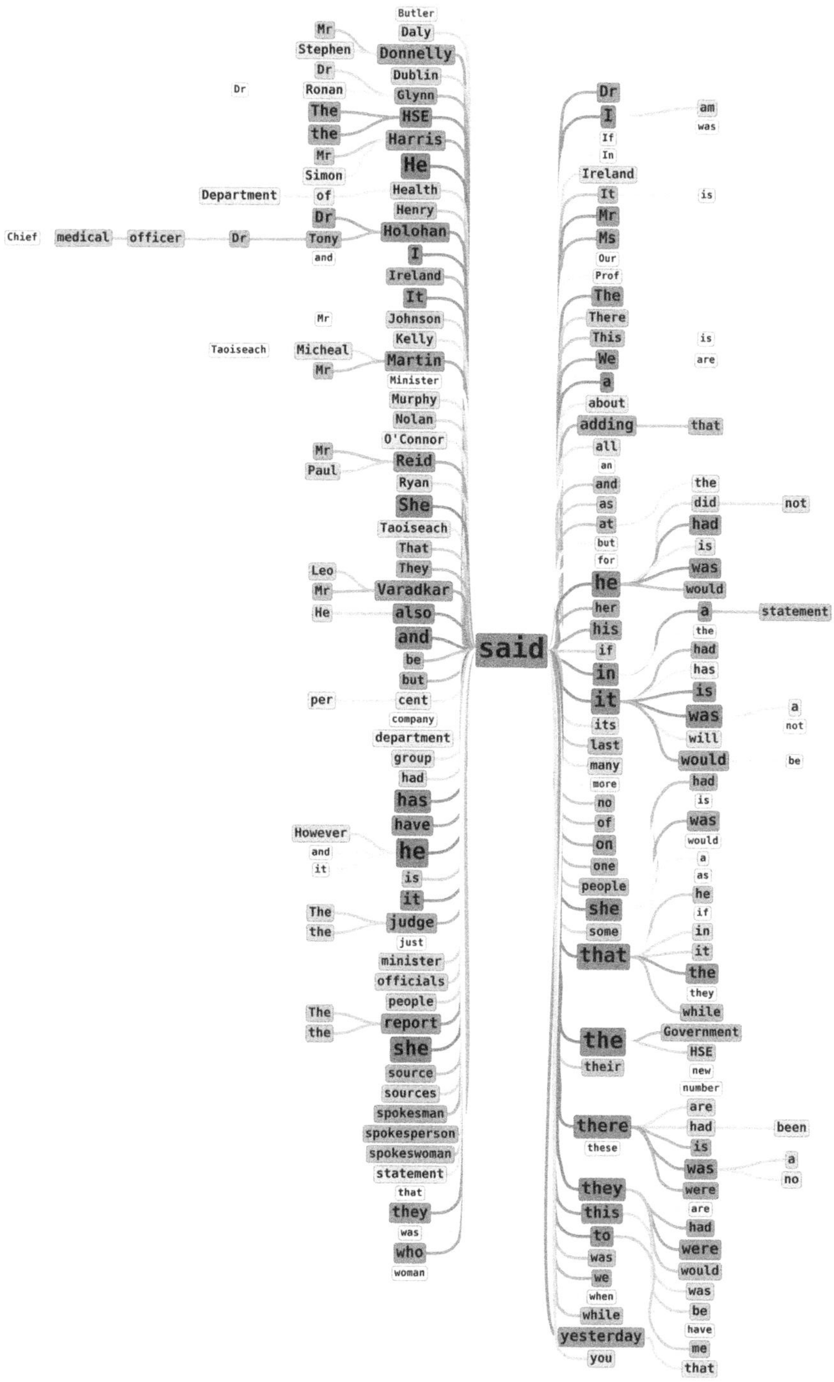

Note: The figure displays a dominance of male and official sources of statements, and a paucity of sources with lived experience of 'mental health', 'underlying conditions', and 'vulnerability'.

repetition. Like well-worn grooves in a stone surface, it is easy to repeat familiar word sequences, harder to say rare combinations, and impossible to speak a novel concept until a new formulation is created to express it.

Case 3: Subjectivity and narrative storytelling

In much of our work, we have also paired statistical word embeddings with more exploratory forms of communication using video-photographic composite artwork. We have exhibited these techniques in analyses of pedestrian flow in public space (Neilson, 2023), tracing bird flight above the city when humans were in pandemic lockdown, and visualising the body language of public personalities. We have noted a significant under-representation of women quoted in print media during the pandemic. There is an under-reporting of the lived experience of single mothers (who greatly outnumber single fathers) juggling blended learning, sick children, and working from home, women in front-line jobs in health care and education, pregnant women faced with frequently changing COVID-19 maternity guidance, and the widespread expectations on women as carers. We processed a set of video sequences to humanise these multifaceted and complex issues in a persuasive image. We selected a convenience sample of three men and three women identified in our newspaper text analysis as often-quoted speakers during the pandemic, and for whom we could locate television interview recordings. All six were interviewed under similar circumstances on the same Irish television programme, RTÉ News. The motion composite exposed the much greater volume occupied by men gesticulating while speaking. The calculated motion intensity visualises spatial volumes symbolic of the political capital accorded to and assumed by men and women in society and in our text corpus.

Discussion: influences and messages

This chapter has outlined how creative approaches to both the analysis *and* representation of big data drawn from word patterns within the media narrative may have utility in supporting readers to see the hidden interplay between influences and messages in the media. The COVID-19 pandemic provided a unique opportunity as large tracts of the media showed a persistent and overwhelming focus on a socially important single topic. Our analysis targets the role degrees-of-freedom decisions play in both the methodological rigour and ensuing findings within quantitative research, using the analysis of big data to explore how such patterns of interest can be made observable within the massive media coverage of the pandemic. We draw from a critical realism framework to support the recognition of both the existing of causal relationships across scientific and public health research. We use this framework to analyse the social construction of our understanding of COVID-19, the seriousness of pandemic outcomes, and what manner of policy response is most appropriate. We have presented the outcomes of our approach across a series of case studies to indicate the value of our approach.

Case Study 1 indicates clear cultural influences regarding how the pandemic was represented within the media data, and clear impacts of government policy on word frequencies within media messaging across countries. This is indicative of a looping effect whereby media narratives impact government policy or responses, with subsequent policy responses, such as lockdowns, also then clearly heavily influencing media narratives in response. Such processes could serve to amplify differences across jurisdictions or media ecosystems over time, leading to increasing polarisation or discord within the populace in response to serious social phenomena, such as the pandemic. Greater awareness among the public of how such feedback loops can develop over time and create a self-sustaining dynamic of their own would be important. Also important would be knowledge of how factors can contribute towards the development of narratives within the media at the outset.

Case Study 2 explored *word embedding* as a visual representation of dominance in media discourse. These embeddings also function as patterns facilitating word choice and repetition, which are linked to particular roles or individuals. As Geena Davis said 'They can be what they see' (Winick, 2013), and vulnerable people can only see themselves as objects of concern, not authors of their own concerns. It is notable, for example, that an initial predominance of 'psychiatric' pathologising terms were replaced in popular discourse by the progressively more fluid (and wellness industry-friendly) phrases 'mental health' and 'mental well-being', marginalising discussion of the scarcity of psychiatric care resources. There may be a range of meanings behind such a choice of terminology or association, particularly regarding communication impact and tractions across as large a segment of the population as possible. Most lay understandings of 'psychiatric' would see the term as denoting more serious psychological or behavioural presentations, which require medical or institutional support in many cases. In other words, they are experiences that impact a relatively smaller subset of the population rather than larger segments of the citizenry. Given the serious and pathological nature of conditions considered as falling under 'psychiatric' categories of psychological diagnosis, they often involve medical diagnosis and specific conditions, presentations, or symptoms. In other words, support for people with psychiatric conditions requires services, support staff, and budgetary spending to put such responses in place.

By comparison, mental well-being, as a concept, has much broader applications across the population, is much less specific or medical in association, and may not require clinical involvement or consultation. As a concept, it is likely to have much greater traction or relatability across a larger subset of the population. In addition, mental well-being does not imply specialist health or psychological services or interventions by the state are required, and can be discussed through the lens of individual actions, choices, or experiences. Such individualistic foci characterise the existing 'well-being industry', which aligns more closely with leisure or lifestyle narratives than with medicalised 'psychiatric' domains.

The *effect* of the increased interest in or concern with 'mental health', 'underlying conditions', and 'vulnerability' in Irish print media was evident in the early reopening of gymnasiums, pubs, and other businesses, the continuous access to alcohol with 'essential retail' throughout the pandemic, and ending both lockdown and all visible restrictions on commerce. Concern has *not* translated into increased funding for psychiatric services, or for social and health care of vulnerable people. Indeed the health budget overall increased by just 4 per cent in 2022 (less than inflation) (Shine et al, 2022), waiting lists for medical procedures have risen by 30 per cent to nearly 900,000 patients (Cullen, 2022), and there is a pending 'cancer timebomb' (Doyle, 2021) of cases undiagnosed and untreated during the pandemic (Cosgrave, 2021). In Ireland, capital spending on public health and the number of staffed hospital beds remains the lowest in the EU (IMO, 2022).

Misinformation and disinformation were widely shared on social media, flooding into mainstream media as both inadvertent misinformation and as quite consciously contentious topics. Marginalisation increases the attractiveness of these alternative sources, further decreasing trust in official supports and access to care, vaccination, and mitigations. All participants identified the media as a catalyst for fuelling Traveller hostility and prejudice throughout the pandemic. Negative representations increased stigma towards vulnerable groups including Travellers (Friel, 2021) and prisoners (User Voice, 2022), who were consistently excluded from media narratives about themselves.

Case Study 3 processed video to expose the difference in volumes occupied by body motion of speakers representative of dominant and non-dominant groups in high-profile media platforms. There was a notably higher frequency of male pundits and experts invited onto media platforms and public discussion programmes on the topic of pandemic-era health policy, leading to greater familiarity and amplification of selected voices or perspectives. In contrast, when women – or Travellers, migrants, or people with underlying conditions – gain access to media platforms, the lower frequency of such opportunities, augmented by their lack of familiarity to viewers and listeners, was associated with noticeable differences in their presentation and behaviour while on air. Case Study 3 processed video into a visual image that contrasts the movement dynamics and physical expressions of male and female panellists on TV discussion programmes. The space the female panellists feel confident occupying is noticeably more constrained than that of the male, despite their expertise, experience, or knowledge being comparable. Frequent panellists develop a level of familiarity, confidence, and experience that is visibly displayed in how they present their views or messages – a profile that undoubtedly impacts how they are perceived by those viewing their contributions. In contrast, less frequent panellists do not have the opportunity to develop this bedrock of expertise, experience, and communications competence, which is unintentionally on display in their non-verbal communications while speaking. Our visualisation extends beyond the quantifiable lack of representation of women, minorities, and vulnerable people quoted in media text or broadcast time to a visceral portrayal of the difference in embodied occupation of media space.

Conclusions: missing voices

Our approach to data analysis creatively visualises the quantitative relationships within language, exploring the cyclic interaction of social media, lobbying, expertise, and media reporting that itself operates like a large language model, with all its attendant oracle-like powers and limitations. We make visible the constant retrenchment of existing biases, amplified by negative and false portrayals of marginalised and vulnerable groups, whose images are appropriated by dominant voices, evident in the visualisation of newspaper reporting about vulnerability.

The people who are the ostensible subjects of concern – mental health patients, Travellers, migrants in Direct Provision, neurodivergent and disabled people, and other vulnerable groups – are visibly missing voices in our text corpus. These are also the groups most affected by the pandemic in measures of illness, mortality, and financial hardship. Identifying these groups is not a remedy, but is the first step on a longer road to expand and rebalance the objective envelope of media representation, to include the first-person contributions of those missing voices, with a lived experience claiming the central role of 'Nothing about us without us' (Charlton, 1998).

Our analysis is limited by our ability to correctly identify relevant vulnerable and powerful voices, and by our own participation in the culture we are analysing. Even seemingly trivial aesthetic choices such as font or colour may carry – perhaps subconscious – cultural meaning.

More broadly, these techniques can be applied to text or video collections that are supposed to represent culture, in which the relative dominance of voices is of interest. The visualisation of word embeddings is relevant to both using large language models as a research tool and as a diagnostic aid to so-called artificial intelligence (AI) generative outputs. There is a paucity of techniques to expose how their output is generated. Visual representations of word embeddings offer a view inside these black boxes, and a measure of their value as analytic tools.

References

Arnall, T. (2014) 'Exploring "immaterials": mediating design's invisible materials', *International Journal of Design*, 8(2): 101–17.

Birhane, A., Kalluri, P., Card, D., Agnew, W., Dotan, R., and Bao, M. (2022) 'The values encoded in machine learning research'. In *Proceedings of the 2022 ACM Conference on Fairness, Accountability, and Transparency*, New York: Association for Computing Machinery, pp 173–84.

Botha, M. (2021) 'Academic, activist, or advocate? Angry, entangled, and emerging: a critical reflection on autism knowledge production', *Frontiers in Psychology*, 12:727542.

Bouchard, G. (2017) *Social Myths and Collective Imaginaries*, Toronto: University of Toronto Press.

boyd, d. and Crawford, K. (2012) 'Critical questions for big data: provocations for a cultural, technological, and scholarly phenomenon', *Information, Communication & Society*, 15(5): 662–79.

Brown, A., Fleetwood, S., and Roberts, J.M. (2002) 'The marriage of critical realism and Marxism: happy, unhappy or on the rocks?', in A. Brown, S. Fleetwood, and J.M. Roberts (eds.), *Critical Realism and Marxism*, London: Routledge, pp 1–22.

Bryan, C.J., Yeager, D.S., and O'Brien, J.M. (2019) 'Replicator degrees of freedom allow publication of misleading failures to replicate', *Proceedings of the National Academy of Sciences*, 116(51): 25535–45.

Caballero, A., Leath, K., and Watson, J. (2020) 'COVID-19 consumer health information needs improvement to be readable and actionable by high-risk populations', *Frontiers in Communication*, 5, 56.

CDC (Centers for Disease Control and Prevention) (2021) 'Understanding health literacy'. Available from: https://www.cdc.gov/healthliteracy/learn/Understanding.html

Charlton, J.I. (1998) *'Nothing About Us Without Us': Disability Oppression and Empowerment*, Berkeley: University of California Press.

Cosgrave, T. (2021) 'After Covexit, what then?', *Irish Medical Times*, 28 May.

Cullen, P. (2022) 'One-quarter of state's population on health waiting lists', *Irish Times*, 6 June.

Doyle, P. (2021) 'Pandemic has created "cancer timebomb"', *Irish Medical Times*, 24 May.

Fletcher, A.J. (2017) 'Applying critical realism in qualitative research: methodology meets method', *International Journal of Social Research Methodology*, 20(2): 181–94.

Friel, D. (2021) 'Life on the margins: an exploration on the impact of COVID-19 on Irish Travellers' lives in the North West Ireland', MA thesis, Institute of Technology, Sligo.

Fuller, S. (2020) 'A post-truth proactionary look at the pandemic', *Postdigital Science and Education*, 2(3): 551–5.

Gravino, P., Prevedello, G., Galletti, M., and Loreto, V. (2022) 'The supply and demand of news during COVID-19 and assessment of questionable sources production', *Nature Human Behaviour*, 6: 1069–78.

IMO (Irish Medical Organisation) (2022) *Pre-Budget Submission, 2023*, Dublin: Irish Medical Organisation.

Jarreau, P.B., Su, L.Y.F., Chiang, E.C.L., Bennett, S.M., Zhang, J.S., Ferguson, M., and Algarra, D. (2021) 'COVID issue: visual narratives about COVID-19 improve message accessibility, self-efficacy, and health precautions', *Frontiers in Communication*, 6: 712658.

Junk, W.M., Crepaz, M., Hanegraaff, M., Berkhout, J., and Aizenberg, E. (2021) 'Changes in interest group access in times of crisis: no pain, no (lobby) gain', *Journal of European Public Policy*, 29(9): 1374–94.

Jurafsky, D. and Martin, J.H. (2000) *Speech and Language Processing: An Introduction to Natural Language Processing, Computational Linguistics and Speech Recognition*, Upper Saddle River: Prentice Hall.

Kickbusch, I., Pelikan, J.M., Apfel, F., and Tsouros, A. (2013) *Health Literacy: the solid facts*, Copenhagen: WHO Regional Office for Europe. Available from: https://www.who.int/europe/publications/i/item/9789289000154

Lima, M. (2011) *Visual Complexity: Mapping Patterns of Information*, New York: Princeton Architectural Press.

Lockie, S. (2016) 'Post-truth politics and the social sciences', *Environmental Sociology*, 3(1): 1–5.

Lubchenco, J. (2017) 'Environmental science in a post-truth world', *Frontiers in Ecology and the Environment*, 15(1): 3.

Neilson, S. (2023) 'The power of an autistic lens: visualising activity in shared public space', in M. Mostafa, R. Baumeister, M. Ramsgaard, and M. Tamke (eds), *Design for Inclusivity: Proceedings of the World Congress of Architects 2023*, Cham: Springer Nature Switzerland, pp 335–46.

Shine, C., Hennessy, M., and Walker, E. (2022) *Spending Review 2022 Health Capital Investment in Ireland*, Dublin: Department of Health.

User Voice and QUB (Queen's University Belfast) (2022) *Coping with COVID in Prison: The Impact of the Prisoner Lockdown*, Belfast: Queen's University Belfast.

Weidinger, L., Mellor, J., Rauh, M., Griffin, C., Uesato, J., Huang, P.S., Chen, M., Glaese, M., Kasirzadeh, A., and Gabriel, I. (2021) 'Ethical and social risks of harm from language models', *arXiv preprint*, arXiv:2112.04359. Available from: https://arxiv.org/pdf/2112.04359.pdf

Winick, K. (2013) 'Geena Davis solves sexism in Hollywood in five minutes flat', *Elle*, 13 December.

Yucel, R. (2018) 'Scientists' ontological and epistemological views about science from the perspective of critical realism', *Science & Education*, 27(5): 407–33.

4

Using discourse analysis to inform content analysis: a pragmatic, mixed-methods approach to exploring how the headteacher role is articulated in job descriptions

Alexandra Morgan, Andrew James Davies, and Emmajane Milton

Introduction

This chapter takes a step-by-step approach to providing guidance and a practical example of using discourse analysis to inform content analysis, in order to gain an understanding of messages being conveyed in an essentially qualitative data set. The focus of this chapter is the creative and methodologically innovative approach taken in combining critical discourse analysis with a detailed quantitative content analysis, to undertake an interpretative examination of headteacher job descriptions in Wales (Milton et al, 2023). We adopted this eclectic and pragmatic methodological approach in order to elucidate understandings of the articulations of professionalism (Evetts, 2009) being conveyed to prospective candidates in a large number of headteacher job descriptions (n = 67) at a very specific point within an ongoing and extensive education policy reform process in Wales (OECD, 2017). It was clear that these reforms were impacting headteacher recruitment, retention, and the lived experience of their roles and identities (Connolly et al, 2018; Davies et al, 2018), and, as such, headteacher job descriptions were useful linguistic artefacts where received accounts of the professional role could be examined.

Generally, critical discourse analysis and content analysis are derived from opposing methodological and epistemological traditions. The first of these being qualitative, offering deep and close analysis of meaning and significance (Hardy et al, 2004), and the other being quantitative, concerned with analysing the frequency with which particular units of data occur within text (Pole and Lampard, 2002). Bryman (1984, cited in Johnson and Onwuegbuzie, 2004, p 15) points to a historical tendency to 'treat epistemology and methodology as being synonymous' in social research. Yet, many researchers have advocated that epistemology should guide, but not necessarily dictate, the methodological approach adopted and, that in certain instances, it may prove useful for researchers to adopt methods that fall outside the paradigm with which their discipline or area of study is historically associated (Teddlie and Tashakkori, 2010). This means that, whilst the particular epistemological position adopted by the researcher (or research team) should be clearly understood and acknowledged, this should not confine them methodologically (Johnson and Onwuegbuzie, 2004, p 15). Adopting this stance of 'methodological eclecticism' (Hammersley, 1996, p 159) means that, for example, an interpretivist epistemological stance should not necessarily stop a researcher (or team) from employing data collection methods more usually associated with quantitative ways of working, where useful. This line of argument supports the approach taken in this study, which adopted an entirely interpretivist epistemology whilst taking a more pragmatic stance on questions of methods and methodology. In this way, our approach aligned with those who suggest that researchers who are judicious, knowledgeable, and intuitive in selecting the optimum technique or approach in terms of methods and methodology for a particular research task are acting as 'connoisseur(s) of methods' (Teddlie and Tashakkori, 2010, p 10).

At an early stage in the project, it was apparent that a purely quantitative deductive approach to analysing the job description statements could have tended to find the obvious and look past opportunities to develop theoretical ideas (Hsieh and Shannon, 2005). However, a purely qualitative inductive approach would not address the scale of analysis required and so would not be suitable for developing meaningful understandings (Campbell et al, 2011). In the final event, the iterative cyclical approach adopted, which involved both inductive and deductive logic, could be viewed as one that was both novel and characteristic of research of this nature (Hesse-Biber, 2010). To understand what was being conveyed in these job descriptions, it seemed important to undertake both an in-depth, critical, and theoretically informed analysis of the discourses, and also to quantify this in a way that helped us to understand the emphasis placed on these across the data set in its entirety. The four phases of our analysis are outlined below, with an explanation of how we went about each phase, along with practical illustrations from our study. In this chapter, we outline how, in our analysis of job descriptions, we navigated the balance between working creatively, iteratively, and yet systematically, and we note the challenges we experienced in adopting this approach (Milton et al, 2023).

This methodology employed aspects of a mixed-methods design, in the sense that both qualitative and quantitative approaches were used to conduct the analysis. However, the methodology discussed here does not sit easily within one of the conventional categories of mixed-methods research that are commonly encountered (Teddlie and Tashakkori, 2010). The design included elements from an exploratory approach, in that the team proceeded sequentially, using a qualitative then quantitative analysis, developing and codifying categories prior to the content analysis taking place (Creswell and Plano Clark, 2011). However, it also deviated from this in that the process was also cyclical, as the research team subsequently revisited codifications during the initial content analysis phases. This was in order to refine working definitions and align understandings of these within the research team, before they were finalised. This approach does not fit neatly into the sequential mixed-method process commonly employed in social science and educational research (Punch and Oancea, 2014), in that no new data was collected during the quantitative phase, but rather a further analysis conducted on the same data, employing a conceptual framework developed during the initial phases. The aim of this was to utilise the affordances of content analysis as a systematic and quantitative approach, in order to develop and extend our understandings of the prevalence of the messages conveyed in the job descriptions and communicated to potential headteachers.

The phases

Advertised headteacher job descriptions in Wales, written in both Welsh and English, were collected over the period of one academic year, from September 2016 through to the end of August 2017. Analysis was undertaken by all six members of the research team working both face to face and at distance via phone on the 67 job descriptions, all those that were fully publicly available. The job descriptions were analysed in the language in which they were written, be that English or Welsh, as members of the research team were bilingual.[1] These job descriptions contained a total of 2,970 individual statements of essential and desirable competences that prospective candidates needed to possess.

Phase one: establishing a shared (and theoretically informed) understanding of categories

Developing shared understandings amongst the research team of key theorisations of professional practice in education was crucial at the start of this analysis, and arriving at this shared understanding constituted a foundational part of the first 'empirical' phase of the project. This was a departure from conventional a priori ways of working in social and educational research, where a conceptual framework and understanding amongst the research team are generally established during the pre-empirical phase, prior to the collection and analysis of data (Punch, 1998). The development of shared understandings was highly collaborative and involved iterative reshaping of ideas, which occurred over time, through a carefully structured set of dialogic protocols (Bakhtin, 1981; Alexander, 2020). This meant scrutinising whole job descriptions systematically in pairs, then triads, and then as a whole team, in light of our theoretical understandings, which later informed the categorisations. The reason this way of working was essential was that the objective was to develop explicitly shared understandings of key concepts related to each of the categories under development – this was necessary in order to achieve consistency of rationale as to why any job description statement might belong to any of the possible categories. These were discussed, negotiated, and gradually became closely aligned and agreed upon as initial categories. This was a highly creative and collaborative stage of the process, and it eventually gave rise to the development of five categories. Therefore, the early allocation of individual competency descriptors to one of the five categories was based on a clear and agreed rationale, informed by a documented theoretical framework.

This project's conceptual framework was grounded in ideas from the sociology of the professions and drew heavily on Evetts' (2009, 2010) distinction between organisational and occupational professionalism. This aligns organisational professionalism with high levels of bureaucracy, external standardisation, technical approaches to practice, and techno-rational approaches to knowledge. Occupational professionalism tends to include as features collegial practices, agency, autonomy, and opportunities for using professional judgement and an appreciation of the importance of context in addressing professional issues (Evetts, 2010).

The framework was further informed by the research team's contextually specific understandings of the headteacher role in practice (Connolly et al, 2018; Davies et al, 2018). We also contrasted competing understandings of accountability within educational research, ranging from Fullan et al's (2015) concept of internal accountability to more organisationally aligned explanations of external accountability (Hoyle and Wallace, 2005). While this particular conceptual framework was appropriate for *our* study (Milton et al, 2020), what is important to remember about this process is that a theoretical framework, based on relevant understandings in the literature related to *your* study, needs to be developed and shared with anyone working on your project (that is, with all members of a research team or between others working on a project, for example, supervisors if you are a student researcher). We would assert that this process of developing a conceptual framework for this specific methodology is perhaps best undertaken as a collaborative endeavour rather than individually. This is because the complex work of developing and agreeing rationales for categorisation requires the significant interpersonal cognitive challenge of other ways of thinking, along with continuous negotiation of interpretation in order to fully articulate and have confidence in the robust nature of the decisions that are made. For our team, this meant many hours together reading out loud the job description statements and grappling with exactly why we felt they belonged in one category rather than another. Carefully considered and well-developed understandings of the rationale for allocation of individual statements or units of text to categories at this point are essential. It is highly likely that the end of the first iteration of coding *your* complete data set will inevitably lead to a process of further dialogue, revision, and refinement.

Fairclough's (2015, pp 129–31) approach to critical discourse analysis involves critical interrogation of the text, scrutinising the formal properties of language, including the experiential, expressive, and relational values that are implicit. In the context of the first phase of our study, this meant posing questions of these textual artefacts such as:

- Does the vocabulary and grammar used reflect occupational or organizational accounts of professional work? Or a mixture of both?

- How are ideologically contested words and concepts deployed? (e.g., accountability, challenge, responsibility, manage(ment)).
- What professional processes and participants dominate the individual statements? For example, considering the extent to which proposed tasks and activities are collaborative, collective and/or collegiate, distributing the locus of control to others; or are they characterized by individual, managerial and/or directed actions?
- To what extent do they promote an ethos of professional judgment *in situ* or are they performative in relation to external drivers?

(Milton et al, 2023, pp 345–7)

The development of the categories themselves for this project was therefore based upon both the conceptual framework and an in-depth reading and re-reading of the individual statements, as outlined. In this way, the process of developing the categorisations went hand in hand with the development and solidifying of shared understandings of the conceptual framework and the underpinning reference points. The abductive development of the categories therefore drew creatively on both inductive understandings emerging from the data and deductive understandings applied from the literature (Tavory and Timmermans, 2014). They were further developed thoroughly, through continually cross-referencing and/or sense-checking the data itself with the emergent categorisations against the theoretical framework, employing a process similar to the constant-comparison method used in grounded theory approaches to analysis (Glaser, 1965). This process is outlined in Figure 4.1.

The final categories that emerged at the end of phase two are listed in Table 4.1 as presented in the original paper (Milton et al, 2023).

Phase two: further defining and iteratively refining categories

During phase two, the categories were further defined and refined through the process of allocating all 2,970 statements to the initial conceptual ideas for the categories. What emerged from this phase were the final categories, as outlined and illustrated in Table 4.1. The team focused on the further iterative defining and refining of categories through the process of allocating every competence statement. This helped to recognise identifying features of statements allocated to each category, and to further clarify category boundaries. Protocols relating to this phase included: any statement only being allocated to a single category; any challenging statements were discussed and their categorisation agreed on by several members of the team; and ensuring all statements were coded. An ongoing process of cross-moderation of the individual statements from the job descriptions improved the interpretive reliability of the categorisation process further.

This process was supported by the use of NVivo 12.[2] The software was used as a tool to support the analytical process, affording the research team the ability to work productively with this large data set, to share analysis transparently and

Figure 4.1: Visual representation of key processes according to phase

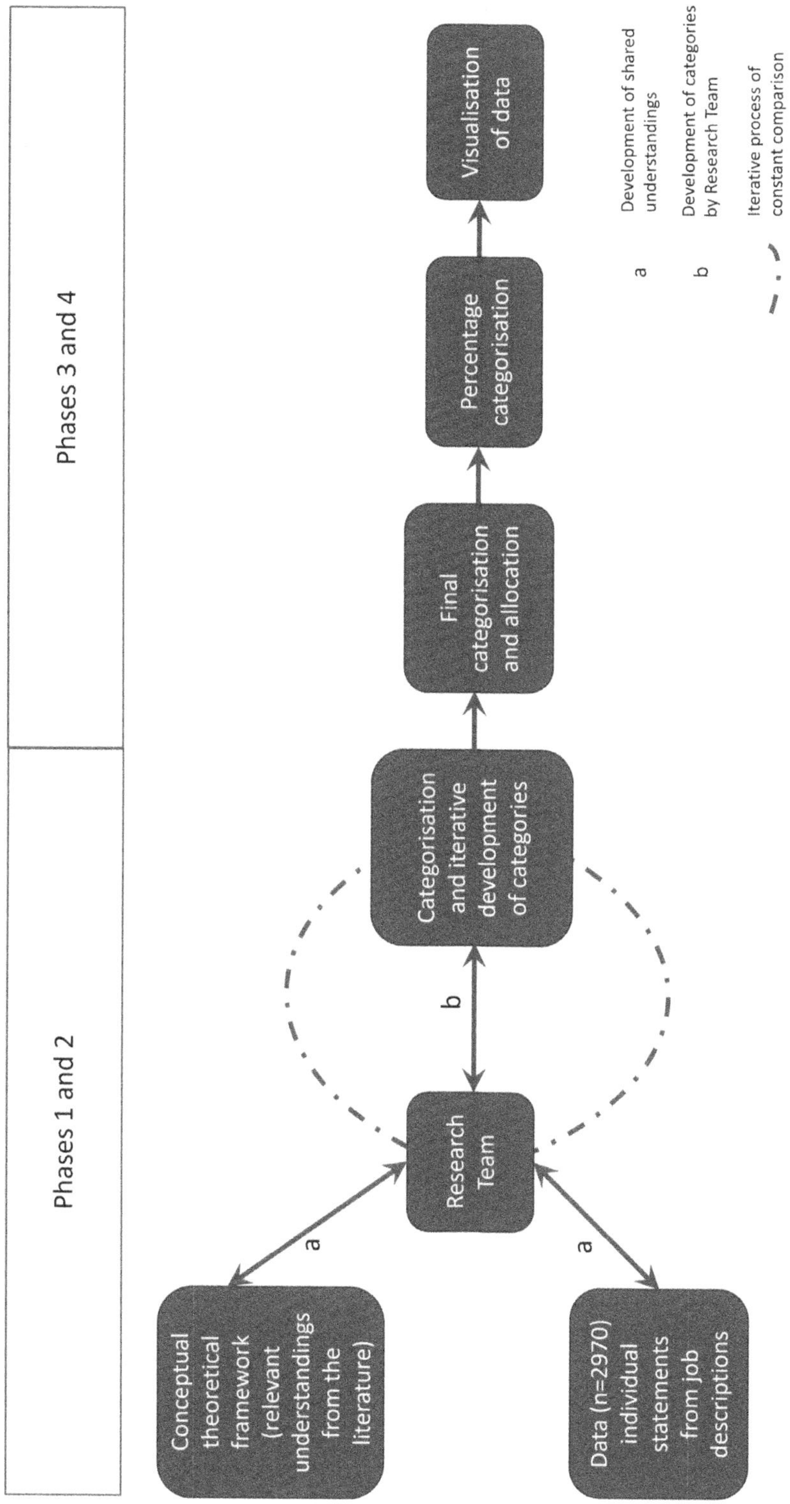

Table 4.1: Definition of categories 1 through 5 with example descriptors

Category	Definition	Role descriptors exemplifying category
1	Competencies and task descriptors grounded in organisational and managerial understandings of professionality and lone 'hero' headship. These are often expressed as being embedded within hierarchical and vertical management structures. Within this conception of organisational professionality high levels of external accountability success is judged by meeting external performance metrics and articulating pupil learning and school development solely against public-facing measures of success and 'mandated outcomes'.	- Manages the school's financial and human resources effectively and efficiently to achieve the school's education priorities and goals. - Challenges underperformance at all levels and ensures effective corrective action and follow-up. - Ensures effective planning, allocation, support, and evaluation of work undertaken by teams and individuals, providing clear delegation of tasks and devolution of responsibilities.
2	As in Category 1, competencies and task descriptors are primarily grounded in organisational and managerial understandings of professionality. However, the articulation of these may involve or make some mention of working collaboratively with others or involvement in shared enterprise. Typically, such shared enterprise and collaboration are deemed to be initiated and controlled by the head teacher.	- Works closely with the governing body and others to ensure that the school vision and strategic plan are effective and pertinent to the school's goals. - Implements successful performance management processes with all staff. - Ensures the vision for the school is clearly articulated, shared, understood, and acted upon effectively by all.
3	Competencies and task descriptors contain expressions of professional practice that are difficult to define, have a mixed or polyphonic (Bahktin, 1981) vocabulary, or hold a negotiated position: often drawn on both organisational and occupational accounts of professional practice.	- To teach regularly as appropriate. - Maintain a high visibility around the school and devote time to interact with learners, staff and parents, and the wider community. - Create an ethos in which all staff recognise their accountability. - Has regard for own workload and that of others to secure satisfactory work/life balance.
4	As in Category 5, competencies and task descriptors are primarily grounded in occupational understandings of professional practice. This encompasses aspects of educative leadership and mentorship – recognising leadership that promotes collegiality, professional autonomy, and contextual awareness. However, these were not as fully expressed as in the references that were coded 5.	- Acknowledges and celebrates the responsibilities and achievements of individuals and teams. - Create an ethos within which all staff are motivated and supported to develop their own skills and subject knowledge and to support each other. - To provide opportunities for dialogue between parents and staff and to encourage their involvement.
5	Competencies and task descriptors are primarily grounded in occupational understandings of professional practice. This encompasses aspects of educative leadership and mentorship – recognising leadership that promotes collegiality, professional autonomy, and contextual awareness.	- To motivate staff and pupils through interest, encouragement, and recognition of their unique value. - Create a culture and ethos where each child is challenged, inspired, and supported to achieve success and become engaged in their own learning. - Builds a collaborative learning culture within the school and actively engages with other educational establishments to build effective learning communities.

Source: Milton et al (2023, p 346).

instantly, and to archive analysis at different points in time. Working to undertake the analysis in this way developed and enhanced what could be known about this data set as a whole, beyond what might have been undertaken by critical discourse analysis alone. The research team were cognisant throughout that the use of NVivo 12 would later afford the calculation of percentages of statements to categories. But the team were also mindful of not using the software as an analytical quick fix, or in a taken-for-granted manner that could potentially lead to an analysis which lacked depth and critical engagement (Coffey et al, 1996).

In terms of the coding process, the individual job descriptions were imported as unformatted text files into NVivo 12, with each created as a file with associated properties such as consortia region, school sector, school language medium, and school denomination. Each of the five categories were created as nodes, and these were used to code all of the text in each file (job description), so that one hundred percent of each file in the data set was coded as one of the five categories. Coding stripes were used to highlight and visualise the coding decisions made in real time, and to check that the entirety of the file had been allocated to one of the categories. Any statement could only be allocated to a single category. Difficult-to-code statements, which were variously indeterminate and/or polyphonic, were allocated to category three. NVivo 12 allows the user to see all the statements coded to each category via the view node tool (as each node represented a category). This enabled scrutiny across the data, and then allowed for the checking and rechecking of accuracy and consistency in allocating statements to categories. Text search proved a useful tool for identifying statements that were similar or the same. The visualisation tool in NVivo 12 allowed the researchers to obtain a clear understanding of the percentage coverage of the job descriptions allocated to each category at this stage of the analysis. The systematic and consistent nature of this process enabled anomalies to be identified and discussed by the team and for further decisions to be made to ensure that the coding approach was being applied consistently and rigorously, throughout this phase and the next.

Phase three: finalising the allocation of job description statements to categories (data cleansing)

During this phase there were further iterations of allocation to categories which took place before all statements were agreed as being finally allocated. This was because it was essential that the entire data set was coded subject to the finalised categories, which only emerged at the end of phase two. This involved a systematic and iterative verification and reverification of the application of the categories. This final check and verification process was undertaken by at least three researchers to eliminate anomalies. This laborious and time-consuming process of data cleansing ensured a high degree of rigour and reliability in allocating statements to categories. The extent to which NVivo 12 automated the calculation of percentage coverage and allowed for the iterative revisiting of the way individual statements had been categorised, served to deepen and extend the level of critical engagement and the rigour of the collaborative analytical process.

Where duplicate, or near duplicate, job descriptions were identified, then a further and final check was undertaken by close examination of the allocation of statements to ensure absolute consistency in coding allocation (Milton et al, 2023). At this point, NVivo 12's 'text search' function provided a particularly useful tool for identifying these, as did viewing via allocation to category (node) to identify similarities in the text used. Once the team had finalised the reliable allocation of the statements to one of each of the five categories, then the percentage coverage of each job description as allocated to each of the five categorisations was viewed graphically using the visualisation tool. This allowed the researchers to obtain a clear understanding of the percentage coverage of the job descriptions allocated to each category.

Phase four: visualisation of the data according to a range of characteristics

Finally, phase four then involved visualising the data graphically, according to a range of case properties (region, school sector, school language medium, and school denomination) in order for the research team to consider if these characteristics seemed to have any relationship with the nature of the articulation of the role in the job descriptions. Figures 4.2 and 4.3 are examples of the type of graphs produced. Figure 4.2 illustrates the percentage coverage of categories 1–5 by geographical region and Figure 4.3 provides percentage coverage of grouped categories by geographical region.

The findings of this analysis highlighted that these job descriptions were highly replicated and contained a dominance of statements that articulated a highly organisationally oriented account of professional practice (Evetts, 2009, 2010). In the paper which reported on this analysis, we argued that this disproportionate emphasis has significance for the practice of headteacher recruitment in Wales

Figure 4.2: Percentage coverage of categories 1–5 in all geographical regions

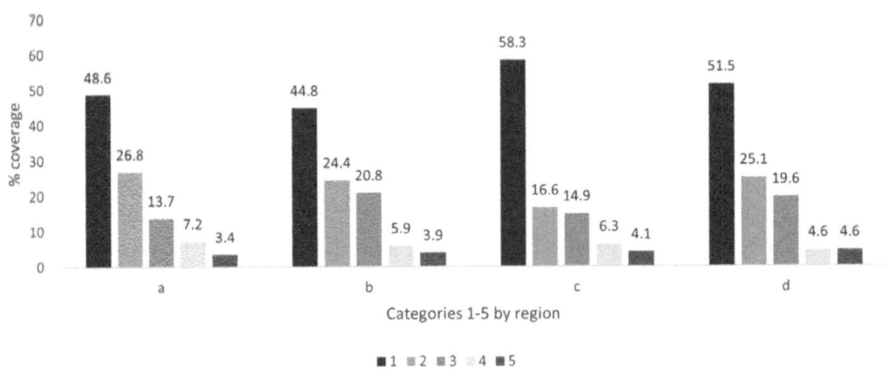

Figure 4.3: Percentage coverage by grouped categories 1 and 2 and 4 and 5 against the region

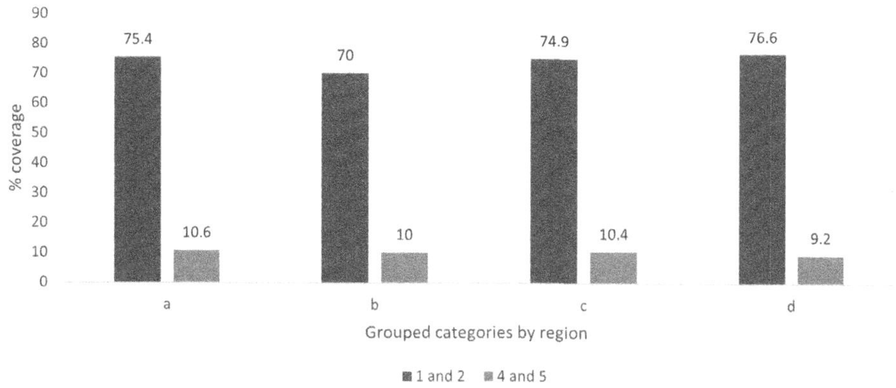

(Milton et al, 2023). If, as this analysis suggests, the articulations of headship in job descriptions at this time were dominated by highly managerial, techno-rational, and organisational ideas, these present a disconnect with the more occupationally oriented ideas of professional practice articulated by policy makers in Wales (Milton et al, 2023). It raises questions about how those responsible need to align, adjust, and contextualise job descriptions with the prevailing policy direction and in support of the ambitious change agenda in Wales (Welsh Government, 2017).

Conclusion

Applying this approach

This epistemologically interpretative, but pragmatically mixed-method, analytical approach is likely to be applicable in a wide range of contexts beyond the illustrative example of headteacher job descriptions. There is scope to utilise this four-phase process to apply a level of scrutiny to a wide range of text-based documents in order to more clearly understand the messages conveyed and the implications of these for policy and practice. Pragmatic, exploratory, and iterative in nature, this mixed-methods approach proved valuable in illuminating and revealing the framing of the headteacher role in job descriptions. It also facilitated exploring both the implicit and explicit meanings embedded in each statement within these. Discourse analysis and content analysis are, epistemologically, conventionally viewed as quite distinct. Our pragmatic stance allowed us to move across these conventionally understood boundaries in order to combine and draw on the strengths of both these well-established analytical methods. Doing this enabled us to develop meaningful understandings of the messages conveyed within job descriptions and to analyse the countable features of these texts.

This approach to data analysis is likely to be of interest to researchers across a range of disciplines who want to develop meaningful understandings of the messages being conveyed in large numbers of documents that may be inconsistent in length but which have a degree of similarity regarding purpose, for example, policy documents, job descriptions, and newspaper articles. Essentially, the approach taken to analysis in this study could be applied to any large data sets of documents consisting of primarily qualitative information.

A collaborative approach

The approach is ideally suited to a research team committed to working together in a way that allows for dialogue, debate, and 'interthinking' (Mercer, 2000, p 141). This is required to develop the pertinent and robust conceptual framework this approach demands. That the development of the shared understandings needed to arrive at a conceptual framework are based both on the data set itself and informed by the literature means that the research team need to be committed to working together in a simultaneously collaborative and creative way. The nature of this work means it cannot successfully be done in a sequential way in the absence of this shared dialogue, because the knowledge that emerges and develops results from the intellectual property of all the members of the research team, and it needs to be shaped and refined in real time.

It is unlikely that this method of analysis might be successfully adopted by an individual researcher. As a minimum this approach requires two researchers dedicated to working on a single data set. Those interested in successfully adopting this approach should consider the following key points:

- Working in this way means adopting an interpretive epistemological stance – whilst quantitative methods are used, this approach does not 'prove' anything.
- This approach requires a significant time commitment for engaging in extended dialogic discussions to help to introduce a duty of doubt in terms of developing the conceptual framework. This is an essential feature of the process and is crucial to the rigour of the work as the team works together to make sense of and align a range of views and interpretations.
- Researchers should be aware of the need to develop confidence and competence with whatever computerised qualitative data analysis software package they utilise to undertake this work. Without a package such as NVivo 12 (as used in this study) it would have proved impossible in practical terms to have iteratively coded and checked statements in order to calculate percentage coverage of each category for each job description. This would have taken so long that it would have been prohibitive.
- At the outset and prior to adopting this approach, researchers should plan their approach with an awareness of the time that will be taken by developing a robust framework, iteratively refining this, coding the documents' content, undertaking a data cleanse, and calculating the final percentage coverage according to

category and particular features of the data set. The time taken will, of course, depend upon how developed ideas are regarding theoretical understandings and the conceptual framework at the start of the process; for example, how large the documents are, how large the entire data set is, and how competent the researcher is with the qualitative data analysis software.

Affordances and constraints

The key benefit or affordance of the approach for large data sets is clear – it allows for a *zooming out* from close discourse analysis and a summative approach to mapping discourse frequency and distribution in a systematic fashion. By working in this way, drawing on both approaches we can understand more about the discourses being communicated and the nature of the variation in these discourses according to features of the data set in any large set of documents of a specified shared type or purpose.

The iterative nature of this approach allows for the incremental development and refinement of the conceptual framework as the data analysis gathers momentum, and avoids confounding cases, forcing a post hoc review and adaptation of the conceptual framework or literature review. Working in this way can guard against the temptation to retrospectively introduce novel theories or ideas to the framework or literature review. This allows for an initially flexible and consequently stronger conceptual base, as well as clear shared understandings among the research team.

This is a pragmatic approach that allows the researcher to build and refine their analytical approach (based on a robust conceptual framework) to respond to the needs of the data and overarching research questions. It is therefore responsive to the data that the researcher *has* and helps to avoid a situation where the data gathered does not align with the conceptual tools at the disposal of the researcher. As such, its flexibility is in keeping with the well-established notion of emergent design (Pailthorpe, 2017).

Engaging with this methodological approach is likely to be an educative experience for researchers. This is because of the open and honest dialogue, which privileges grappling with ideas and is inherent to the process.

There are inevitably constraints and limitations to this way of working and the approach outlined. These include the need for the data set to have a systematic collection rationale, with underpinning inclusion and exclusion criteria. The data set therefore needs to be a large data set comprising of documents with a degree of similarity regarding purpose. Additionally, care needs to be taken to articulate exactly why documents have been selected or excluded and to avoid strategic inclusion.

The biggest risk is that this approach could be misunderstood and misinterpreted by researchers keen to identify statistical predictive validity, generalisability, and reliability. These are not (and were never) the intention of the research or the method – an interpretivist epistemology was maintained throughout, even though quantitative analysis was used in this process.

In conclusion, this creative approach enables teams to answer previously unanswerable research questions focused on the discourses communicated in documents that are drawn from large numbers of comparable but different texts intended for a similar purpose. In this way it offers an important vehicle to develop meaningful understandings to bring about change within the contexts to which their focus is orientated.

Notes

1 The Welsh Language Measure (2011) and the Welsh Language Act (1993) mean that Welsh language has an official status equal to English language in Wales. Therefore, in the context of this project all publicly available job descriptions (n = 67) were analysed in the language in which they were written.

2 QSR International Pty Ltd (2018) NVivo (version 12 pro). [Software].

References

Alexander, R.J. (2020) *A Dialogic Teaching Companion*, London: Routledge.

Bakhtin, M. (1981) *The Dialogic Imagination: Four Essays*, trans M. Holquist and C. Emerson, Austin: University of Texas Press.

Bryman, A. (1984) 'The debate about quantitative and qualitative research: A question of method or epistemology?', *British Journal of Sociology*, 35:78–92.

Campbell, R., Pound, P., Morgan, M., Daker-White, G., Britten, N., Pill, R., et al (2011) 'Evaluating meta-ethnography: systematic analysis and synthesis of qualitative research', *Health Technology Assessment*, 15(43). Available from: https://www.journalslibrary.nihr.ac.uk/hta/hta15430/#/abstract

Coffey, A., Holbrook, B., and Atkinson, P.A. (1996) 'Qualitative data analysis: technologies and representations', *Sociological Research Online*, 1(1): 80–91.

Connolly, M., Milton, E., Davies, A.J., and Barrance, R. (2018) 'Turning heads: the impact of political reform on the professional role, identity and recruitment of head teachers in Wales', *British Educational Research Journal*, 44(4): 608–25.

Creswell, J.W. and Plano Clark, V.L. (2011) 'Choosing a mixed methods design', *Designing and Conducting Mixed Methods Research*, 2: 53–106.

Davies, A.J., Milton, E., Connolly, M., and Barrance, R. (2018) 'Headteacher recruitment, retention and professional development in Wales: challenges and opportunities', *Wales Journal of Education*, 20(2): 204–24.

Evetts, J. (2009) 'The management of professionalism: a contemporary paradox', in S. Gewirtz, P. Mahoney, I. Hextall, and A. Cribb (eds), *Changing Teacher Professionalism: International Trends, Challenges and Ways Forward*, London: Routledge, pp 19–30.

Evetts, J. (2010) 'Organizational professionalism: changes, challenges and opportunities', paper presented at the DPU Conference: Organizational Learning and Beyond, Copenhagen, 20 October.

Fairclough, N. (2015) *Language and Power* (3rd edn), London: Routledge.

Fullan, M., Rincon-Gallardo, S., and Hargreaves, A. (2015) 'Professional capital as accountability', *Education Policy Analysis Archives*, 23(15): 1–17. Available from: http://dx.doi.org/10.14507/epaa.v23.1998

Glaser, B. (1965) 'The constant comparative method of qualitative analysis', *Social Problems*, 12(4): 436–45. Available from: https://doi.org/10.2307/798843

Hammersley, M. (1996) 'The relationship between qualitative and quantitative research: paradigm loyalty versus methodological eclecticism', in J.T.E. Richardson (ed), *Handbook of Qualitative Research Methods for Psychology and the Social Sciences*, Leicester: BPS Books, pp 159–74.

Hardy, C., Harley, B., and Phillips, N. (2004) 'Discourse analysis and content analysis: two solitudes?', *Qualitative and Multi-Method Research*, 2(1): 19–22. Available from: https://doi.org/10.5281/zenodo.998649

Hesse-Biber, S. (2010) 'Feminist approaches to mixed methods research: linking theory and praxis', in C. Teddlie and A. Tashakkori (eds)' *SAGE Handbook of Mixed Methods in Social and Behavioral Research*, Thousand Oaks: SAGE Publications, pp 1–42.

Hoyle, E. and Wallace, M. (2005) *Educational Leadership: Ambiguity, Professionals and Managerialism*, London: SAGE Publications.

Hsieh, H.F. and Shannon, S.E. (2005) 'Three approaches to qualitative content analysis', *Qualitative Health Research*, 15(9): 1277–88.

Johnson, R.B. and Onwuegbuzie, A.J. (2004) 'Mixed methods research: a research paradigm whose time has come', *Educational Researcher*, 33(7): 14–26.

Mercer, N. (2000). *Words and Minds*, London: Routledge.

Milton, E., Morgan, A., Davies, A.J., Connolly, M., Donnelly, D., and Ellis, I. (2023) 'Framing headship: a demand-side analysis of how the headteacher role is articulated in job descriptions', *International Journal of Leadership in Education*, 26(2): 339–58. Available from: https://doi.org/10.1080/13603124.2020.1811898

OECD (Organisation for Economic Co-operation and Development) (2017) *The Welsh Education Reform Journey: A Rapid Policy Assessment*, Paris: OECD.

Pailthorpe, B.C. (2017) 'Emergent design', in C. Davis, J. Mattes, and R. Potter (eds), *The International Encyclopaedia of Communication Research Methods*, Hoboken: Wiley-Blackwell, pp 1–2.

Pole, C. and Lampard, R. (2002) *Practical Social Investigation: Qualitative and Quantitative Methods in Social Research*, Harlow: Pearson.

Punch, K. (1998) *Introduction to Social Research: Quantitative and Qualitative Approaches*, London: SAGE Publications.

Punch, K. and Oancea, A. (2014) *Introduction to Research Methods in Education* (2nd edn), London: SAGE Publications.

Tavory, I. and Timmermans, S. (2014) *Abductive Analysis: Theorizing Qualitative Research*, Chicago: University of Chicago Press.

Teddlie, C. and Tashakkori, A., (2010) 'Overview of contemporary issues in mixed methods research', *Sage Handbook of Mixed Methods in Social and Behavioral Research*, Thousand Oaks, SAGE, pp 1–44.

Welsh Government (2017) *Education in Wales: Our National Mission. Action Plan 2017–21*, Cardiff: Welsh Government.

5

Word clouds as creative data analysis – what can they tell us about student views of learning something new?

Louise Gascoine, Kate Wall, and Steve Higgins

Introduction

This chapter focuses on the application of word clouds as a method of creative data analysis. Word clouds are a method of visualising textual data, where the frequency of words in the analysed text is related to their size within the visualisation or word cloud (McNaught and Lam, 2010). It is important to note that this is a collaboratively written chapter, all three authors have a practitioner/teacher background as well as research experience. However, neither our perspective nor position is singular. We would describe ourselves as relatively pragmatic, fluid, and creative in relation to research design and methods to address different research questions appropriately. Our positionality in relation to the creative data analysis using word clouds in this chapter is a shared one and was reached through collaborative discussion of the data. The holistic nature of pragmatic approaches, which acknowledge and engage with complexity and practicability (Clarke and Visser, 2019), are cognisant with the complexity of the research focus (metacognition) and data in the research that will be described in this chapter. Figure 5.1 is an example of a word cloud that provides a visual summary of this chapter by illustrating the frequency of the words used, acting as a visual abstract.

We see word clouds as an inherently mixed method due to the statistical algorithm used to generate the words in the cloud, which are then presented visually and can be subjectively interpreted. Aspects of the presentation of the 'cloud' are configurable, such as the relative size and direction of the words and the minimum frequency for inclusion. Word clouds have been described as increasing in popularity since the early 2000s (Viegas et al, 2009; Henderson and Segal, 2013), although research literature about their utility is relatively sparse. In terms of how they have been used, word clouds have been described or used as a 'supplementary' method of analysis (Dietz, 2016), as a means of engagement with content (Viegas et al, 2009), and, pertinent to the example in this chapter, as a means of visualising qualitative data (Henderson and Segal, 2013). The purpose of this chapter is to describe and explain how we used word clouds as a means of creative data analysis.

Figure 5.1: Word cloud summary: a visual abstract of this chapter

The word clouds as creative data analysis example in this chapter comes from doctoral research (Gascoine, 2016) that focused on exploring developmental trends in metacognition using data collected with a visual research tool – pupil views templates (PVTs) (Wall and Higgins, 2006).[1] As Figure 5.2 illustrates, PVTs are a visual tool developed in the mid-2000s, they are both a research and pedagogical tool that can be used by both teachers and researchers to facilitate discourse with students about their learning (Wall et al, 2012). PVTs are creative because they tell stories, both visually, if students choose to engage with or annotate the image on the PVT, and textually, within the speech and thought bubbles. The doctoral research in this example built on existing research that had used PVTs and found evidence of developmental trends in the textual data recorded on them (Wall, 2008; Wall et al, 2012). A total of 374 PVTs were analysed from students aged 4–16 years across three schools (one secondary school and two primary schools); they were completed in focus groups of between four and six students with the first author. Over 10,000 words in total were transcribed from these individual PVTs (Table 5.1). In line with the existing deductive approach to coding PVTs for evidence of metacognition (see Wall, 2008; Wall et al, 2012), they were coded using Moseley et al's (2005, p 315) 'integrated model for understanding thinking of learning' and Veenman et al's (2005) definitions of metacognitive knowledge and skilfulness. This allowed a statistical analysis of the codes across the age groups, exploring developmental trends using a series of one-way analyses of variance (ANOVA) with post hoc testing to explore relationships between age groups and the individual dependent variables from the deductive coding.

Word clouds were introduced as part of the analysis process to facilitate visualisation of the PVT data. They allowed us to make language and contextual-focused comparisons across participant age groups. Essentially, word clouds enabled us to retrieve information about the stories that the data told in

Figure 5.2: PVT used for data collection

Note: PVT completed by a 7-year-old primary school student.

Source: Adapted from Wall et al (2007, p 22).

Table 5.1: Age range of participants and words transcribed from PVTs

School year group	Age (years)	Total number of participants (PVTs completed)	Total number of words transcribed
Reception	4–5	32	411
Year 1	5–6	32	427
Year 2	6–7	32	507
Year 3	7–8	32	577
Year 4	8–9	32	857
Year 5	9–10	32	937
Year 6	10–11	32	910
Year 7	11–12	34	1,249
Year 8	12–13	32	1,090
Year 9	13–14	30	1,076
Year 10	14–15	31	1,245
Year 11	15–16	23	1,100
–	–	**374**	**10,386**

relation to the frequency and use of different words that students had included in their PVTs in relation to the prompt of 'think about a lesson when you learnt something new …'. Word clouds complemented the statistical analysis and allowed us to make links between the quantitative results (see Gascoine, 2016) and the textual data and stories within the PVTs. We used what Feilzer (2010, p 12) described as a pragmatic approach to mixed-methods research: 'moving back and forth between the data sets with the knowledge produced by each one'. The inherently creative and visual nature of word clouds embraced this back and forth in the analysis of the PVT data. Accordingly, the deductive and word cloud analysis maintained and stayed true to the visual integrity of the PVT template itself in the analysis.

To contextualise the creative approach to data analysis using word clouds, it is important to situate this research by reflecting on our understanding of the core concept – metacognition. Metacognition is a complex phenomenon that has been widely researched and has been defined in different ways since Flavell's (1976) conceptualisation of it as 'one's own knowledge concerning one's own cognitive processes and products or anything related to them' (p 232). It was described by Wellman (1983), less than a decade after Flavell, as fuzzy, and it continues to be described, defined, and applied in a range of different ways within education and other disciplines. This wide-ranging description, definition, and use of the term metacognition means that it can lack clarity (see Hofer, 2004; Schunk, 2008), further supporting, firstly, the need to be clear about exactly what we were exploring within the research and, secondly, the need for creative approaches to data analysis to ensure we included this 'fuzziness'. The decision to use PVTs to gather data in this research was supported by the results of a systematic literature review (Gascoine et al, 2017), which indicated that PVTs were one of very few tools suitable to collect data about metacognition across a broad age range such as 4–16 years. We collected and analysed the data in this research with explicit consideration of the complexity of metacognitive development and sought to be non-reductive in presenting and synthesising analysis of the stories of the data collected using PVTs. This approach is cognisant with both the creativity inherent in PVTs and the complexity of metacognition.

Complexity and creativity in data analysis using word clouds

We did not use word clouds as a method of analysis to provide simple frequency statistics or to engage only at surface level with the data visualisations that they generated. Rather, we used them to provide a structural and strategic means of eliciting and comparing the stories of the data collected using the PVTs. The research described in this chapter adopted a mixed-method research approach to pragmatically explore a complex concept (metacognition) within a complex context (education). The approach to data analysis using word clouds described in this chapter considered four key stages:

1. *The data was collected using PVTs – an inherently qualitative and creative research tool that produces complex data.* In the analysis process there was a need to avoid transforming this in-depth qualitative PVT data reductively to something that was solely quantitative. Creative data analysis using word clouds stayed true to the creativity inherent in the visual of PVTs. Word clouds translate frequencies to visual clusters by displaying word frequencies as relative size relations. This helped to emphasise the relational nature of the words and their meanings.
2. *The research focused on student views of their own learning.* There are power relations to consider when researching with children and young people, such as the adult-centrism prevalent in research (McLaughlin, 2015). The PVT as data collection method goes some way to mediating this by seeking to elicit students' perspectives and it was important to continue to consider these issues in the data analysis. Word clouds facilitated a simultaneous distance from and closeness to the data and its stories.
3. *The complexity (of concept and data) and value of a nuanced approach to data analysis.* Word clouds provided a valuable statistical mnemonic, both structural and strategic, that allowed us to compare and explore the stories of the data from the different age groups of participants (that is, different school year groups and key stages). We were concerned with the importance of the data and its stories, and keen to avoid reductionism – word cloud analysis enabled the relations between different aspects of the stories told by the data to be maintained, foregrounding the whole rather than separating into discrete units.
4. *Word clouds as allowing visual comparisons.* Word clouds provided a way of visually stratifying the transcribed PVT data for comparison; in this sense the word clouds had a clear purpose and link to the comparisons in the statistical analysis. For example, statistical analysis looked at the different deductive codes across age groups, as did word clouds, offering two ways of looking at and working with the same data.

Teddlie and Sammons' (2010) description of mixed-methods research resonates with the rationale for the approach to data analysis adopted in this research, focusing on the 'synergy' (p 116) between numbers and story. Word clouds of the PVT data enabled an analysis that appreciated the presence of both numbers and stories, and the synergy between them. Our approach to both data collection and analysis was grounded in our belief that the complexity of metacognition necessitated a creative approach to data analysis and synthesis. It was an iterative approach that acknowledged the unavoidable complexity and, we hope, stayed true to the individual and collective stories that the participants provided. The research process and analysis are illustrated in Figure 5.3. The linearity of this diagram represents the time from start to finish of the research rather than linearity in the analysis process itself. The analysis process, as represented by the double-ended and dashed arrows, was iterative and there was interplay between the statistical analysis and word cloud analysis. This allowed us to hold central the stories of the data as whole parts as well as individual deductive codes that were statistically analysed and compared.

Transcribed textual data from PVTs was inputted into the analysis software NVivo 10.[2] This software was used for both the deductive coding and the word cloud analysis, and this was important in enabling the synergy between analyses. We note that word clouds can be created using a variety of different computer-assisted qualitative data analysis software (CAQDAS) packages and that there are several free online word cloud generators. An initial reading of the data for the purpose of data familiarisation preceded the analysis. As illustrated in Figure 5.3, clouds were generated both with the aim of unpicking and exploring the codes from deductive coding as well as the different ages and stages represented in the completed PVTs. For example, PVT data was coded as 'strategic and reflective thinking' (Moseley et al, 2005, p 312) and then this code was reanalysed for evidence of metacognitive knowledge and skilfulness (Veenman et al, 2005). The word clouds were generated to represent each of these individual codes, allowing comparison of the stories they contained. Word clouds for different age groups (see Table 5.1) and for the different deductive codes were generated in NVivo with consistent parameters, namely the same font sizes, layout, colours, and a minimum word length of three letters. We adopted a creative approach to data analysis, but it was necessary for limits to be set in place so we could make appropriate comparisons in a consistent way.

Statistical analysis was an important means by which to understand if there was evidence of developmental trends in metacognition, as had been explored in previous PVT research (see Wall et al, 2012). However, without simultaneous word cloud analysis there was potential for reductivity in the analysis – discounting the stories of the PVT data and the roots of PVTs in pupil voice within research and practice (Cremin et al, 2011). The word cloud analysis helped us to unpick and exemplify the stories within the data. It would have been easy, in the analysis of the deductively coded data, to get lost in the numbers and statistical analysis. Triangulating the statistical analysis with the word cloud analysis served to bring the data to life and as a visual reminder that the PVTs were completed by students in schools.

Word clouds and foregrounding stories within the data

When looking at the PVTs there was often a clear link to specific lessons or topics that students were reflecting on in relation to learning something new. This level of detail and story in the data could easily have been missed in the analysis if it was not for the visual reminder provided by the word clouds. For example, reflecting on the content of the lessons that the reception year group students (aged 4–5 years old) included on their PVTs – sharing and remembering within the context of talking with each other and telling the researcher about some of the content (life cycles). We can see in Figure 5.4 that some of the youngest student

Figure 5.3: Process of research: from data collection to analysis

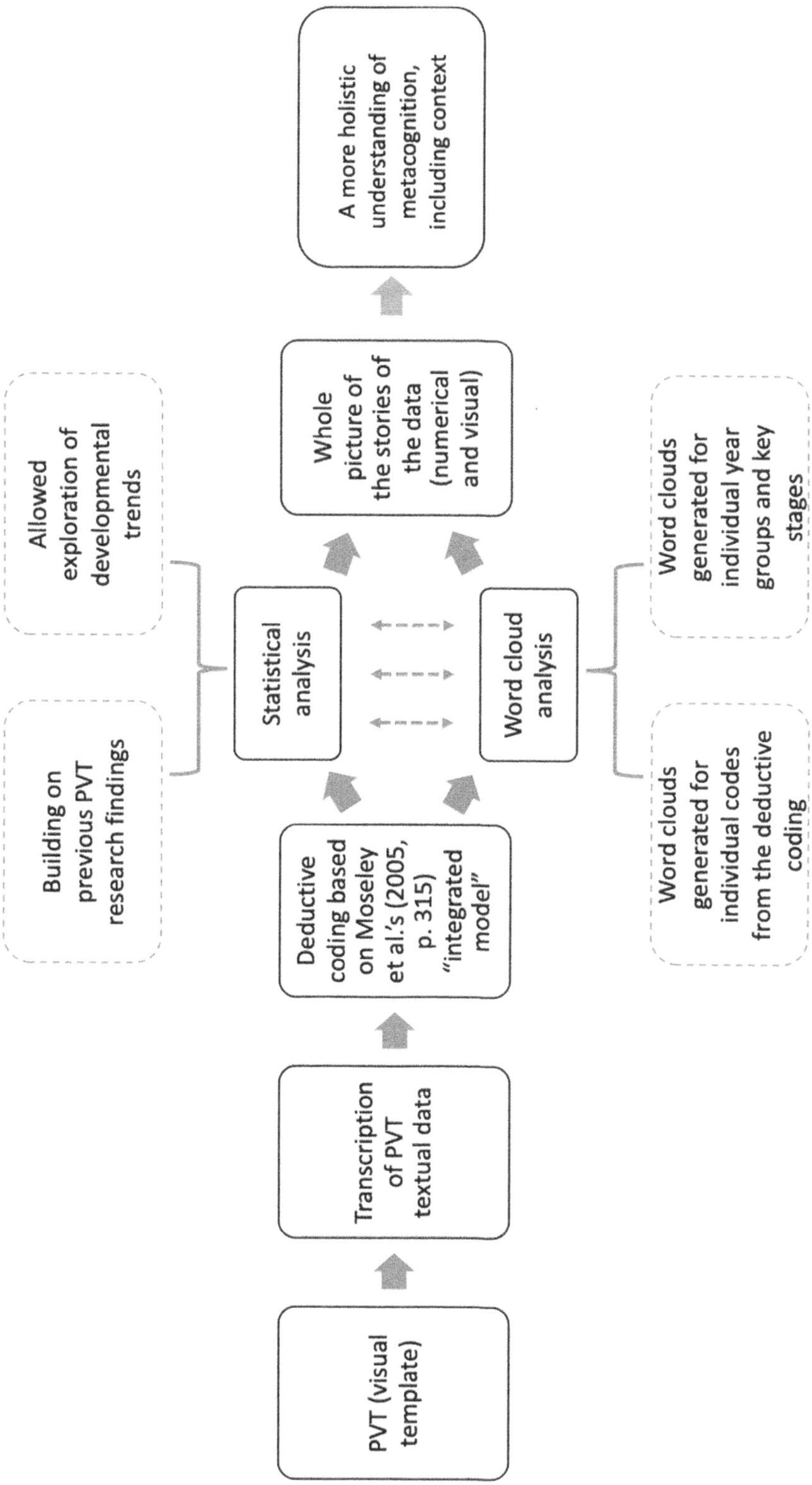

participants recorded information about different stages of chicks hatching – the word chicks is prominent in the centre of the word cloud and also to the left of this, and, less prominent, we can make connections to the words 'yolks' and 'egg'. To make a comparison to Figure 5.4, Figure 5.5 shows a word cloud created from PVT textual data of the students in year 8 (aged 12–13 years). The older student's PVT data tells a different story that focuses less on content and more on describing learning experiences, for example, 'it being hard', and processes such as 'understand' and 'help'. Therefore, we used word cloud analysis to understand the contexts in which there were, or were not, evidence of cognitive skills and metacognition in the PVT data. Word clouds allowed us to oscillate between a consideration of the data both in relation to the statistical analysis and separately from it – simultaneous close engagement, overview, and contextualisation of the PVT data in a visual way; therefore, making a loop back to the visual way in which the data was collected using PVTs.

The statistical analysis of the deductively coded data showed evidence of some developmental trends in cognitive skills and metacognition. While this was an important finding, it did not tell us what these trends looked like in the data. For example, in the statistical analysis we were able to refute the null hypothesis that 'the distribution of productive thinking is the same across the category of age group' (Gascoine, 2016, p 205). Productive thinking was defined as 'planning what to do and say, imagining situations, reasoning, solving problems, making decisions …' (Moseley et al, 2005, p 313). We knew that there was evidence of an age-related trend in the PVT data for productive thinking, but we wanted to explore what this looked like, what stories the data told us about productive thinking. Looking at word clouds of individual codes, like productive thinking, allowed us to begin to illustrate the stories of the data. Word clouds generated for the data of all age groups included in the statistical analysis for productive

Figure 5.4: Word cloud for PVT data from students aged 4–5 years (n = 32)

Figure 5.5: Word cloud for PVT data from students aged 12–13 years (n = 32)

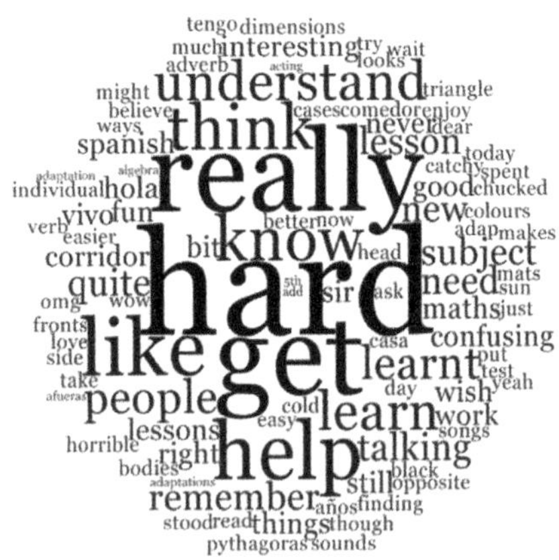

thinking contained the word 'remember'. The examples in Figure 5.6 comprise PVT data coded as productive thinking from students in the four age groups in the statistical analysis: Reception and Key Stage 1 (age 4–7 years, n = 96), Key Stage 2 (age 7–11 years, n = 128), Key Stage 3 (age 11–14 years, n = 96), and Key Stage 4 (age 14–16 years, n = 54).

The valuable visual oversight provided by word clouds of what productive thinking looked like for each age group enabled a consideration of the contexts in which the word 'remember' was used across the data set. For example, a 4-year-old student wrote: 'thinking about it helps me to remember to spell my name'. Returning to the PVT completed by this student we can see that they are referring to spelling their name out using the individual letters that are visible

Figure 5.6: Word cloud of text from PVTs coded as productive thinking for students aged 4–7 years (top left), 7–11 years (top right), 11–14 years (bottom left), and 14–16 years (bottom right)

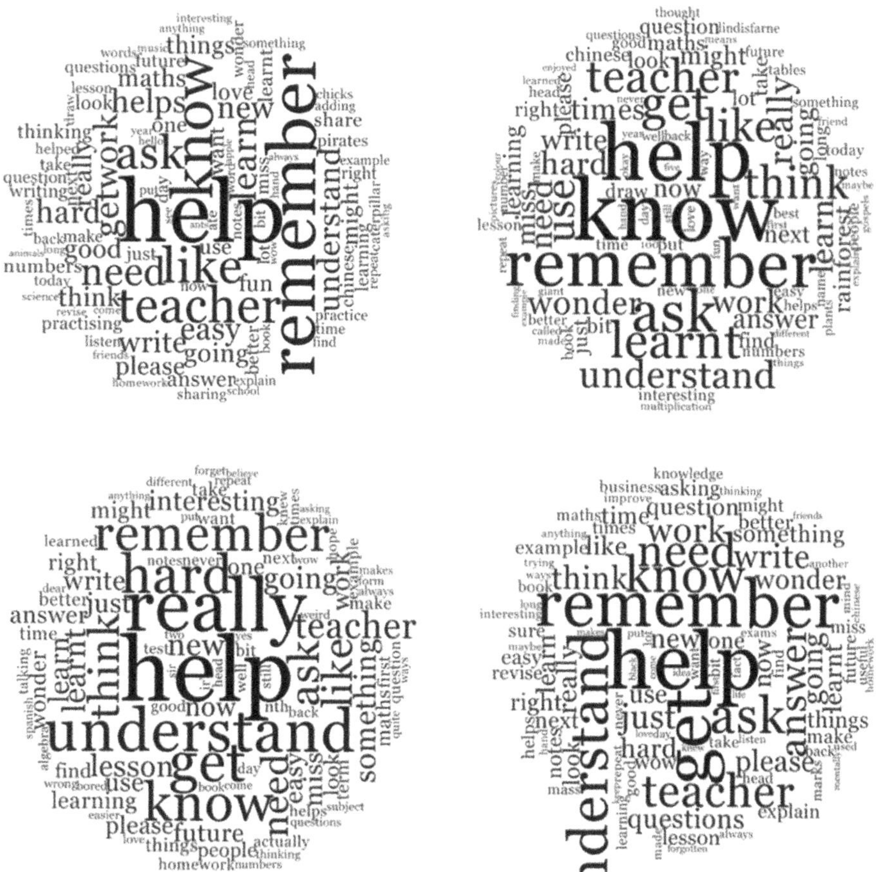

on their PVT. Another student, who was aged 7, wrote: 'Practising the word helps me remember it.' We noticed development here from the quite non-specific language of 'thinking' to the mention of a specific strategy (practising), which the student describes as aiding remembering. In addition, word clouds illustrated the context in which the word 'remember' was used. This developed further for students aged 10–11 years to encompass multiple different and specific strategies, including drawing things, writing things down, and using number patterns. For example, 'Oh I remember I can use the pattern of numbers'; here we can see reflection on past learning in terms of remembering a strategy and applying it to a situation that is occurring at a different point in time (as illustrated within that student's PVT reflections).

Word clouds facilitated a strategic unpicking and digging deeper into the textual data, where identified developmental trends could be linked to specific words via word cloud analysis, which then allowed us to consider the context and stories of specific words like 'remember' within the PVT data. The frequency of the word 'remember' was distinctive across the different year groups – without the prompt of the word cloud we think we would have been less likely to pick up on this finding. 'Remember' in the 4–5 years age group is the 14th most frequent word; by 6–7 years it is the third most frequent word. PVTs had not previously been used to unpick the specifics of metacognitive vocabulary and 'mental verbs', which can be comprehended as young as age 4 (Peskin and Astington, 2004, p 254), with words like 'remember' suggesting prior knowledge. A solely quantitative approach to the analysis of the qualitative textual data gathered using the PVTs would not have done justice to what was a profoundly creative method of data collection.

The triangulation of data analysis using word clouds alongside statistical analysis offered a simultaneous closeness to and distance from the data. We undertook an iterative process of making sense of the statistical analysis of the deductive coding and what this told us about the presence of developmental trends in metacognition within the analysed data. Alongside this we could maintain, through the word clouds, a broader view of the data. Word clouds as a method of creative data analysis facilitated an increased transparency of analysis – we used them to make complicated data analysis more accessible. Word clouds also provided a means of analysis in themselves (context and frequency), and foregrounded the visualisation and interpretation of specific deductive codes from within the statistical analysis. Where the statistical analysis could identify developmental trends, word cloud analysis could be used to contextualise and interpret these. Accordingly, the word clouds acted as a catalytic tool to aid the transparency of the data analysis, increasing the accessibility of the analysis and presenting complex statistical analysis alongside visual representations of specific aspects of the analysis and the broader stories of the data. Word clouds allowed us to look at the data in different ways and provided a rapid way of generating different visualisations of the data to explore and compare both within and across the PVTs.

Word cloud analysis: some thoughts

It is important to consider the benefits, drawbacks, and challenges of different methods of analysis, and be reflexive about this at all stages of research. The main focus of this chapter has been the use of word clouds as an approach that allows more creative data analysis. However, there are both interesting developments and cautions that should be considered. Henderson and Segal (2013) presented a framework to aid understanding of different types of data visualisation for qualitative textual data. Their framework classifies data visualisation in two dimensions: level of display (word, sentence, narrative) and complexity (simple to complex). In this framework, word clouds can be seen as relatively simple (in terms of complexity of use) and at the level of 'word' in terms of their 'level of display' (word – sentence – theme/narrative). We found that they are easy (and quick) to create but that the consequences in our data analysis and in terms of engagement with our data set showed they were far from simple. We have demonstrated the utility of word clouds as a 'fast and visually rich way to enable researchers to have some basic understanding of the data at hand' (McNaught and Lam, 2010, p 630). Word clouds were also useful in facilitating distance from PVT data, which was particularly advantageous for the first author, who was immersed in the data collection process.

The speed at which word clouds can be generated could be both beneficial and risky; this is where consistency (alongside creativity) becomes important. The word clouds were created within the same parameters to aid consistency of comparison across the data set, and they were visually rich and provided useful summaries of textual data. However, there can be a risk that they could be used to illustrate frequency alone and not include an exploration into the rich stories that make up the textual data from which the word clouds were created. Word cloud analysis was used in this research example to support and validate trends identified in previous analysis, specifically the trends identified from the PVT data through the statistical analysis. Yet, we also knew that it was important not to discount the stories of the data on the PVTs themselves prior to transformation via transcription. This involved picking out individual words to explore the meanings that word clouds highlighted. They also facilitated a valuable link back to the original PVT data. We were able to actively explore and make links between the deductive coding, associated statistical analysis, and the metacognitive vocabulary that students were using in talking about their learning. The depth of consideration of the PVT data at a level of metacognitive vocabulary that word clouds provided enabled us to think about the data in a different way. This was initiated by word frequency (a simple count) but the considerations were not simple; for example, the in-depth exploration of how the word 'remember' was used by different age groups.

It is interesting to consider whether word clouds, as a method of analysis, would have been enough on their own, and to contemplate the value of synergy with the statistical analysis. The synergy certainly provided an opportunity for this research to build on previous research that had explored developmental trends in metacognition using PVTs (Wall et al, 2012). Word clouds help to resist reductive data analysis and can go some way to mitigating the power relations inherent in transforming the PVT data from a visual tool to transcribed text for deductive analysis of the text content. This is mindful of the mediating role PVTs had in the focus groups in which the data was collected. They mediated between the researcher (an adult) and the participants (students within school settings) (Wall and Higgins, 2006). Word clouds facilitated a heightening of the character and meaning of the textual data by foregrounding the meanings that were inherent in the PVTs themselves (visually) prior to transcription. This felt like a neat 'closing of the feedback loop' within the data set, providing a connection to the pragmatic nature of the intent with which the data was collected.

An added potential value of using word clouds is a link to the co-production of research knowledge, as an accessible way of sharing data gathered with participants. For example, drawing on the work of Jenkins (2006), Viegas et al (2009, p 1138) described the utility of word clouds in a 'participatory culture' where in 'a culture of convergence ... consumers are encouraged to seek out new information and make connections among dispersed media content'. This does not speak directly to educational research but there are important parallels if we think about stakeholders and participants instead of 'consumers'. The visual accessibility of word clouds could provide novel pathways in educational research that are less rigid and more 'impactful' (Mannay, 2020, p 3), by making it easier to collaborate and to share research outcomes with stakeholders and participants, for example, practitioners and students of different ages. Accessibility of research outcomes also links to our belief that, in this research, word clouds helped to retain the meaning and character of the data by making it visually accessible in line with the way it was collected. We also think it allowed us to stay true to the value of PVTs as both a pedagogical and research tool and respected the meanings that the students were communicating through their use.

We have aimed to show that word clouds offer a useful analysis lens for qualitative data. They have sufficient structure and consistency in their production to support helpful comparisons and, when used alongside other methods, they can offer a complementary and insightful view. Their inherent creativity and accessibility are likely to make them attractive to the researcher. However, researchers planning to use word clouds to analyse data need to think about the question of intent. What are word clouds adding – is it that they provide visual accessibility for research outcomes and impact to be shared? Do they allow visual comparison that aids stepping into and away from your data to allow you to engage with both the finer detail and the bigger picture? Clear strategic use and appropriate selection of data and word cloud parameters are paramount, as without this word clouds could become decorations rather than useful elements of the research.

Notes

1 The doctoral research (Gascoine, 2016) received ethical approval from the Durham University School of Education Ethics Subcommittee.

2 NVivo (2012) NVivo qualitative data analysis software (version 10), QSR International Pty Ltd.

References

Clarke, E. and Visser, J. (2019) 'Pragmatic research methodology in education: possibilities and pitfalls', *International Journal of Research and Method in Education*, 42(5): 455–69. Available from: https://doi.org/10.1080/1743727X.2018.1524866

Cremin, H., Mason, C., and Busher, H. (2011) 'Problematising pupil voice using visual methods: findings from a study of engaged and disaffected pupils in an urban secondary school', *British Educational Research Journal*, 37(4): 585–603. Available from: https://doi.org/10.1080/01411926.2010.482977

Dietz, S.S. (2016) 'Using Wordle in qualitative research: a supplemental tool for case studies', *Instructional Technology*, 13(8): 23–35.

Feilzer, M.Y. (2010) 'Doing mixed methods research pragmatically: implications for the rediscovery of pragmatism as a research paradigm', *Journal of Mixed Methods Research*, 4(1): 6–16. Available from: https://doi.org/10.1177/1558689809349691

Flavell, J.H. (1976) 'Metacognitive aspects of problem solving', in L.B. Resnick (ed), *The Nature of Intelligence*, Hillsdale: Lawrence Erlbaum Associates, pp 231–5.

Gascoine, L. (2016) 'Investigating developmental trends in metacognitive knowledge with school-aged children using pupil views templates', PhD thesis, Durham University. Available from: http://etheses.dur.ac.uk/11555/

Gascoine, L., Higgins, S., and Wall, K. (2017) 'The assessment of metacognition in children aged 4–16 years: a systematic review', *Review of Education*, 5(1): 3–57. Available from: https://doi.org/10.1002/rev3.3077

Henderson, S. and Segal, E.H. (2013) 'Visualizing qualitative data in evaluation research', *New Directions for Evaluation*, 2013(139): 53–71.

Hofer, B.K. (2004) 'Epistemological understanding as a metacognitive process: thinking aloud during online searching', *Educational Psychologist*, 39(1): 43–55.

Jenkins, H. (2006) *Convergence Culture: Where Old and New Media Collide*, New York: New York University Press.

Mannay, D. (2020) 'Revisualizing data: engagement, impact and multimodal dissemination', in L. Pauwels and D. Mannay (eds), *The SAGE Handbook of Visual Research Methods* (2nd edn), London: SAGE Publications, pp 2–26.

McLaughlin, H. (2015) 'Ethical issues in the involvement of children and young people in research', in H. McLaughlin (ed), *Involving Children and Young People in Policy, Practice and Research*, London: National Children's Bureau, pp 13–25.

McNaught, C. and Lam, P. (2010) 'Using Wordle as a supplementary research tool', *The Qualitative Report*, 15(3): 630–43.

Moseley, D., Baumfield, V.M., Elliott, J., Gregson, M., Higgins, S., Miller, J., and Newton, D.P. (2005) *Frameworks for Thinking: A Handbook for Teaching and Learning*, Cambridge: Cambridge University Press.

Peskin, J. and Astington, J.W. (2004) 'The effects of adding metacognitive language to story texts', *Cognitive Development*, 19(2): 253–73. Available from: https://doi.org/10.1016/j.cogdev.2004.01.003

Schunk, D.H. (2008) 'Metacognition, self-regulation, and self-regulated learning: research recommendations', *Educational Psychology Review*, 20(4): 463–7. Available from: https://doi.org/10.1007/s10648-008-9086-3

Teddlie, C. and Sammons, P. (2010) 'Applications of mixed methods to the field of educational effectiveness research', in B. Creemers, L. Kyriakides, and P. Sammons (eds), *Methodological Advances in Educational Effectiveness Research*, Oxford: Routledge, pp 115–52.

Veenman, M.V.J., Kok, R., and Blöte, A.W. (2005) 'The relation between intellectual and metacognitive skills in early adolescence', *Instructional Science*, 33(3): 193–211. Available from: https://doi.org/10.1007/s11251-004-2274-8

Viegas, F.B., Wattenberg, M., and Feinberg, J. (2009) 'Participatory visualization with Wordle', *IEEE Transactions on Visualization and Computer Graphics*, 15(6): 1137–44.

Wall, K. (2008) 'Understanding metacognition through the use of pupil views templates: pupil views of learning to learn', *Thinking Skills and Creativity*, 3(1): 23–33. Available from: https://doi.org/10.1016/j.tsc.2008.03.004

Wall, K. and Higgins, S. (2006) 'Facilitating metacognitive talk: a research and learning tool', *International Journal of Research and Method in Education*, 29(1): 39–53. Available from: https://doi.org/10.1080/01406720500537353

Wall, K., Higgins, S., and Packard, E. (2007) *Talking about Learning: Using Templates to Find out Pupil's Views*, Crediton: Southgate Publishers.

Wall, K., Higgins, S., Remedios, R., Rafferty, V., and Tiplady, L. (2012) 'Comparing analysis frames for visual data sets: using pupil views templates to explore perspectives of learning', *Journal of Mixed Methods Research*, 7(1): 22–42. Available from: https://doi.org/10.1177/1558689812450211

Wellman, H.M. (1983) 'Metamemory revisited', in M.T. Chi (ed), *Trends in Memory Development Research*, Basel: Karger, pp 31–51.

Section II

Creative embodied analysis

6

Analysing creative multimodal data for a scientific audience

Jennifer S. Leigh, Jennifer R. Hiscock, Sarah Koops, Anna J. McConnell, Cally J.E. Haynes, Claudia Caltagirone, Marion Kieffer, Emily R. Draper, Anna G. Slater, Kristin M. Hutchins, Davita Watkins, Nathalie Busschaert, and Larissa K.S. von Krbek

Introduction

This chapter will discuss analysis of multimodal creative and qualitative research that has been undertaken by the International Women in Supramolecular Chemistry Network (WISC, 2020). WISC's overarching aim is to create a community to support the retention and progression of women and other marginalised genders within the field of supramolecular chemistry. WISC use an ethos that 'calls in' the community and embeds equality, diversity, and inclusion (EDI) expertise (Caltagirone et al, 2021a, 2021b). We intentionally use reflective methods as part of an Embodied Inquiry (Leigh and Brown, 2021) to capture and share their 'invisible, embodied, emotional experiences' (Leigh et al, 2023, p 1) as a way to raise awareness and effect change (Leigh et al, 2022a).

Context

Science is not known for its diversity and inclusion, whether that is regarding gender (Rosser, 2012), race (Prasad, 2021; RSC, 2022), sexuality (Smith, 2019), or disability (CRAC, 2020). Women are subject to resistance in academia generally (Shelton et al, 2018; Murray and Mifsud, 2019), and this is intensified within the science, technology, engineering, and mathematics (STEM) disciplines (Rosser, 2012). Despite investment in programmes designed to address the gender imbalance, women remain a minority, particularly in more senior roles (Rosser, 2017). This imbalance is more pronounced in some STEM disciplines than others. Women are one of several marginalised groups in the chemical sciences (RSC, 2018, 2022), with the period post-PhD being when they are most likely to leave (RSC, 2019a). Although there is not one easy explanation for this, the combination of the so-called chilly gender climate and challenges in planning families and pregnancy in a lab are certainly contributing factors (Stockard et al, 2018; Leigh et al, 2022b; Slater et al, 2022). In 2019, the Royal Society of Chemistry (RSC) stated that the rate of attrition is such an issue that the current rate of change would never result in gender parity despite half of undergraduate chemistry students identifying as women (RSC, 2019a). The RSC recognised that women are subject to multiple small barriers (or micro-aggressions) (Ahmed, 2017), resulting in them publishing less, in lower impact journals, and in their publications receiving fewer citations (RSC, 2019b). This, combined with intersectional factors due to disability, class, and race, mean that there is a significant under-representation of women and other marginalised groups in the chemical sciences.

WISC was launched in November 2019 to address the need for support to retain and progress women and other marginalised groups within the specific field of supramolecular chemistry. Jean-Marie Lehn described supramolecular chemistry as 'chemistry beyond the molecule' with the aim of 'developing highly complex chemical systems from components interacting by non-covalent intermolecular forces' (Lehn, 2002, p 4763).

> Supramolecular chemistry can simply be defined as the study of the non-covalent interactions between molecules. This is often referred to as a host-guest scenario in which reversible interactions temporally hold together two or more chemical species, thus forming a supermolecule. In essence you can think of these supramolecular reversible interactions that hold molecular species together as a type of atomic hook and loop fastening, similar to those developed by VELCRO®. If you pull the hook and loop partner strips apart, they remain unaltered, bring them back together and they will stick to each other, holding two pieces of material in place. This is a process that can be repeated, depending on the quality of your hooks and loops, an infinite number of times, and is the key principle behind all areas of supramolecular chemistry. (Leigh et al, 2022a, p 69)

WISC takes an intersectional approach (Crenshaw, 1989), recognising that gender is not binary, intersects with other protected characteristics such as race and ethnicity, and cannot be considered in isolation. While WISC uses the word 'women', we explicitly include trans women, genderqueer women, and non-binary people who are significantly female-identified. Rather than take a generalist approach to the long-standing (Fara, 2018) or wicked problem of marginalisation of women and other groups in STEM that is seemingly resistant to initiatives and funding (Rosser, 2004; Colwell and Bertsch McGrayne, 2020), we chose to build and extend a network to create a sense of kinship and responsibility so that the supramolecular chemistry community looked after its own.[1] The power of networks and community to support marginalised groups has been long recognised (O'Leary and Mitchell, 1990). Since its launch, WISC has established a mentoring programme, a website,[2] and bank of resources, collaborated on a series of webinars with an RSC Special Interest Group (RSC, nd), held events, created specialist support clusters (for parents, disabled/chronically ill/neurodivergent supramolecular chemists), and undertaken both community and funded research projects. These clusters have generated projects around aspects of parenting (Leigh et al, 2022b, 2022c; Slater et al, 2022) and accessible laboratories (Egambaram et al, 2022). WISC have also published on lived experiences of managing research through COVID-19 (Leigh et al, 2022a), and the more general lived experiences of women principal investigators (Leigh et al, 2022d), combining community research such as qualitative surveys with more reflexive projects incorporating creative research and collaborative autoethnography. The collaborative autoethnography was utilised to explore topics such as balancing work and family life, supporting research groups, career challenges, and life milestones in greater depth. Autoethnography (Bochner and Ellis, 2016) and collaborative autoethnography (where a group collaborates to combine experiences and insights, see Chang et al, 2016) are qualitative research techniques that use autoethnographic data to gain rigorous understanding of an issue set within a wider context. This chapter will primarily discuss the analysis of data collected as part of WISC's work with research groups and ongoing reflexive collaborative autoethnography into the lived experiences of women leading research in supramolecular chemistry.

Ethical and theoretical concerns

WISC unashamedly undertakes feminist research that does its best to capture and share the voices of those who are marginalised (Evans-Winters, 2019). Our research with the wider supramolecular chemistry community utilises conventional online qualitative surveys (Silverman, 2006). However, in order to capture and explore the emotional and embodied reality of lived experiences of women principal investigators working in a field in which they are marginalised due to their gender, we chose to take a phenomenological approach (Merleau-Ponty, 2002) and conduct an Embodied Inquiry (Leigh and Brown, 2021). This resulted in our approaches and data including open-text survey responses and transcripts together with objects and images used as metaphors for experiences (Lakoff and Johnson, 2003), video as method (Harris, 2016), rhythmanalysis (Lyon, 2019), and evocative collaborative autoethnography (Bochner and Ellis, 2016; Chang et al, 2016). Our analytical processing of these data was also embodied and creative (Leigh and Brown, 2021).

In terms of ethical approvals, the process was quite complex. The lead university in the funded research project did not require formal ethical approval for a collaborative autoethnography, as it was understood that the research was being undertaken with and by the group who would be conducting the analysis and publishing from the process. However, this is not the case for every institution, and it is wise to clarify ethical requirements before beginning a collaborative autoethnography research project (Bochner and Ellis, 2016). It is vital that any collaborative autoethnography group establish their agreed ground rules of how they intend to go about their research in terms of meetings, expectations, and publications at the very beginning of the process (Chang et al, 2016). Formal ethical approvals were needed for the research carried out using online surveys with the wider supramolecular chemistry community, the work with specific research groups, and to film in laboratory spaces. In addition to the formal ethical approval process (which was conducted through a social science department), which discussed methods, anonymity, data protection, and informed consent, WISC also had to gain approval and consent from the science department to bring cameras and recording equipment into the laboratory. In terms of participant recruitment, the online survey participants were recruited through social media and at in-person and online supramolecular chemistry conferences and events. The collaborative autoethnography participants were recruited through word of mouth from WISC board members, and the specific research groups who took part were led by collaborative autoethnography participants who volunteered.

In both the collaborative autoethnography and specific research group meetings multimodal data were gathered in the form of meeting recordings and automatic transcripts, images, drawings, and presentations created and shared by participants. The analytic process was iterative and collaborative. The creative outputs showcased and discussed here formed part of the analysis process and are not simply the product of creative methods of data generation. We used a process of identifying embodied responses and reactions to data (Leigh and Brown, 2021), generated through and by the collaborative autoethnography group, to identify 'hotspots' (MacLure, 2003), which were then discussed. We combined this process together with a reflexive thematic analysis of data generated through the online community surveys and worked with specific supramolecular chemistry research groups led by collaborative autoethnography participants (Braun and Clarke, 2006, 2019) to triangulate data of different types and from different sources (Leigh et al 2022a, 2022d).

When it came to wider ethical and axiological issues of avoiding harm (Kara, 2018), and given the close-knit nature of the supramolecular chemistry community, WISC decided to use a creative approach to ensure that there was no risk to any participant by inadvertently allowing them to be identified, even after a process of anonymisation or pseudonymisation. WISC has an ethos of 'calling in' rather than 'calling out' (Caltagirone et al, 2021b). The issues that WISC was exploring included experiences of harassment, discrimination, and marginalisation (Leigh et al, 2022d). Just as we did not want any potential 'victims' to worry that they might be identifiable, we did not want any potential 'perpetrators' to be recognised either. The collaborative autoethnography group had concerns that if it was possible, or even if audiences believed it was possible, to identify individual stories from participants of the group, then these participants would become vulnerable to being accused of complaining (Ahmed, 2021) or suffer the consequences of whistle-blowing (Bjørkelo, 2013). While we were keen to use a narrative approach, many draw on autoethnographic experiences and allow the author(s) to be identified (Shelton et al, 2018; Caine et al, 2022). We addressed this by using a process of fiction as research (Leavy, 2016) to construct a series of fictional narratives or vignettes. The use of narratives within research has a long history (Clough, 2002). WISC used the process of creative writing as part of the analytic process as well as to produce a means of disseminating 'true' stories that were not 'real', in part to safeguard research participants from harm (Bakan, 1996). The vignettes were created from all the research data available to us (see example Second Vignette Adi. All vignettes are available in the Open Access book *Women in Supramolecular Chemistry: Collectively Crafting the Rhythms of Our Work and Lives in STEM* (Leigh et al, 2022a)). The vignettes were shared with the collaborative autoethnography participants and with the specific research groups involved in the study before a process of editing and sharing iteratively to ensure that the vignettes did not inadvertently allow any individual to be identified and that they represented the themes and stories that were important to participants. The iterative process allowed participants and creators to reflect on the stories themselves, how the stories were shared, and how they as participants and their stories were represented within the vignettes, and allowed for edits and changes at every stage.

SECOND VIGNETTE ADI, 23, INTERNATIONAL PHD STUDENT

I don't think I want to stay in academia. I don't know what I want to do yet; there's a big part of me that wants to use my degree as I have worked so hard and spent so much time on it, but I don't think academia is for me. For one thing I don't see any faculty who look like me – there aren't any Black women in my department. I look at my supervisor and see all the hours she works.

Last year I know she was in the lab until 11 or 12pm every night. She tells us that she has to work that much just to get things done. She's not even a professor yet. There never seems to be any time to take stock of where we are and everything we've done – as soon as something works we just move on to the next thing. Before COVID we used to celebrate in the pub if the group published a paper, but these days it just doesn't happen. It can't happen.

I don't think it's only my supervisor either. I remember how burnt out all my lecturers looked when I was an undergrad. Always rushing from one thing to another. It's not that they didn't help – they did or I wouldn't be here now – but I don't think it's what I want for myself. Someone I know got a job as a lecturer straight after a post-doc. Almost unheard of right? But he has to do so much teaching and with everything online I can see that it's almost breaking him. At least my group isn't as bad as some. One friend I know is expected to work 11 hours a day 6 days a week and her supervisor regularly sets meetings at 8pm on a Friday night. I don't think I could cope with that; I want to have a life!

Even doing this is stressful. I cannot tell you the levels of stress that I just seem to cope with on a day-to-day basis and see as normal now! Worrying about not getting enough data, worrying when things don't work, worrying that I am not living up to the huge sacrifices I made to come here. So much

> worry and fear. Leaving my family and boyfriend and only seeing them a few times a year is horrible. I worry that when things aren't working I'm letting my supervisor down. I have to live up to all their expectations of me and make this worthwhile. My whole support network is in another country. There are wellbeing services and things at uni – but they are really general and the waiting lists are so long it's like I'll graduate before I get any real help. My mental health is definitely suffering. As an international student the worry about the financial and emotional cost of my PhD is huge. And every time I want to apply for any kind of financial help they ask for a million things and really intrusive information so I just give up.
>
> Even if I did want to stay, it's not as though getting a job is easy. There are so few positions out there, and everywhere people are being made redundant. I heard somewhere that a CV that won't get you into a post-doc these days would have been better than one that would have got you tenure 30 years ago! It's as though expectations just keep going up and up and up and the pressure goes up with them.
>
> I also know I want to have a family and I just don't see how that's possible in academia. None of the really senior women I see in the field have children. I know that there are more younger ones coming through who have had kids or who are having kids, but who knows if they will make it to professor? There is so much holding women back in the field, doesn't being a mom just make it harder? How could I be the kind of parent I want to be and work that hard?
>
> That said, I don't know what I am going to do. Industry seems to be almost as bad. Maybe publishing or editing is the way forward. Something where I can use my science but also have a life. (Leigh et al, 2022d, p 50)

WISC's primary audience is the wider scientific community, more particularly those working within the chemical sciences, and specifically those within the field of supramolecular chemistry. One of the main challenges for us as researchers has been documenting, disseminating, and justifying the analytic processes in a way that is understandable and deemed rigorous enough for a scientific audience. Scientific ideas of objectivity and meritocracy dominate within STEM disciplines (Popper, 1959). Meritocracy implies that success is awarded by merit, and undermines the evidence seen in the data and lived experiences of those who are marginalised and face additional barriers to success in science (Rosser, 2017; McGee and Robinson, 2020). The idea of objectivity means that qualitative research can still be viewed with suspicion by some more used to the idea that there is an 'objective truth' (Bakan, 1996), and this is particularly true for creative qualitative approaches (Leigh, 2023). Even conventional qualitative approaches such as data saturation do not always apply to reflexive thematic analysis (Braun and Clarke, 2021) or phenomenological research methods (Moustakas, 1994). This does not mean to say that we shied away from using creative and embodied methods of data gathering, analysis, or dissemination despite the fact these were unfamiliar to the participants and co-researchers. In fact, we intentionally brought these into our analytic process. Reflecting on how it felt to be, in the words of Sarah Franklin (2015, p 158), a 'wench in the works', one collaborative autoethnography participant wrote:

> so aware of heart pumping and expanding
>
> not sure whether terror or pride.

Realities of researching and analysing data in the real world

WISC's original plan was to set up a collaborative autoethnography project, where the participants would meet once a month for the duration of the funding, and use arts-based and reflective methods to explore aspects of their work and lives. Jennifer Leigh, once-chemist turned creative and qualitative researcher (Leigh, 2020), led this social science aspect and was to also work closely with one supramolecular chemistry research group, meeting them weekly and introducing them to reflective and arts-based methods to facilitate communication, group dynamics, and, hopefully, moments of creativity and inspiration. She would film in the laboratory, using time-lapse photography, to conduct a rhythmanalysis of the space and see how the researchers utilised the space within the lab. This work with the research group and the rhythmanalysis would feed back into the collaborative autoethnography, and throughout the period of funding there would be visits to the laboratories and research groups of other members of the collaborative autoethnography, to conduct shorter ethnographic studies, rhythmanalysis, and in-person workshops with each group. In addition, we originally planned to use a mixed-methods approach, to collect data on indicators of achievements at fixed time points, and to see if we could show quantitatively whether our creative and reflective approach had facilitated more and/or higher quality science. However, the project began in 2020 during a period of national lockdown due to COVID-19. The members of the collaborative autoethnography and their individual research groups were for the most part shut out of their labs. The collaborative autoethnography meetings became a site of peer support and connection as well as research. Work with the research group had to be online, which changed the shape of sessions from exploratory and creative workshops using arts materials and Lego© to online meetings on a virtual platform for group video calls (Peabody and Noyes, 2017). Leigh was not allowed into the labs when they reopened, as space was at a premium, with maximum numbers of bodies in a room, and it was imperative that the chemistry research be prioritised. A small amount of footage was filmed when access to the labs was less restricted. Travel was curtailed, and so planned visits to other collaborative autoethnography members' labs could not take place.

Figure 6.1: 'Care packages' of art materials sent to participants

Collaborative autoethnography

The issues that people were sharing in the collaborative autoethnography and reflective research group meetings went beyond the expected trials and challenges of chemistry research and navigating academia to a broader sense of collective trauma; loneliness, isolation, helplessness, and burnout. As illustrated in Figure 6.1, participants were sent out 'care packages' to facilitate self-care and participation in reflective and artistic activities.

The value of the meetings as a safe space to vent, share, explore, and process became very clear to the participants as they reflected on their own and others' experiences and explored various themes and ideas about how they could be a chemist during COVID-19 with no access to a laboratory, or how to have the courage to take their own advice.

The project's timing meant that the collaborative autoethnography participants discussed the benefits and challenges of working from home, and how they experienced the continual pressure to work and not let life get in the way while being 'brilliant' within the neoliberal 'hustle culture' of academia:

> What is hustle culture anyway? In a nutshell huStLe cUlTurE (as the name implies) means constant working <she writes before 9 am>
>
> Bombarded with stuff
>
> Brain overload
>
> Never resting
>
> Never sitting down
>
> Never good enough. (Collaborative autoethnography participant)

Other topics included the importance of communicating transparently in order to build a healthy research group culture or community, broader issues such as abortion and women's rights – 'I'm worried we have less rights now than in our mothers' generation' (collaborative autoethnography participant based in the US) – and acceptable clothes to wear in the lab:

> Conventionally, in a chemistry laboratory it is imperative for health and safety reasons to ensure that along with wearing a lab coat, gloves, and safety goggles, all skin is covered up. … In practice, 'lab-safe' clothes can equate to being told to wear a typically masculine uniform of baggy cotton jeans, trainers, and the like. The argument for this is that clothes in the lab need to be able to be removed easily in case of spills. Adding into the equation oversized lab coats and hair that has to be tied back can mean that for young women being in the lab is somewhat equivalent to removing their femininity, if they have been used to expressing that femininity through their clothes. Although some may say that a lab is not a place to express personality and identity, this perspective homogenises lab workers, and expects them to all tend towards the 'average' – which of course is skewed by imbalanced representation. We all have personality and identity, but only some are told that this should not be expressed, or that it is the wrong sort of personality. In the labs where we have been, typical 'blokey' behaviour raised very few eyebrows, but overtly feminine behaviour had a different impact depending on the group.
>
> The goal of equality is not to remove femininity in order for women to be considered on a par with men. This does not put people on a level playing field. Instead, it reinforces the idea that only some bodies (male) are allowed to exist within the space. Equality and inclusivity mean that the space needs to genuinely reflect diversity and be comfortable to all people who might inhabit it. (Leigh et al, 2022a, p 101)

The collaborative autoethnography group quickly expanded from the original and intended six members to 14 across the UK, Europe, US, and Japan, as the original participants reached out to others within the supramolecular community who were struggling. They found support knowing that other people were feeling the same: 'It feels so good to speak about my feelings with you! Thanks a lot for let me be part of this [sic]' (collaborative autoethnography participant).

Reflective research groups

The sessions with the supramolecular research groups were more structured and boundaried than the collaborative autoethnography meetings, as they were delivered to support the students and their development as well as gather or create data (Leigh, 2023). The two groups that Leigh worked with had different needs, and the focus of sessions was adapted in response. For example, one research group required more explicit input to establish a cohesive and supportive environment. In line with the research study's aims to improve scientific outputs through creative and reflective methods, work with this research group aimed to develop a group culture, improve communication, and create a supportive group dynamic to facilitate an environment more conducive to productivity. Towards the beginning of working with them, one group member said about the culture, 'it's not very team driven' (man 24, postgraduate research student (PGR)). The sessions explored a lot of themes including expectations of self and others, creativity, productivity, and communication. On reflection, the participants described the sessions as 'a bit more like group therapy when we started' (man 21, PGR). Leigh has a background in therapeutic work (Leigh, 2020), and she was able to contain and boundary the work as it became more therapeutic (Leigh, 2023). At the end of the project participants reflected:

> Lots of changes for the better … and that has improved my research. I have improved as a chemist in my ability to be an independent researcher. (man 23, PGR)

> The last year has seen a significant amount of change and progress. I struggled to take responsibility and ownership of my PhD. … A lot of things needed to click into place and I have now done 2/3 [of the work necessary to complete] in 8 months. … The culture of the group has changed – more accountable for themselves and also willing to help. Adds to a community and makes it more enjoyable and we can be more productive. (woman 22, PGR)

> The first year/year and a half were difficult. … Lots of stress and anxiety and imposter syndrome. … [Now I'm] much happier with what I'm doing, how I work, more positive feelings. (man 26, PGR)

> The group in general have come a long way. Everybody seems more involved and feeling the ownership. (man 21, postdoc)

While some meetings were less cathartic (Finlay, 2022), they still addressed issues such as pressure, expectations of themselves and others, and what success might look like or feel like (Figure 6.2):

> Success is when I don't feel overwhelmed at the start of the day. (woman 14, PGR)

> There are two parts to it – professionally and in academia it's about being acknowledged and acknowledging what I've done. … Success I want to attain is just being happy. (woman 13, PGR)

Both supramolecular chemistry research groups used images and objects as metaphors for their experiences (Lakoff and Johnson, 2003), for example a rollercoaster ride as their PhD journey, or an 'I ❤ Jesus' bracelet as a symbol of gratitude and a reminder to enjoy every second of life. They reflected that they found the sessions very helpful, valued feeling part of a community, and that it was beneficial to hear how people were feeling as well as to talk about research. The meetings gave them a confidence boost, and helped them appreciate the things that went right, and to celebrate the small things more.

> 'Makes it easier to get on with the actual science' (woman 18, PGR)

This comment encapsulates the wider aim of the research project, which was to use reflective and creative methods to enable individuals and groups to do better science.

Strengths, challenges, and further developments

A strength of WISC's creative data analysis process was that it allowed us to implement transdisciplinary perspectives (Stone, 2015), and move beyond scientific and disciplinary norms and expectations of research (Aikenhead and Ogawa, 2007). Qualitative research is very different from the cutting-edge chemical research the members of WISC and the supramolecular chemistry community engage in day to day. Qualitative research is not, and nor should it be, generalisable (Braun and Clarke, 2013). Instead, our approach was designed to capture and record lived experiences, and, at the same time, allow time and space to process these experiences (Evans-Winters, 2019). This approach continued through analysis. The images and accounts shared here are products of analysis as much as of a data capture process. The aim was to create a multilayered account, and at the same time allow the analytic process to 'be soul work that serves to heal thyself' (Evans-Winters, 2019, p 7). The fictional vignettes were created to capture the lived and embodied realities of women working within a discipline in which they are marginalised. Their purpose was to evoke emotions from the audience (Bochner and Ellis, 2016) so that they might better understand and

Figure 6.2: What 'success' looks and feels like

empathise, without being distracted by attempting to identify *who* a story was about. WISC have shared these vignettes at international conferences as a way of highlighting lived experiences at different career stages. The response we have had has been unequivocally positive. The mode of sharing images created by participants, or reading out a fictional vignette or 'story-time', is very different from conventional scientific conference papers based around figures, graphs, and numbers. However, every time we have taken this approach we have been overwhelmed with positive feedback from the audience, expressing to us that it was as though we were telling their personal story. For example, each time after sharing Adi's vignette, international students, postdocs, and early career researchers have approached us to say that before hearing it they had thought they were the only one that felt that way. As a consequence of hearing our fictional vignette they felt less alone and more connected to the supramolecular community. Similarly, those who had not directly experienced those emotions were more able to recognise them in others, and see where they could bring about change. By disassociating the stories within the vignettes from 'realness' we are allowing the community space to take stock and enact change to more proactively include those who might otherwise be marginalised.

One of the biggest limitations of our creative analytic process has been finding where to situate our work generally, for scientific audiences in particular. Our analytic process resulted in objects, free-form writing, images, and drawings, including those shared here, which can be challenging to incorporate into conventional academic outputs. The content of our initial paper on WISC's ethos and approach to EDI (Caltagirone et al, 2021a) might have 'fit' better into a journal on higher education; however, it would not have reached the scientific audience we wanted to reach. For the most part, disciplinary scientific journals are set up to publish and disseminate scientific data.[3] Reviewers for chemistry or science journals are versed in methodological techniques appropriate for their specialisation, and are not generally experts in EDI issues, social science, or qualitative or creative methods. This has meant that some chemistry and science journals and funders, while stating the importance of WISC's work and aims for developing an inclusive and diverse research culture, are unable to 'fit' us into their standard criteria for publication or funding. Another limitation connected to this is the reputation and standing of EDI work generally (Ahmed, 2012). EDI work is undervalued, and if a chemist or scientist is seen to focus on EDI work this can in fact have a negative impact on their career. They can be perceived as being less dedicated to their science, with the time and energy spent on EDI work detracting from the research that 'counts' for matters of prestige, reward, or recognition. We need to change this within the wider research culture, and begin to value and reward EDI work, recognising that those who are marginalised or face barriers are often those most engaged with EDI (Ahmed, 2012). WISC members and others passionate about addressing EDI and research culture risk potential damage to their careers by devoting time to this work rather than their chemistry or scientific research.

Our aim is to continue with our approach of combining creative data creation *and* creative data analysis with more conventional data collection and analytic methods to triangulate our findings so that we can disseminate our work to a wide audience. Our choice to undertake an Embodied Inquiry (Leigh and Brown, 2021) through design, redesign, and data creation and analytical processes resulted in rich, evocative outputs such as those shared here, which are able to 'haunt' our audience (Wilson, 2018). This evocation of emotion is necessary to persuade people to effect change and address the barriers faced by those who are marginalised. In addition, our creative analytic method allowed the collaborative autoethnography participants to make sense of and process their own lived experiences beyond the catharsis of sharing their stories. We believe that our approach of using creative data analysis and creating *true* but not *real* stories has many possible applications, highlighting marginalised voices and experiences within research culture and other scenarios.

Notes

1 A wicked problem is a problem whose solution requires a great number of people to change their mindsets and behaviour. For more on wicked problems see Rittel and Webber's (1973) original paper defining wicked problems in planning, and Lönngren and van Poeck's (2021) mapping of the literature.

2 www.WomenInSupraChem.com

[3] Please see Helen Kara's *Qualitative Research for Quantitative Researchers* (Kara, 2022) for exceptions to this in STEM disciplines, such as qualitative physics, as developed by de Kleer and colleagues in the 1980s (de Kleer and Brown, 1984; de Kleer, 1993).

References

Ahmed, S. (2012) *On Being Included: Racism and Diversity in Institutional Life*, Durham: Duke University Press.

Ahmed, S. (2017) *Living a Feminist Life*, Durham: Duke University Press.

Ahmed, S. (2021) *Complaint!* Durham: Duke University Press.

Aikenhead, G.S. and Ogawa, M. (2007) 'Indigenous knowledge and science revisited', *Cultural Studies of Science Education*, 2(3): 539–620. Available from: https://doi.org/10.1007/s11422-007-9067-8

Bakan, D. (1996) 'Some reflections about narrative research and hurt and harm', in R. Josselson (ed), *Ethics and Process in the Narrative Study of Lives*, London: SAGE, pp 3–8.

Bjørkelo, B. (2013) 'Workplace bullying after whistleblowing: future research and implications', *Journal of Managerial Psychology*, 28(3): 306–23. Available from: https://doi.org/10.1108/02683941311321178

Bochner, A. and Ellis, C. (2016) *Evocative Autoethnography: Writing Lives and Telling Stories*, London: Routledge.

Braun, V. and Clarke, V. (2006) 'Using thematic analysis in psychology', *Qualitative Research in Psychology*, 3(2): 77–101.

Braun, V. and Clarke, V. (2013) *Successful Qualitative Research: A Practical Guide for Beginners*, London: SAGE.

Braun, V. and Clarke, V. (2019) 'Reflecting on reflexive thematic analysis', *Qualitative Research in Sport, Exercise and Health*, 11(4): 589.

Braun, V. and Clarke, V. (2021) 'To saturate or not to saturate? Questioning data saturation as a useful concept for thematic analysis and sample-size rationales', *Qualitative Research in Sport, Exercise and Health*, 13(2): 201–16. Available from: https://doi.org/10.1080/2159676X.2019.1704846

Caine, V., Clandinin, D.J. and Lessard, S. (2022) *Narrative Inquiry: Philosophical Roots*, London: Bloomsbury.

Caltagirone, C., Draper, E., Hardie, M.J., Haynes, C., Hiscock, J.R., Jolliffe, K.A., et al (2021a) 'An area specific, international community-led approach to understanding and addressing EDI issues within supramolecular chemistry', *Angewandte Chemie International Edition*, 60(21): 11572–9. Available from: https://onlinelibrary.wiley.com/doi/10.1002/anie.202015297

Caltagirone, C., Draper, E., Hardie, M.J., Haynes, C., Hiscock, J.R., Jolliffe, K.A., et al (2021b) 'Calling in support', *Chemistry World*, 8 March. Available from: https://www.chemistryworld.com/opinion/supramolecular-community-calls-in-support-for-gender-equity/4013248.article

Chang, H., Ngunjiri, F. and Hernandez, K.-A. (2016) *Collaborative Autoethnography*, Abingdon: Routledge.

Clough, P. (2002) *Narratives and Fictions in Educational Research*, Maidenhead: Open University Press.

Colwell, R. and Bertsch McGrayne, S. (2020) *A Lab of One's Own: One Woman's Personal Journey through Sexism in Science*, New York: Simon & Schuster.

CRAC (Careers Research & Advisory Centre) (2020) *Qualitative Research on Barriers to Progression of Disabled Scientists*, London: Careers Research & Advisory Centre.

Crenshaw, K. (1989) Demarginalizing the intersection of race and sex: A black feminist critique of antidiscrimination doctrine, feminist theory, and antiracist politics. *University of Chicago Legal Forum*, 1989(1): Article 8. Available from: http://chicagounbound.uchicago.edu/uclf/vol1989/iss1/8.

de Kleer, J. (1993) 'A view on qualitative physics', *Artificial Intelligence*, 59(1–2): 105–14.

de Kleer, J. and Brown, J. (1984) 'A qualitative physics based on confluences', *Artificial Intelligence*, 24(1–3): 7–83.

Egambaram, O., Hilton, K., Leigh, J., Richardson, R., Sarju, J., Slater, A., et al (2022) 'The future of laboratory chemistry learning and teaching must be accessible', *Journal of Chemical Education*, 99(12): 3814–21. Available from: https://doi.org/10.1021/acs.jchemed.2c00328

Evans-Winters, V. (2019) *Black Feminism in Qualitative Inquiry*, Abingdon: Routledge.

Fara, P. (2018) *A Lab of One's Own: Science and Suffrage in the First World War*, Oxford: Oxford University Press.

Finlay, L. (2022) *The Therapeutic Use of Self in Counselling and Psychotherapy*, London: SAGE.

Franklin, S. (2015) 'Sexism as a means of reproduction', *New Formations*, 86: 14–33.

Harris, A. (2016) *Video as Method: Understanding Qualitative Research*, Oxford: Oxford University Press.

Kara, H. (2018) *Research Ethics in the Real World: Euro-Western and Indigenous Perspectives*, Bristol: Policy Press.

Kara, H. (2022) *Qualitative Research for Quantitative Researchers*, London: SAGE Publications.

Lakoff, G. and Johnson, M. (2003) *Metaphors We Live By*, Chicago: University of Chicago Press.

Leavy, P. (2016) *Fiction as Research Practice: Short Stories, Novellas, and Novels*, Abingdon: Routledge.

Lehn, J.M. (2002) 'Toward complex matter: supramolecular chemistry and self-organization', *Proceedings of the National Academy of Sciences*, 99(8): 4763–8.

Leigh, J. (2020) 'What would a longitudinal rhythmanalysis of a qualitative researcher's life look like?', in B.C. Clift, J. Gore, S. Gustafsson, S. Bekker, I.C. Batlle, and J. Hatchard (eds), *Temporality in Qualitative Inquiry: Theory, Methods, and Practices*, Abingdon: Routledge, 72–92.

Leigh, J. (2023) *The Boundaries of Qualitative Research: With Therapy, Education, Art, and Science*, Bristol: Bristol University Press.

Leigh, J. and Brown, N. (2021) *Embodied Inquiry: Research Methods*, London: Bloomsbury.

Leigh, J., Hiscock, J., Koops, S., McConnell, A., Haynes, C.J.E., Caltagirone, C., et al (2022a) 'Managing research throughout COVID-19: lived experiences of supramolecular chemists', *CHEM*, 8: 299–311. Available from: https://doi.org/10.1016/j.chempr.2022.01.001

Leigh, J., Busschaert, N., Haynes, C.J.E., Hiscock, J.R., Hutchins, K.M., von Krbek, L.K.S., et al (2022b) 'Planning a family', *Nature Reviews Chemistry*, 6(10): 673–5. Available from: https://doi.org/10.1038/s41570-022-00427-0

Leigh, J., Smith, D., Blight, B.A., Lloyd, G.O., McTernan, C., and Draper, E.R. (2023) 'Listening to fathers in STEM', *Nature Reviews Chemistry*, 7(2): 67–8. Available from: https://doi.org/10.1038/s41570-022-00459-6.

Leigh, J., Hiscock, J., McConnell, A., Haynes, C., Caltagirone, C., Kieffer, M., et al (2022d) *Women in Supramolecular Chemistry: Collectively Crafting the Rhythms of Our Work and Lives in STEM*, Bristol: Policy Press.

Leigh, J., Hisock, J.R., McConnell, A.J., Haynes, C.J.E., Caltagirone, C., Kieffer, M., et al (2023) 'Women in supramolecular chemistry: narratives of resilience and community building in a gender-constrained field', in M. Ronksley-Pavla, M. Neumann, J. Manakil, and K. Pickard-Smith (eds), *Academic Women: Voicing Narratives of Gendered Experiences*, London: Bloomsbury.

Lönngren, J. and van Poeck, K. (2021). 'Wicked problems: a mapping review of the literature', *International Journal of Sustainable Development & World Ecology*, 28(6): 481–502. Available from: https://doi.org/10.1080/13504509.2020.1859415.

Lyon, D. (2019) *What is Rhythmanalysis?* London: Bloomsbury.

MacLure, M. (2003) *Discourse in Educational and Social Research*, Buckingham: Open University Press.

McGee, E. and Robinson, W. (eds) (2020) *Diversifying STEM: Multidisciplinary Perspectives on Race and Gender*, New Brunswick, NJ: Rutgers University Press.

Merleau-Ponty, M. (2002) *Phenomenology of Perception*, trans C. Smith. Oxford: Routledge.

Moustakas, C. (1994) *Phenomenological Research Methods*, London: SAGE.

Murray, R. and Mifsud, D. (2019) *The Positioning and Making of Female Professors*, ed R. Murray and D. Mifsud, Cham: Palgrave Macmillan.

O'Leary, V. and Mitchell, J. (1990) 'Women connecting with women: networks and mentors', in S. Stiver Lie and V. O'Leary (eds), *Storming the Tower: Women in the Academic World*, London: Kogan Page, pp 58–73.

Peabody, M.A. and Noyes, S. (2017) 'Reflective boot camp: adapting LEGO SERIOUS PLAY in higher education', *Reflective Practice: International and Multidisciplinary Perspectives*, 18(2): 232–43. Available from: https://doi.org/10.1080/14623943.2016.1268117

Popper, K. (1959) *The Logic of Scientific Discovery*, London: Routledge & Kegan Paul.

Prasad, A. (2021) 'Why are there still so few Black scientists in the UK', *The Observer*, 10 April. Available from: https://www.theguardian.com/science/2021/apr/10/why-are-there-still-so-few-black-scientists-in-the-uk

Rittel, H.W.J. and Webber, M.M. (1973). 'Dilemmas in a general theory of planning', *Policy Sciences*, 4(2): 155–69. Available from: https://doi.org/10.1007/BF01405730

Rosser, S. (2004) *The Science Glass Ceiling: Academic Women Scientists and the Struggle to Succeed*, New York: Routledge.

Rosser, S. (2012) *Breaking into the Lab: Engineering Progress for Women in Science*, New York: New York University Press.

Rosser, S. (2017) *Academic Women in STEM Faculty*, Basingstoke: Palgrave Macmillan.

RSC (Royal Society of Chemistry) (2018) *Diversity Landscape of the Chemical Sciences*, London: Royal Society of Chemistry.

RSC (Royal Society of Chemistry) (2019a) *Breaking the Barriers: Women's Retention and Progression in the Chemical Sciences*, London: Royal Society of Chemistry.

RSC (Royal Society of Chemistry) (2019b) *Is Publishing in the Chemical Sciences Gender Biased? Driving Change in Research Culture*, London: Royal Society of Chemistry.

RSC (Royal Society of Chemistry) (2022) *Missing Elements: Racial and Ethnic Inequalities in the Chemical Sciences*, London: Royal Society of Chemistry.

RSC (Royal Society of Chemistry) (nd) 'Macrocycllic and supramolecular chemistry special interest group', *Royal Society of Chemistry*. Available from: https://www.rsc.org/membership-and-community/connect-with-others/through-interests/interest-groups/masc/

Shelton, S.A., Flynn, J.E., and Grosland, T.J. (eds) (2018) *Feminism and Intersectionality in Academia: Women's Narratives and Experiences in Higher Education*, Basingstoke: Palgrave Macmillan.

Silverman, D. (2006) *Interpreting Qualitative Data*, London: SAGE Publications.

Slater, A.G., Caltagirone, C., Draper, E., Busschaert, N., Hutchins, K., and Leigh, J. (2022) 'Pregnancy in the lab', *Nature Reviews Chemistry*, 6: 163–4. Available from: https://doi.org/10.1038/s41570-022-00362-0

Smith, D. (2019) 'A personal matter?', *Cell Press*, 1439–42. Available from: https://www.cell.com/matter/pdf/S2590-2385(19)30291-7.pdf

Stockard, J., Greene, J., Richmond, G., and Lewis, P. (2018) 'Is the gender climate in chemistry still chilly? Changes in the last decade and the long-term impact of COACh-sponsored workshops', *Journal of Chemical Education*, 95(9): 1492–9. Available from: https://doi.org/10.1021/acs.jchemed.8b00221

Stone, D. (2015) 'Transdisciplinary approach to interdisciplinary communication and collaboration', in J. O'Rourke, M. Crolwey, S. Eigenbrode, and S. Wulfhorst (eds), *Enhancing Communication and Collaboration in Interdisciplinary Research*. Thousand Oaks: SAGE, pp 82–102.

Wilson, S. (2018) 'Haunting and the knowing and showing of qualitative research', *The Sociological Review*, 66(6), 1209–25. Available from: https://doi.org/10.1177/0038026118769843

WISC (Women in Supramolecular Chemistry) (2020) *The International Women in Supramolecular Chemistry Network*. Available from: https://www.womeninsupramchem.com

7

Object-work as a creative approach to data analysis in Embodied Inquiry

Nicole Brown

Introduction

In introducing my contribution to this edited collection, I feel I need to outline my background and what has brought me to where I am now. I have always been interested in the details of how research is carried out, especially in settings that may not allow for conventional research approaches and methods. As such, I have always engaged deeply with questions of power, hierarchy, and ownership as well as the location and generation of knowledge. Research for me became a process of active doing and making in collaboration with participants with the aim to foster social justice and pursue activist work. It is this philosophical outlook on research that has led to my substantive research relating to the lived experience of disabled, chronically ill, and/or neurodivergent staff and students in higher education, predominantly in the UK. It is also this philosophical outlook that has led me to focus on object-work in the context of analysis.

The methodological and theoretical framing for my work is Embodied Inquiry (Leigh and Brown, 2021), an approach to research that foregrounds the role of the body and seeks to transcend the body–mind dualism. As such, Embodied Inquiry may focus on any or all of the following four typologies (pp 22–6):

1. the lived experience
2. the researcher's body in the field
3. the body as communicator
4. the body in interaction

The exploration of these research foci draws on phenomenology and hermeneutics, thereby recognising knowledge as produced and relational, contextual, and multimodal (pp 39–40). Embodied Inquiry is based on the principle that human language is insufficient and inexact, that human understanding is embodied and that therefore human communication relies on metaphors (pp 28–31). Consequently, analysis within Embodied Inquiry requires engagement with data that goes beyond conventional coding and theme finding in order to account for embodiment within analysis. For researchers to be able to make sense of data and generate knowledge as active agents Embodied Inquiry asks that they attend to their bodies. This attention to the body may feel unnatural and even uncomfortable, as through socialisation in academia we have learned to distance our selves and our bodies from the processes in which we engage. In Embodied Inquiry we are now tasked to genuinely listen to our gut to identify and respond to what excites or affects us in our data (Brown, 2019b). Instead of approaching analysis as a cerebral activity that requires rational logic, we should attend to our natural, embodied reactions and the 'moments of nausea, vertigo, disgust, embarrassment, guilt, fear, or fascination' as they indicate important phenomena and 'hot spots' in our research (MacLure, 2011, p 1003f). It is in this context that the use of objects is particularly conducive. For the remainder of this chapter, I shall focus on how I used objects as a creative form of analysis constituting a systematic analytical process within Embodied Inquiry by focusing on my research 'The "I" in fibromyalgia: the construction of academic identity under the influence of fibromyalgia' (Brown, 2020).

Research context

For my research project 'The "I" in fibromyalgia' I explored how academics make sense of their academic identity if they must navigate the realities of fibromyalgia (Brown, 2020). Fibromyalgia is a complex, often contested condition that is diagnosed by exclusion, as there is no definitive medical test for it. It is characterised by widespread pain, psychological disorders, sleep disturbances, chronic fatigue, cognitive dysfunctions, such as issues with word retrieval, memory, and sequencing, along with a wide range of symptoms from irritable bowels, restless legs, and migraine-type headaches (White and Harth, 2001; Yunus, 2008, 2009; Wolfe and Walitt, 2013). What adds to the complexity of fibromyalgia is the fluctuation in the severity and nature of symptoms, so that many people diagnosed with fibromyalgia experience each day differently. I therefore approached data collection with a very open mind and effectively negotiated with my participants how they would best share their knowledge and express their experiences. The data set I ended up with was particularly varied and ranged from social media contributions to interview transcripts, from essay-type narratives to written responses to questions, from objects and photographs of objects to collages. I was determined to identify an approach to analysis that would allow the creation of a more holistic representation of the lived experience of fibromyalgia. The aim was to recognise the individuality of experiences, but to connect these individual specificities to new entireties and, more generally, applicable representations of experiences to foster understanding and enable activist work. This is where the use of objects comes in. In the following sections, I will outline how I as a researcher analyse my data using objects before showing how objects may also be used when analysis is participatory and co-constructed with participants.

Using objects as analysis in Embodied Inquiry

When I analyse my data, I use objects to make sense of that data by creating installations, some of which are intended for public sharing, whereas others are merely a stepping stone towards theorisations and deeper understanding. The process from data to installation follows the common principles of thematic analysis. The stages of thematic analysis consist of familiarisation with the data, the identification of initial codes, the generation of themes, and the naming of themes (Braun and Clarke, 2006, 2019). The difference here lies with how, in Embodied Inquiry, I respond to what the data tells me. When I engage in thematic analysis, I can explicitly describe and justify how I have moved from code to theme. Embodied Inquiry analysis feels more intuitive in that I listen to my body and the emotional connections I make from the data. This is why I draw on the process of fictionalisation, where authors select information and details from their worlds and then combine them to create a storyworld that is then recognised as a fictionalised account of the authors' truths (Iser, 1997). The selection and reshuffling of details represent a creative process that results in new knowledge although the explicit, individual steps for this cannot be retraced. Despite the focus on the specificities of the setting, in which the selection and reshuffling have occurred, this new contextualised knowledge generated in this process is still generally applicable because of how I focus on the quality criteria for arts-based research (see the section 'A penny for your thoughts').

In practice, I familiarise myself with the data to then begin data analysis by answering the question 'What do I see?' This first stage of analysis focuses on noticing and describing. As my data usually consists of a variety of different sets, the noticing and describing relates to textual as well as visual elements. For example, I notice and describe what kinds of words and phrases participants use to express their experiences in texts, which colours, compositions, or items are used in the visual materials. I then ask myself what my data is telling me at a holistic level. Rather than getting lost in the details of codes and themes, I look for the bigger picture, a grand narrative.

In my fibromyalgia research, for example, across visual and textual data sets many participants expressed their experiences in terms of 'when I have a flare-up …' or 'at work …', 'at home', or 'if I am at University …'. The 'hot spot' here was the divergence between what participants experience in public versus how they deal with their illness experience in private. I effectively selected significant and identifiable details from the research data, as per the fictionalisation process (Iser, 1997), while also identifying the theme 'public versus private' as per the thematic analysis (Braun and Clarke, 2006, 2019). In order to better understand, but also to be able to better communicate that duality of the public and the private self, I drew on creativity and the use of objects. I asked myself what constitutes my public and private self and how I would portray them both. I experimented with that duality by collecting objects found in my home and by creating artefacts.

For my public self (see Figure 7.1), I chose the objects that were most meaningful to me at the time: a loom monster I had created with my son to

Figure 7.1: Collection of objects to represent my public self

represent my motherhood; an owl I had made as a bookmark to represent my academic, scholarly, and creative work; a broken GPS system to represent my determination and commitment even with obstacles in the way; and an essential oil although I no longer remember what that was supposed to represent.

To express my private self (Figure 7.2), I created an image of my hand with bandages around my fingers to indicate the pain management I tend to undertake in my private sphere, but not in public.

I would like to emphasise key issues related to this kind of analysis with objects. The fact that I cannot remember what the essential oil stood for highlights the intuitive nature of the work (Figure 7.1). At the time, the object was important for me, but as time has moved on, and as I have moved on, that detailed element has become less relevant. That does not make the collection any less meaningful because the core of who I am and how I see myself, the 'essence' of my experience, are still reflected. Analysis through object-work allowed me to put phenomenology, the study of the lived experience, into practice (Heidegger, 1996; Husserl, 2001). I considered the entirety of the experiences of public and private selves, condensed those to what is at the core, what constitutes the essence of my private and public self, and then represented that in my collected objects and curated photograph. Engaging with that 'essence' of an experience through the object-work in Embodied Inquiry is very personal and also emotionally taxing. For me, it felt much more difficult to curate or create my private self than my public self. This is probably an indication of the depths I was going to in this reflexive analytic process and the vulnerability to which I opened myself up. Such emotion work and emotional labour are not common in academic and research contexts (Butler-Rees, 2021), so I was not particularly used to that either.

Figure 7.2: Curated photograph to represent my private self

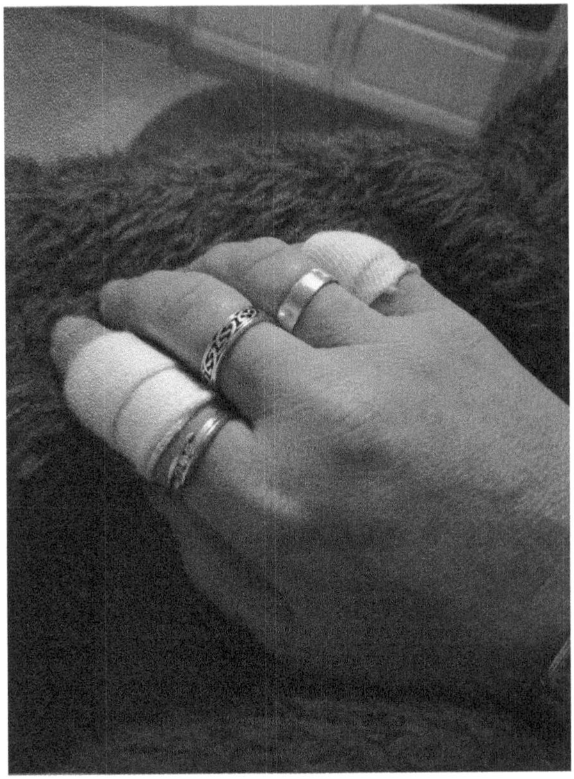

Having considered my own public and private self, I returned to the data and sifted it for specific information participants offered on their view of what constitutes a public academic self. When asked about academic identity participants mentioned 'a male professor stood at a microphone holding their 98th publication', 'an academic, these days will usually be white', and 'if you're like in senior levels you will be a man'. Participants also reflected on how 'it has now become sort of main stream that you are expected to have a PhD' and that typically academics would be in 'a big wooden lecture hall'. The general image created in such snippets of information would be that of an academic that is male, White, a bit older, and engaging in traditional 'scientific' research. Again, I gathered objects and artificially curated a photograph that would represent that picture painted by participants (see Figure 7.3).

The analytical process did not end with the collected objects and curated photograph. This first phase of selection was followed up with combination, as per the fictionalisation process (Iser, 1997). Thus, I brought these creations together. I combined them. In the context of fictionalisation, the combination or rearrangement of details allows for the creation of a storyworld through which the researcher represents truth in such a way that it may affect the reader. The process with

Figure 7.3: Curated photograph to represent the public academic self

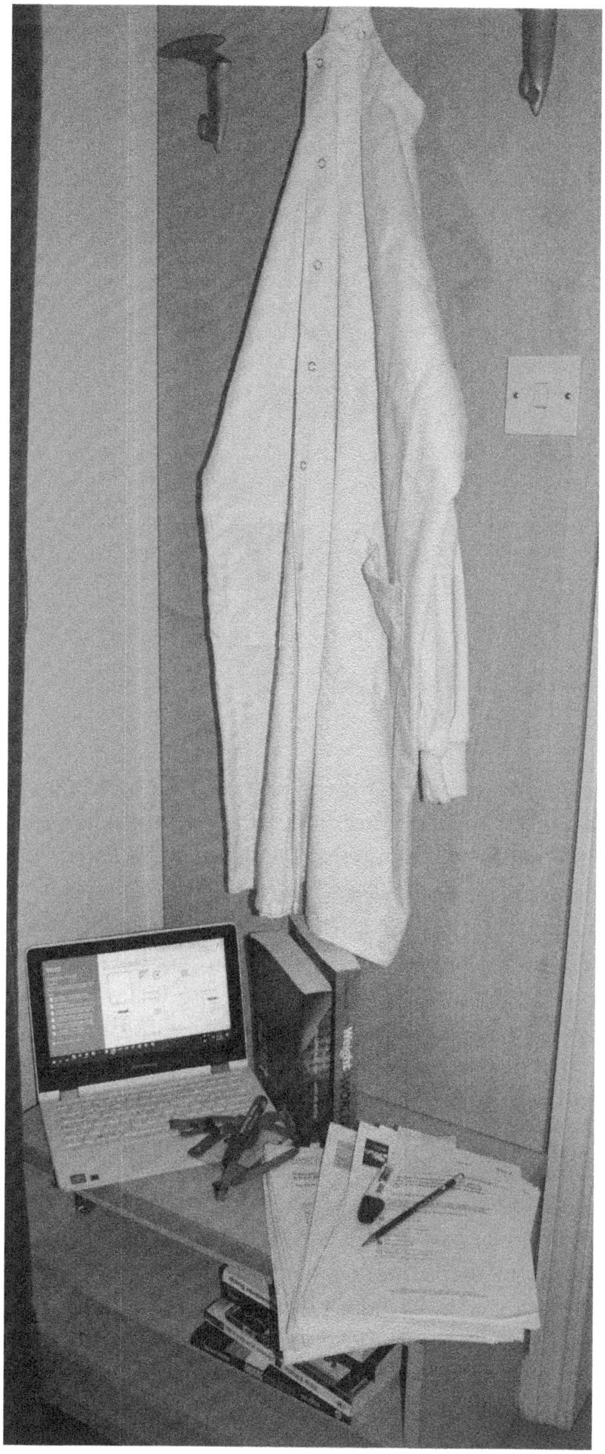

Figure 7.4: Painting *Two Parts of a Whole #2*

the object-work is not dissimilar. Through the process of selecting and combining, I transcended the immediacy and specificity of my data and transformed them into a more generically applicable experience. Following the above steps of engaging with the public and the private self, and to move from the specific to the generic, I created the painting *Two Parts of a Whole #2* (Figure 7.4). The painting depicts the difference between public and private self in the form of the happy, smooth, friendly exterior as opposed to the pixelated, rough, cool interior, which together represent the lived experience of higher education for people with fibromyalgia.

Through this analytic process – from the collection of objects and curated photographs through to the painting – a theme can be conceptualised: how individuals with fibromyalgia navigate different identities and selves in private and public spheres. Of course, the navigation of multiple identities and selves is ubiquitous, with differences only existing in how people make sense of their beings (Josselson and Harway, 2012). In subsequent research with academics and doctoral students with disabilities, chronic illnesses, and/or neurodivergences I found that this duality of public and private self applied too. As such, the theme now no longer relates to one specific experience of a particular participant in one research project. Instead, it represents a more generally applicable truth of the lived experience of disabilities, chronic illnesses, and/or neurodivergences in higher education.

'A penny for your thoughts'

The presentation of my analytic process in this chapter is not entirely sufficient. It appears that the process is linear, tidy, and chronological. In truth, the process is messy and iterative, with much toing and froing between the conventional analysis by coding for themes and the creative analysis by engaging in the embodied, physical process of doing and making. My way of working often raises questions relating to systematicity and robustness.

Although the analytic process through the curation of objects in the fibromyalgia research was messy and iterative, the process was also systematic. Systematicity was achieved by regularly interconnecting the specifics of the research data with the generic of the creative representation. I familiarised myself with what participants were saying and identified the specifics of 'at work', 'when I'm at home', and 'at University'. I then pushed analysis forward creatively with objects to create the painting of *Two Parts of a Whole*, before returning to the specific research data to look for consolidation or contradiction. Equally, the themes generated in the conventional coding processes enabled further engagement at an intuitive, creative level, thereby deepening my own understanding to support conceptualisation. Just like the constant comparative method in grounded theory (Glaser, 1965; Hallberg, 2006), this constant interconnection between codes and creative work made my approach systematic and robust.

Another strategy to achieve systematicity and robustness was to engage participants in the meaning-making process. For example, I shared my preliminary collages and curated images, such as those in Figures 7.1 to 7.4, to further explore experiences and representations with my participants and to check for realistic representation, authenticity, and resonance. Member checking in this way is crucial to achieve verisimilitude, which is often mentioned as a key criterion for good-quality arts-based research. Ultimately, the decisions I take concerning what I include and what I exclude are rooted in the data but intuitive, and I could potentially create several different artefacts, collages, and curated objects from one and the same data set. The choices that I make are therefore based on purpose and intent. If the purpose and intent of the creative analysis are to make sense of data, I remain as close as possible to the details from the research data. The creations presented in this chapter so far, for example, were never meant for public sharing. They clearly link to the data, as they were developed from the data and in response to the data, with the themes recurring in and across data sets. Yet, I felt they lacked aestheticism, artfulness, incisiveness, evocation, and illumination. When the purpose and intent are to foster empathy, raise awareness, develop understanding, and enable emotional connections – thus, the creation is to be shared publicly – then I take liberties and use my artistic licence to work towards a representation that will meet the criteria for good-quality arts-based research. Let me offer a specific example from my fibromyalgia research here. From the collected objects, the curated

Figure 7.5: Objects collated to represent the lived experience of fibromyalgia

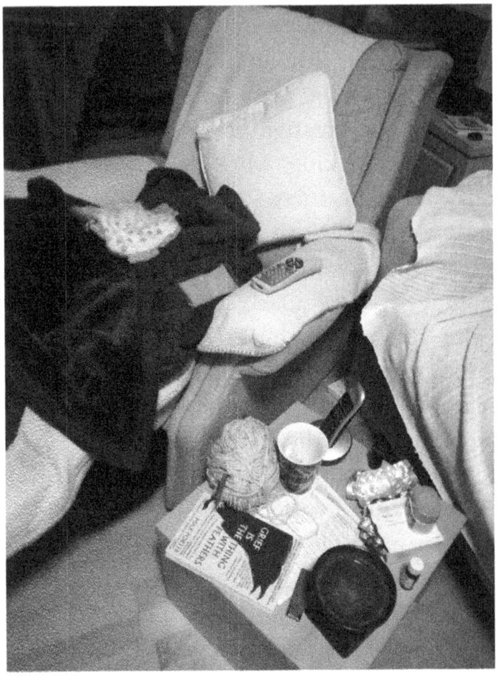

Figure 7.6: *Peace Treaty* as it was exhibited in the gallery

photographs, and the painting, I continued developing representations of the lived experience of fibromyalgia (Figure 7.5). Please note that for copyright and property rights matters, an object had to be blurred. It is the book *Grief is the Thing with Feathers* by Max Porter.

This was still very much a private endeavour to make sense of experiences. But as I recognised the potential impact an installation could have in showing what it feels like to have fibromyalgia, I continued to pursue the idea further with the purposeful intent to create an installation to be exhibited publicly. *Peace Treaty* (Figure 7.6) was exhibited in an art gallery and a part of the installation travelled to an academic conference about equality in neoliberal times.

The difference between the object-work in Figures 7.1 to 7.4 to the installation in Figure 7.6 lies in the application of the criteria for good-quality arts-based research (Barone and Eisner, 2012; Leavy, 2015). The basic principles of arts-based research are not too dissimilar from good-quality criteria in qualitative research (see Tracy, 2010) in that they relate to the robustness and ethics of research, its significance and relevance, its credibility and resonance. However, the evaluative frameworks for arts-based research also emphasise the need for considering evocation and illumination (Barone and Eisner, 2012) and audience response, aesthetics, artfulness, and creativity (Leavy, 2015). The decision of whether I share a creation publicly or not is therefore embedded in my critical reflections and evaluations of my own work. It could well be argued that the painting in Figure 7.4 meets those criteria, but in my mind the painting does not meet the criteria for aesthetics and artfulness, and so it will not elicit a strong audience response. For me, a creation only meets the criteria for sharing publicly if the picture truly is worth a thousand words. I felt I had achieved that powerful message in the object-work in Figure 7.5, which is why I then prepared *Peace Treaty* for public sharing. If I am overly critical of my other object-work, then maybe that tells us something about my own public, academic, and private selves.

Continuing the development of object-work in analysis

I have shown that the use of object-work for analytic purposes allows us to get closer to our data, to develop deeper understanding of our participants' experiences in order to then foster empathy and raise awareness. Further, I have shown that object-work is just one option. Embodied Inquiry offers a wealth of opportunities in engaging with different creative processes, with the only limitations set by the researcher's skill, techniques, and craft. Despite my own shortcomings as a painter, I still engage in the process of painting in the privacy of my office/art studio in order to make sense of data in ways I would not be able to with conventional coding only. Sometimes, a particular form or medium feels more adequate than others, and it is then that I pursue that analysis behind closed doors. Yet, for anyone who may be a proficient painter, performer, visual artist, games coder, videographer, the opportunities just keep opening up, as shown elsewhere in this book. Analysis and dissemination then often converge. *Peace Treaty* from Figure 7.6 started out as a sense-making process, but eventually became a means of sharing experiences, a dissemination of research findings, and a public engagement activity aimed at helping the general public understand what it feels to be isolated in your own home because of fibromyalgia, a chronic condition that is often misunderstood and misinterpreted as malingering and laziness. In which way and when exactly the shift from private analysis to public dissemination happens is difficult to ascertain, especially because the consistent going backwards and forwards between the various data sets and the analytical object-work blurs the lines between analysis and dissemination. This is further compounded when considering that some of the data sets themselves consist of analytic object-work.

So far in this chapter I have not explored the fact that this analytic object-work is not limited to the researcher engaging with data. But research participants, too, can make sense of their experiences through working with objects. For example, for my research projects, 'The "I" in fibromyalgia' and 'Disclosure dances in doctoral education', I asked my participants to curate their own experiences. For the fibromyalgia project I developed five specific questions for the participants, which they were tasked with answering by placing objects into a box. Once a participant had answered a question, they sent me an email with the photograph of the box containing the object(s) and a brief statement of what the objects were and what they stood for. At this stage, the next question would be released to repeat the process. Over the course of several weeks, participants essentially curated their experiences through the objects in their boxes (for more details see Brown, 2019a, 2019b, 2022). For the disclosure dances project, I developed this initial idea further in that I directly asked participants to curate their experiences. I asked for photographs displaying the contents of their handbags and rucksacks that would exemplify and highlight through the objects what it feels like to be disabled, chronically ill, and/or neurodivergent as a doctoral researcher. In Figures 7.7 and 7.8 I share two examples from the research project. Again, some objects had to be blurred for property rights and copyright reasons. These are branded medications, like Advil and Tylenol, or branded muesli bars, for example.

Even without any further details on the conditions and disabilities experienced by these two participants, both images are detailed representations of how individuals manage their lived experiences. Although it appears like a data generation task, this object-work represents a first, tentative interpretation on the part of the participants; therefore, it is in effect both data generation and analysis. In this process, participants explored the contents of their handbags, investigated their bedside tables, and reflected on their experiences before being able to identify the objects they would need for the final picture. The staged pictures represent analysis in that the objects supported participants in the sense-making processes and deepened their understanding of what it means to navigate academia with a disability, chronic illness, and/or neurodivergence.

Figure 7.7: Picture of an experience as curated by a research participant (1)

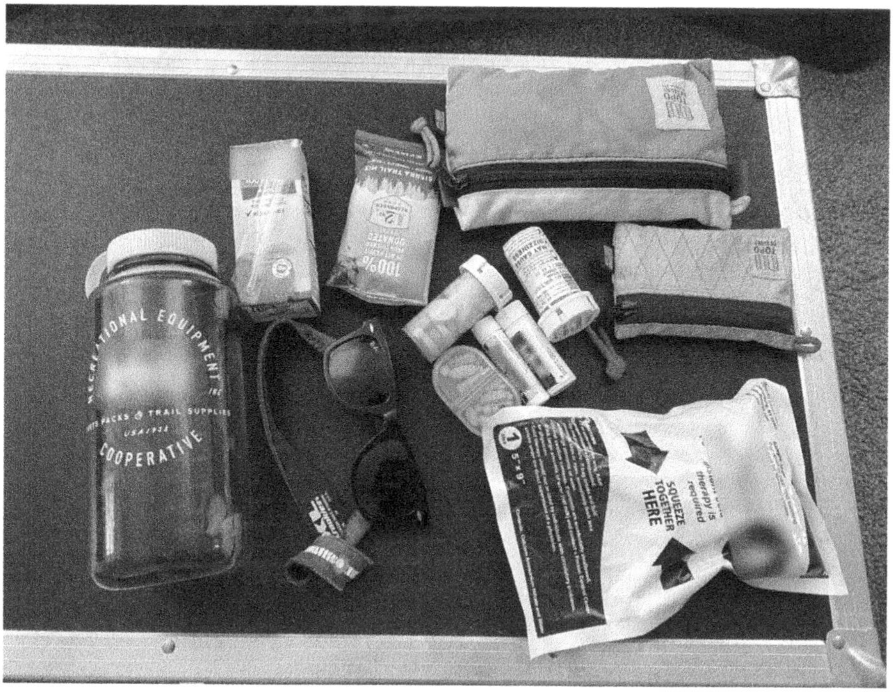

Strengths, limitations, challenges, and opportunities of object-work

Naturally, as with all approaches to data collection and analysis, this approach of creative analysis through object-work within the context of Embodied Inquiry also has its strengths, limitations, challenges, and opportunities.

A major strength of the embodied approach to analysis is that it enables the researcher and participant to paint a clearer picture for the audiences of what it means to live with particular experiences. In a sense, it is the very criteria for good quality of arts-based research, such as illumination and evocation, that are at play here (Barone and Eisner, 2012; Leavy, 2015). Audiences are emotionally and emotively drawn into the work of art and the data upon which the artwork is based. In turn, this close connection at an emotional level creates the context for empathetic understanding and transformation.

Additionally, through communicating research findings in a creative manner, researchers can make their research accessible for a much wider audience. In the contemporary academy, with its focus on impact and engagement as well as gaining research funding, public scholarship plays an ever-increasing role. For research institutions it is no longer sufficient to demonstrate impact and engagement within the academy itself. Instead, researchers and institutions are tasked with making sure that stakeholders in research as well as the general public have access to research

Figure 7.8: Picture of an experience as curated by a research participant (2)

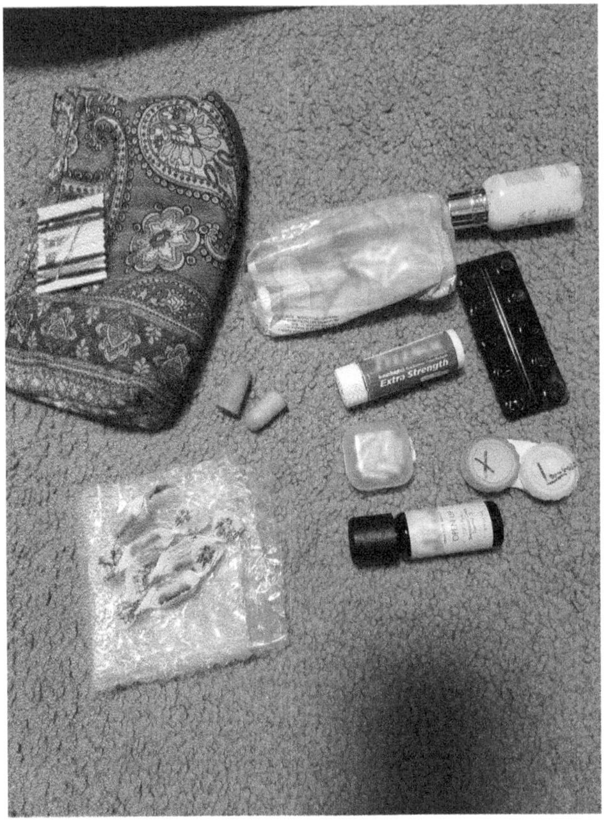

findings (UKRI, 2022a, 2022b). In this context, creative analysis through object-work is particularly effective.

The most significant limitation in engaging with analysis from an Embodied Inquiry viewpoint is probably the researcher's own limitations regarding skills and techniques. As we have seen, it is not sufficient to create an artefact, curate a box of objects, or produce a painting for these to be of worth in communicating research findings. The piece created needs to meet certain standards of aestheticism and artfulness, for which skills and techniques are required. Whether the standards of arts-based research need to be the same as the standards of art for art's sake is disputed, but there is consensus that the arts-based craft needs to be developed and the crafter trained (see Phillips and Kara, 2021; Watson, 2022). Although it is possible to hone a craft, it is quite likely that most researchers will not achieve the specialist techniques and skills required across all media. Work with objects therefore feels like a meaningful compromise. There is significant materiality and physicality involved in gathering objects to produce an installation, but rather than this being a process of creation, it is a process of curation (see Figures 7.5 and 7.6).

Challenges of embodied analyses relate to questions of quality and truthfulness, as the creative process is somewhat difficult to describe and share, and therefore attracts criticism of being self-indulgent and artistic rather than scholarly. It is therein that the greatest opportunity lies for developing object-work as an embodied analysis. Through analysing data conventionally as well as creatively and then using objects to represent experiences it is possible to record and explain the steps that have been taken to move from data to analysis. In the course of painting, mark making, or performing, as is also customary within Embodied Inquiry, the analytic process becomes more blurred, which raises suspicions about its robustness. In truth, however, the researcher makes sense and learns through doing. Within the social sciences, writing as a way of meaning making is already well established (see Richardson, 2000; Faulkner, 2009; Wolcott, 2008). It is now on us to demonstrate that other artistic creation is not too dissimilar from writing.

References

Barone, T. and Eisner, E.W. (2012). *Arts Based Research*, London: SAGE.

Braun, V. and Clarke, V. (2006) 'Using thematic analysis in psychology', *Qualitative Research in Psychology*, 3(2): 77–101.

Braun, V. and Clarke, V. (2019) 'Reflecting on reflexive thematic analysis', *Qualitative Research in Sport, Exercise and Health*, 11(4): 589–97.

Brown, N. (2019a) 'Identity boxes: using materials and metaphors to elicit experiences', *International Journal of Social Research Methodology*, 22(5): 487–501.

Brown, N. (2019b) '"Listen to your gut": a reflexive approach to data analysis', *The Qualitative Report*, 24(13): 31–43.

Brown, N. (2020) 'The "I" in fibromyalgia: the construction of academic identity under the influence of fibromyalgia', PhD dissertation, University of Kent.

Brown, N. (2022) '"It is ..., it stands for ..., it shows ..." : arts-based representations in data generation and analysis', in K. Hinsliff-Smith, J. McGarry, and P. Ali (eds), *Arts Based Health Care Research: A Multidisciplinary Perspective*, Cham: SpringerNature, pp 15–29.

Butler-Rees, A. (2021) '"There's no place for emotions in academia": experiences of the neoliberal academy as a disabled scholar', in N. Brown (ed), *Lived Experiences of Ableism in Academia: Strategies for Inclusion in Higher Education*, Bristol: Policy Press, pp 37–52.

Faulkner, S.L. (2009) *Poetry as Method: Reporting Research through Verse*, Walnut Creek, CA: Left Coast Press.

Glaser, B.G. (1965) 'The constant comparative method of qualitative analysis', *Social Problems*, 12(4): 436–45.

Hallberg, L.R. (2006) 'The "core category" of grounded theory: making constant comparisons', *International Journal of Qualitative Studies on Health and Well-Being*, 1(3): 141–8.

Heidegger, M. (1996 [1953]) *Being and Time: A Translation of Sein and Zeit*, trans. J. Stambaugh, New York: State University of New York Press.

Husserl, E. (2001 [1901]) *Logical Investigations: Volume 2*, trans and ed D. Moran, M. Dummett, and J.N. Findlay, London: Routledge.

Iser, W. (1997) 'The significance of fictionalizing', *Anthropoetics*, 3(2): 1–9.

Josselson, R. and Harway, M. (eds) (2012) *Navigating Multiple Identities: Race, Gender, Culture, Nationality, and Roles*, Oxford: Oxford University Press.

Leavy, P. (2015) *Method Meets Art: Arts-Based Research Practice* (2nd edn), New York: Guilford Publications.

Leigh, J.S. and Brown, N. (2021) *Embodied Inquiry: Research Methods*, London: Bloomsbury.

MacLure, M. (2011) 'Qualitative inquiry: where are the ruins?' *Qualitative Inquiry*, 17(10): 997–1005.

Phillips, R. and Kara, H. (2021) *Creative Writing for Social Research: A Practical Guide*, Bristol: Policy Press.

Richardson, L. (2000) 'A method of inquiry', in N.K. Denzin and Y. S. Lincoln (eds), *The Handbook of Qualitative Research*, London: SAGE, pp 923–48.

Tracy, S.J. (2010) 'Qualitative quality: eight "big-tent" criteria for excellent qualitative research', *Qualitative Inquiry*, 16(10): 837–51.

UKRI (UK Research and Innovation) (2022a) 'Guidance on engaging the public with your research'. Available from: https://www.ukri.org/what-we-offer/public-engagement/guidance-on-engaging-the-public-with-your-research/

UKRI (UK Research and Innovation) (2022b) 'Open research'. Available from: https://www.ukri.org/about-us/policies-standards-and-data/good-research-resource-hub/open-research/

Watson, A. (2022) 'Writing sociological fiction', *Qualitative Research*, 22(3): 337–52.

White, K. and Harth, M. (2001) 'Classification, epidemiology, and natural history of fibromyalgia', *Current Pain and Headache Reports*, 5: 320–29.

Wolcott, H.F. (2008) *Writing Up Qualitative Research* (3rd edn), London: SAGE Publications.

Wolfe, F. and Walitt, B. (2013) 'Culture, science and the changing nature of fibromyalgia', *Nature Reviews Rheumatology*, 9: 751–5.

Yunus, M.B. (2008) 'Central sensitivity syndromes: a new paradigm and group nosology for fibromyalgia and overlapping conditions, and the related issue of disease versus illness', *Seminars in Arthritis and Rheumatism*, 37(6): 339–52.

Yunus, M.B. (2009) 'Central sensitivity syndromes', *Journal of Musculoskeletal Pain*, 17(4): 400–8.

8

A composite approach to movement analysis for embodied methodology

Rachel Kurtz and Laura Mazzoli-Smith

Introduction

This chapter will present a composite method of analysis for embodied movement data, developed in the first author's doctoral project (in progress at time of writing), which considered attitudes towards the body in English primary schools. This project sought to understand embodied experiences and their pedagogical implications, suggesting that heightened embodied awareness might support a more perceptive and responsive pedagogy. It drew heavily on the first author's career as a dancer and aerialist specialising in creative movement. Residencies in three schools over the course of nine months consisted of weekly, one-hour sessions of creative movement play with at least two classes in each school, over the course of six weeks in the pilot school and the first enquiry school, and ten weeks in the second enquiry school.

Children were mainly aged 7 to 11 and attended in whole-class groups. Class teachers supported, sometimes joining in and sometimes observing the movement. Data generated included: interviews with staff, video of the children moving using both high-quality video and infrared cameras, and drawings and voice-recorded comments made by the children at the end of the sessions. These different data types were used to triangulate and cross reference during later stages of analysis, which are not discussed in this chapter.

Theoretical foundations

Extending and adapting Rhythmanalysis (Lefebvre, 2004), we used creative dance practice for non-verbal interpretation of the intangible experience of the learner. Analysing primary, embodied, movement data through a series of dance-led lenses avoids immediately grappling with the limitations of translating movement into language, focusing instead on the perceptual level of these experiences and opening the possibility of deeper understanding. However, this approach does then consider what is entailed in the move to reflect and define through words, mindful that this translation will necessarily be occurring in research on embodiment, but also that words can usefully enrich and then invoke as data the movement language that is generated. The question is not then whether reflection and analysis through words should happen at all, but rather how and at what stage.

In keeping with Dewey's transactional constructivist understanding of the situated learner (Vanderstraeten and Biesta, 1998), data generation and analysis methods equally value physical, emotional, and intellectual foci, in order to understand embodiment as an integral aspect of experience. The approach is therefore composite, both at the level of analytical lenses that can be brought to bear on the analysis of movement but also then through turning to other forms of representational and symbolic construction; in this case, children's drawings and reflections. The focus of this chapter is the collection and analysis of the movement and the value of this composite approach in terms of providing a structure of complementary frames for observation. The method draws on dance movement psychotherapist Helen Payne's (2019) work, in emphasising the importance of movement for development, social connection, belonging, and emotional well-being.

Overview of the composite method

The four approaches within this composite method have been adapted to be accessible to researchers with no prior experience of embodied research and are as follows:

- 5Rhythms®
- Ways of Seeing (Tortora, 2005)
- Laban Movement Analysis (LMA)
- Open Floor

While each analytic lens offers a distinct map or method in its own right, they may be combined to fit the needs and focus of analysis. Maps aid transparency, an important criterion in qualitative research (Lincoln and Guba, 1985), and we aim for transparency in both the embodied researcher's activities and in the overall process as apprehensible to others. Each lens is described in Table 8.1, followed by an example of their practical application as a composite method in the case example.

Video data: supporting rationale

'The body is rich in meaning when understanding another's actions' (Thorburn and Stolz, 2020, p 99); it is not merely a physical object, but also a subject that is integrated with mind and in constant relationship with other (van Manen, 1997). The body communicates what remains pre-conscious and unsaid (Dewey, 1958; Brown et al, 2011) and the subtlety of embodied experience in perception and communication may ordinarily go unnoticed on a conscious level (Stanton-Jones, 1992). There is a lack of focus on pre-reflective embodied experience within education and research in favour of discursive accounts, in part because of the challenge of apprehending and capturing fluid embodied experiences

Table 8.1: Analytical lenses

Analytical lens	Insight	Description of analysis
5Rhythms® (Roth, 2011)	Overall mood	Framework of related qualities, elements, and emotions for mapping experience within the perpetual rise and fall of energy
Ways of Seeing (Tortora, 2005)	Individual, subjective experience	Three-stage process for understanding observed movement through the experience of the witness
LMA (Davies, 2007)	Objective description	Method for translating movement into words – aims to describe all aspects of movement with accuracy and detail
Open Floor curriculum*	Potential motivations, needs, and desires	Provides a structure and language for mapping and articulating motivational forces and modes of expression

* https://openfloor.org/our-curriculum/intro-and-overview/

without immediately resorting to metaphor and other linguistic frames (Brown et al, 2011). The use of video data foregrounds embodied experience and, used in educational settings, responds to evidence that well-considered embodied teaching is both beneficial and effective (Horn and Wilburn, 2005; Thorburn and Stolz, 2020).

Use of a filmmaker

This project drew on the expertise of a specialist filmmaker to produce high-quality video footage. While video is not always possible to obtain, being able to draw on this enabled repeated, detailed observation of salient moments and the development of an analytical framework, scaffolding multilayered consideration of the data and therefore a deeper level of analysis than in-the-moment observation might afford.

However, it should be remembered that the camera invites performance and, accordingly, several performative responses were evident. This self-consciousness was mostly countered by the filmmaker's experience and skill. Using a small, handheld camera with a zoom lens, shots were taken from down on the floor, among the children, so that in most of the footage they seemed unaware of being filmed. Furthermore, the movement explorations absorbed the children's attention so the camera often produced little more than a quick smile when noticed. Children were increasingly unself-conscious and spontaneous as the weeks passed.

A further limitation is that filming is necessarily selective and narrows what can be studied later, therefore we suggest that video be used as a prompt to stimulate recollection of multisensory cues and contextual information, and that additional types of data should be generated to widen perspective and capture what the camera might miss.

Movement sessions

The first session began with a guided exploration of moods and movement qualities. Children lay on the floor listening to short bursts of emotive music. They were invited to imagine scenes or stories and move their hands to express whatever feelings were invoked. Building on this we examined emotions conveyed through static poses. Sitting in a circle, volunteers were invited to make a shape to convey a particular emotion while the rest of the class offered observations on how emotion was communicated. These 'emotion shapes' were then animated into a movement improvisation to music to enrich the volunteers' embodied exploration. The work developed slowly over the weeks using adaptations of these and other movement games to facilitate increasingly nuanced explorations of emotional experience. In the final session children were asked to apply their movement vocabulary to express elements of school experience in groups of six, while the class guessed the emotions being expressed.

Subjective and objective attention to embodied cues – breath, heartbeat, and muscle tension – developed children's interoception – the ability to track inner experience including both conscious (for example, heartbeat) and unconscious (for example, non-specific feelings of health or ill-health) sensations. Interoceptive skill is linked to better emotional regulation and can be improved through training (Garfinkel et al, 2015). Heightened interoceptive awareness offered insight into relational experience that informed children's assessment of emotional expression by their peers in later sessions. This highlighted individual differences as well as similarities in embodied expression, reinforcing the importance of affording time to observation of embodied movement as a foundation for perceptive and responsive pedagogy.

Multistage, iterative analysis

Creative data analysis is not only the analysis of creative data, but also a creative approach to data analysis, and while a systematic process should not be rejected, it is not the over-riding aim of this composite method. In practice, analysis often begins with what appears most salient and the maps presented here support exploration and route finding through various lenses onto the data, rather than dictating a course that needs to follow this series of approaches in a linear fashion.

During the movement sessions, initial, spontaneous analyses (an embedded part of the first author's established teaching practice) often occurred in the moment and influenced facilitatory choices, thus shaping project development and direction. For example, when asked how they knew a shape expressed a particular emotion most children created explanatory narratives instead of physical cues. While this was interesting in itself, for the first author this generated frustration as it shifted attention away from the body. Knowledge that frustration (anger) is a *staccato* energy in 5Rhythms® led her to the *staccato* qualities of clarity and precision. Sharpening instruction accordingly, she guided attention back to body position, muscle tension, gaze direction, and breath, thus developing children's interoceptive awareness. Observations and analytical conversation with staff also helped shape sessions and uncover initial themes, with the process kept on track by regularly revisiting underpinning research questions.

In-depth analysis using the composite method on selected, thematically organised excerpts followed. The deeper analysis of the composite method, facilitated by the ability to slow down and pause movement, often uncovered further themes, necessitating a return to the complete footage to check for previously missed instances of the new theme.

This iterative and responsive process relies heavily on the first author's experience as a movement practitioner and introduces the possibility that the work might be overly shaped by implicit interpretations about the qualities of movement built up through many years of practice. This approach has therefore been developed to promote vigilant self-reflection, whether the observer is new to movement or an experienced movement practitioner, but we also accept some normative assumptions about the qualities inhering in movement as inevitable, not only in the case of the first author given her long background in dance, but also in any observer. In this chapter we both draw on this insider perspective as offering insight in the areas of focus, while benefiting from conversations with the second author whose non-dance perspective offers a challenge to some of the normative assumptions about movement that a trained dancer might have. Our analysis does not attempt to avoid norms inhering in interpretations of movement, but rather to aid transparency and reflexivity with respect to the interpretation of movement.

Thematic analysis by the first author

I watched more than six hours of footage in its entirety several times, until key moments generated interest due to emotional impact, incongruence between words and actions, uniqueness, or repetition. For example, in the first session the children laughed at my rolling demonstration and later took time out of their own play to remark on their teacher participating in the movement games. These and similar instances indicated that the usual norms of adult behaviour had been transgressed and revealed children's sensitivity to behavioural norms, a rich theme within the project. I added notes and time stamps to emerging themes, identified the most important ones, then extracted, labelled, saved, and catalogued relevant clips a few seconds long (see Table 8.2).

Table 8.2: Clip catalogue example

Time	Clip name	Children's names	Description	Themes
00:5:06	Tummy slide 1		Girl and boy sliding together on tummies, smiling, stopping to look around and tell friends 'Mrs. [*Teacher*] is doing it!'	Gender, normalisation of movement (teacher joining in)

Figure 8.1: 5Rhythms® demonstration

Video link

Elements of the composite method

5Rhythms®

5Rhythms® is an improvised conscious dance practice that can open a doorway to altered states of consciousness (Roth, 2011).[1] The rhythms are: *flowing*, *staccato*, *chaos*, *lyrical*, and *stillness*, articulated as distinct stages in the natural arc of energetic motion, building gradually to a crescendo and then subsiding. The photographs in Figure 8.1 intimate something essential about each rhythm; however, viewing the full video (1) is recommended for a better understanding.

Together the rhythms form a '*wave*' of energy that provides a framework of connected associations as described in Figure 8.2. At the centre we see each rhythm, followed by its associated element, emotion, qualities, and shadow as we move outward. The element describes the overall energy, for example, *lyrical* (air) is light, 'emotion' indicates a category of emotions, and 'qualities' includes those most useful for our purposes. This shorthand describes the tone of each rhythm. In reality the 'emotion' of *flowing* (fear) might manifest as apprehension, anxiety, or terror. The shadow of a rhythm arises when its expression is obstructed. In the case of *flowing* the shadow is inertia (lethargic and resistant to movement) in which the dancer might appear to be bored.

While the order of the rhythms within a 'wave' does not change, they might be thought of as a palate of colours that can be mixed so that we can have *flowing* fear, mixed with *lyrical* joy producing excitement or anticipation or with shadow *chaos* (confusion) resulting in panic.

For a researcher concerned with embodiment we suggest 5Rhythms® can support and develop observational capacity by understanding the inherent qualities of each rhythm. Its inclusion is also important as an act of transparency regarding an unavoidably implicit lens, since the first author was apprenticed to its creator, Gabrielle Roth, and has been a qualified 5Rhythms® teacher for more than two decades.

The structure of the rhythms informed session planning. For example, activities early in the session included opportunities for children to increase body awareness

Figure 8.2: 5Rhythms® maps as defined by Gabrielle Roth

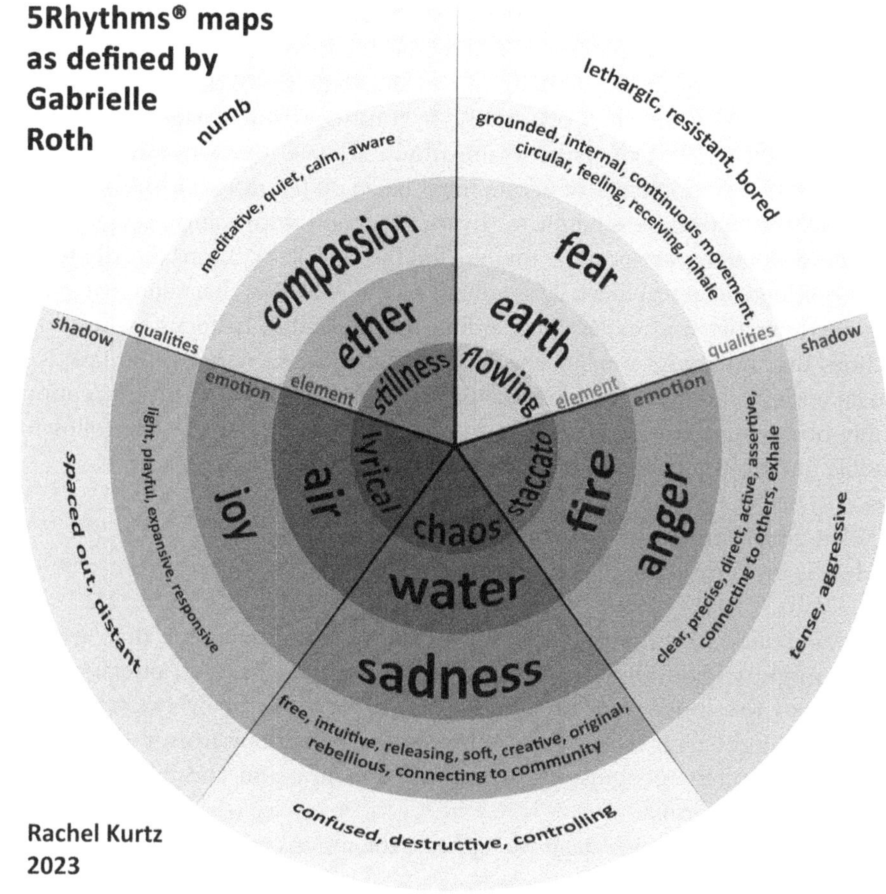

to encourage qualities such as internal focus, groundedness, and tuning into the weight of the body. This accommodated the possibility of fearful emotions such as shyness, self-consciousness, and caution or the flowing shadow – boredom.

During analysis, if a particular quality or emotion is felt or observed by the researcher, the 5Rhythms® maps can suggest possible associations that might offer deeper understanding of the dynamic, as described in the section 'Case example by the first author'. Due to space limitations, we direct the curious to further reading on 5Rhythms® (Vargas-Gibson et al, 2017).

Ways of Seeing

Tortora's (2005) method centres on building and exploring relationship as a foundation for development. It combines dance movement psychotherapy, Authentic Movement, and LMA (Tortora, 2005). Developed for therapeutic settings, notably with non-verbal children with additional needs, it has been adapted here for ease of use in research. While Tortora integrates LMA into the overall approach, we have separated this for clarity and to enable the substitution of LMA with less complex observational methods where appropriate. The process is embodied and empathetic, seeking to recreate elements of the inner landscape of the observed within one's own body. It consists of three stages: witnessing, kinaesthetic seeing, and kinaesthetic empathy.

Tortora's 'observation stage' of witnessing is based on (extracted) LMA; therefore, we bypass this and move straight to Tortora's second witnessing stage – that of the observer's internal response – thereby linking our observations directly to our own experience as researchers. This attends to the relational dynamics that situate embodied practice and experience, aiding the researcher to focus on relational qualities that otherwise remain elusive and inchoate. As witness, the task is to remain curious and open, avoiding the projection of narratives or preconceptions, simply observing one's inner state in the presence of the mover. The following possible questions provoked by Tortora (2005) are useful tools:

a. What does witnessing this stir in me?
b. How is my breath/muscle tension/emotions?

Observing may not stir empathy; for example, a movement that appears defiant might trigger anger, solidarity, or fear in the observer, offering clues not only to the inner state of the mover, but also the likely reactions their behaviour might trigger in others. Untangling behaviour from response can help the researcher recognise the relational dynamic and consciously choose a constructive response.

In the kinaesthetic seeing stage we replicate the movements to experience them at first hand, asking:

- What is the sensory effect?
- What information is being received through this action?
- How does this movement influence my perception?

Thus, we gain insight into the sensations of the observed. For example, in the first pilot session the teacher reprimanded a group of boys for being 'silly'. They were lying on the floor, giggling, and shaking their legs so that the ends of their oversized sports trousers flopped about in the air over the top of their feet. On trying the action later, the first author felt an interesting sensation in her legs, although, dressed differently, she imagined feeling the momentum of the trouser-ends created by the shaking.

In the kinaesthetic empathy stage, we notice emotions that arise in us in the *kinaesthetic seeing* stage that can lead to insight and more compassionate facilitation. Coming back to the trouser example, the autonomous-looking 'dance' performed by the excess fabric was undeniably funny and the shaking action was an effective way of dispersing energy.

There had been some resistance to 'dance' sessions from some of the boys early on and therefore some nervousness about joining in. Given the freedom for creative interpretation within the sessions, the boys' actions were an effective way of dispersing feelings of awkwardness and uncertainty. Perhaps there was also delight in sharing and expressing emotion without the need for words. This possibility guided facilitation towards group work exploring shared and contrasting emotions through movement and in a short time both boys and girls were fully engaged and incidents of 'silliness' subsided.

Ways of Seeing was also practised after sessions in the first researcher's shared office. Analysis was undertaken alone but could be a collaborative endeavour, which may be preferable where the researcher is new to movement analysis. In Figure 8.4 the first author can be seen in the video practising kinaesthetic empathy to the left of the screen (0:27 seconds), tucking her hand under her knees and rocking backwards in the same way as the child.[2]

LMA

LMA is widely employed for recording dance choreography and by movement therapists (Davies, 2007). Elements of movement are teased apart and analysed according to type, providing a detailed picture of the whole movement that can then be communicated to other professionals.

While becoming a Laban Movement analyst requires in-depth training, its maps are nonetheless helpful for researchers new to movement observation. Varied, often complex, LMA maps are available. The first author developed a mind-map version (Figure 8.3), suggesting this is a congruent visual support for a creative approach to analysis, based on Tortora's (2005) succinct summary of LMA. An advantage is the ability to visually connect items on the mind map, allowing multiple starting points and possible pathways to follow the flow of the inquiry. It is intended as an evolving tool with room for adaptation as appropriate and is readily adoptable by a researcher relatively new to this field.

The LMA map has five main categories of analysis: body, time (phrasing), effort, shape, and space, centrally connected by elements of relationship. Arrows radiating from these categories connect to subcategories in ellipses (time, flow, space actions, etc) and, from these, further arrows connect to cloud-shaped bubbles (eg rhythm, free/bounded, rise/sink). The arrows signify the direction of movement from coarser to finer detail; however, the connection can be followed in either direction, supporting thorough analysis. Tortora's (2005) prompts may also support observation:

- What are the movement qualities?
- How long does it last?

- How much does it expand into the surrounding space?
- Does it have a clear direction?
- Which body parts are involved?

Example LMA map use

The following analysis relates to the case example video below and can be tracked using Figure 8.3, with the categories from the map being marked in bold and the transitions from one to another indicated by arrows.

The mover is kneeling with her hands beneath her legs, therefore we might start at **'curled'** as a subcategory of → **'shape'** with → **'central'** (rather than **'peripheral'** involvement of) → **'body parts'** (legs, hands, face – smiling when making the shape). From here we might move to → **'relationship'** (Who is she smiling at? How is she positioned in relation to them?). Beyond this we might move to → **'phrasing'**, which would lead us to notice her rocking backwards later in the clip, therefore **'time'** and → **'rhythm'** draw attention. We can explore further connected subcategories and qualities, continuing until all relevant connections are exhausted or begin again from another category or quality. Through this process we arrive at a detailed description of the movement. Thus, a sustained inquiry

Figure 8.3: LMA mind map

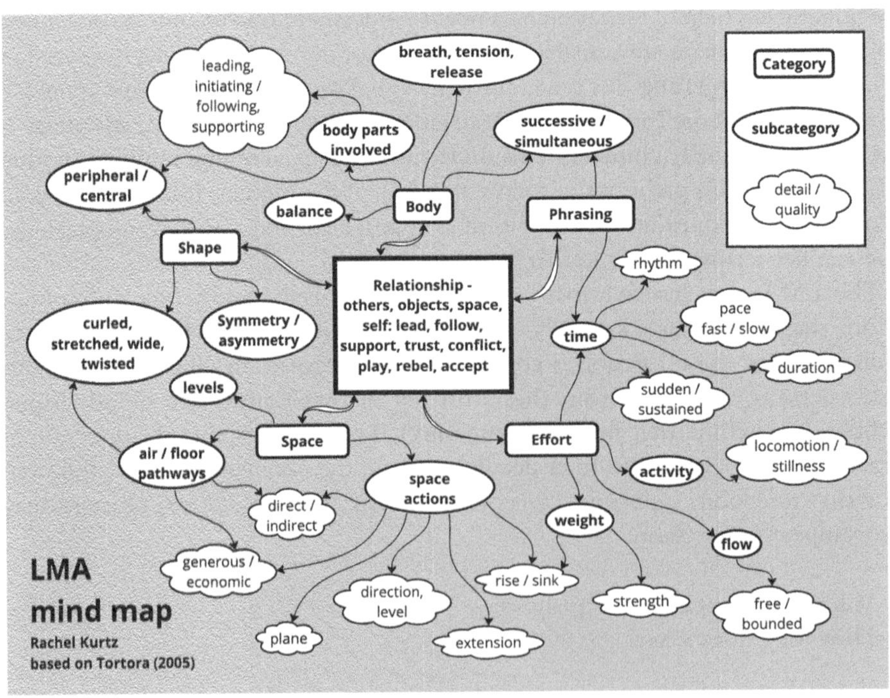

is possible, with the choice of entry points enabling foregrounding of the most salient information.

Open Floor

> When we move it is not just our physical body that we are moving.[3]

Open Floor was collaboratively developed from 5Rhythms® and 'offers a dynamic, creative tool kit of resources to move through life, fully embodied'.[4] Its curriculum consists of ten core movement resources, four dimensions of embodiment – physical embodiment, emotional embodiment, embodied mind, and embodied soul – and four relational hungers – solitude, connection, belonging, and spirit. In the section 'Case example by the first author' we draw on the relational hunger for belonging and the core movement resources, which are:

1. ground
2. centre
3. spatial awareness
4. expand and contract
5. release
6. towards and away
7. activate and settle
8. dissolve
9. vector
10. pause

The core movement resources describe dynamic attributes or 'energies' and are therefore related to intention or tone. For example, a stance might feel very different for both mover and observer depending on whether the energy is grounded (yielding and connected to the floor) or ungrounded ('uptight' and resistant to gravity). While a posture or movement may be described through detailed observation of individual elements (as with LMA) this is not always necessary and the overall energy may be lost in this dissection. While it is not suggested that LMA be discarded completely, Open Floor may be more manageable for some applications – in the case example Open Floor might lead us straight to the most salient movement resource, perhaps the 'towards and away' of the girl's slight backward rock.

Case example by the first author

The following analysis demonstrates how I applied these approaches to an interaction during a session in school during my doctoral project. The drawings in Figure 8.4 are adapted from a short, animated video clip (2).

Figure 8.4: Case example

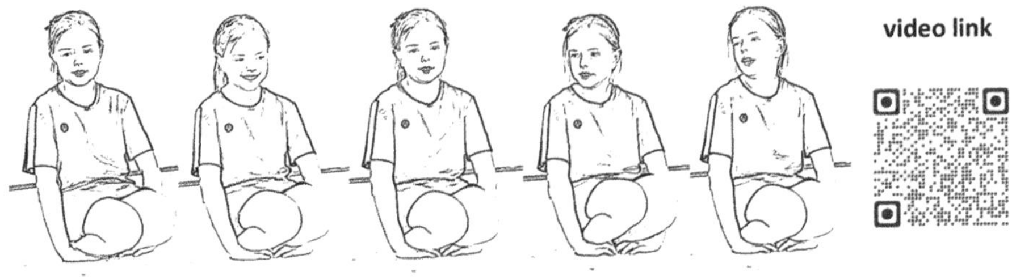

The clip speaks to the research question 'What are children's embodied experiences of school and how might understanding them inform pedagogical practice?' The footage contributes to one of the major emerging findings – how much children's physical self-expression is shaped by often unquestioned, embodied social norms. Whatever children's inner experiences (which we cannot infer from observation alone), these are expressed physically through a series of delimited normative behaviours; accordingly, physical expression of emotional experience is modified by social expectation. Such behaviour modification undoubtedly supports social integration; however, it is also important to question whether these forms of behaviour modification harm the individual. Where modification becomes suppression, tension or anxiety may result, as we know from autistic children who engage in stimming – repetitive movements or behaviours for emotional self-regulation – when expressions are curtailed. Many of the adults I have worked with in non-judgemental, non-verbal connection through dance, express enormous relief at being accepted and not feeling that their body is too big or small, 'wrong' or 'in the way'.

I ask the subject, a girl of approximately 8, how she feels (proud) about a lesson of her choice (Art). Modes and stages of analysis are in brackets in bold after the descriptions as follows:

- 5Rhythms® **(5R)**
- Ways of Seeing:
  - Kinaesthetic Seeing **(WoS:KS)**
  - Kinaesthetic Empathy **(WoS: KE)**
- Laban Movement Analysis **(LMA)**
- Open Floor **(OF)**

> The girl is kneeling, hands under her legs, smiling. Still at first, then rocking backwards and forwards (**OF** – towards/away) with a steady rhythm, moderate effort (**LMA**) emphasising the backward movement (**OF**). The Rhythms I perceive are 'stillness' (composure, quiet, aware) with some shadow 'staccato' (slightly tense, punctuated motion of rocking, particularly

when speaking or spoken to directly, which is a staccato act (connecting to/communicating with others (**5R**)).

I find myself holding my breath as I focus (**WoS:KS**).

We do not immediately identify the girl's emotion. Another child suggests that she might feel awkward, which feels congruent to me, however, she reveals that she is expressing the emotion 'proud'.

When I try on the movement to see how it feels in my own body (**WoS: KE**), I am struck by how disabling it is to sit on my hands. I am curious. The hands express, give and receive. By suppressing their movement, I make myself passive (**5R** – flowing) – I can neither give nor receive actively, and to some degree I can hide what free, expressive movement might otherwise reveal. It speaks to me of control. The rocking, on the other hand speaks of restlessness and a degree of discomfort (**WoS:KS**). It moves me towards and away from the audience (**LMA; OF**). It suggests a body that is filled with more energy than is comfortable being static, and yet the face is smiling, which feels incongruent when I do it – I use rocking to soothe (**WoS:KS**).

Watching the video later, I try on the movement again and notice that I connect more with this sense of pride on my inhale than on my exhale (**WoS:KS**). The inhale is one focus of *flowing* and the associated emotional category is fear, which includes emotions anticipation and excitement (**5R**), both of which are congruent with the girl's movement and could be encompassed within her concept of pride. Her rocking tension around the towards/away dynamic and the emphasis on the backward rock suggests retreat from the intensity of being in the limelight (**OF**).

The girl's action appears to be motivated by a need for social recognition which sits in tension with the assimilated social value of modesty. Since both of these are connected to the *hunger for belonging* (**OF**) 'pride' becomes a complex emotional blend when expressed publicly and the suggestion 'awkwardness' resonates. The girl may be unconsciously expressing a routinely British cultural norm of discomfort around 'pride'. Through this analysis I arrive at an interpretation of the movement that the girl herself may not have been aware of and that speaks of the social nature of embodied communication (**LMA; OF**) – in private the girl's pride may have been embodied entirely differently.

Method in practice

We suggest that it is not simply the application of the four analytic lenses for apprehending movement as data that draws out meaning, but the interplay between them. The initial observation of 'stillness' combined with 'shadow staccato' suggests a tension between different types of energy. The application of LMA and *kinaesthetic seeing* also highlights incongruence – between the smiling face and the bounded, restless posture. By combining 5Rhythms® with *kinaesthetic empathy* the perception of increased congruence is discovered between movement and inhale, and the door opens to the possibility of fear in the movement, an association strengthened by the passive, disabling posture. Open Floor then guides us to the understanding of the social element through the *hunger for belonging*. Using one method alone we would not have arrived at the same detail in an interpretation. This interpretation then resonated with several more explicit examples pointing to how normative assumptions and behaviours about physicality are pervasive and what movements are delimited, such as seen in the aforementioned delight children expressed at adults joining their physical play.

Ethics

Strict ethical protocols are necessary to safeguard children, including parental and child permission for video and sensitive filming to preserve anonymity. The use of multiple types of data required consent to be specified individually for each, making the parental consent form quite complex. To minimise misunderstandings videos were made explaining both the project and the form. These were accessed by up to a quarter of parents whose children were involved in the project and therefore successfully increased accessibility. Given that data is commonly returned to for analysis many months after it is generated, children's names and faces can be forgotten, and good channels of communication must be maintained with schools to ensure that published material is compliant with parental wishes. Consent was given for anonymised data to be stored by the university for ten years, with anonymisation including changing or omitting names, but not disguising voices or video images. For the purpose of this chapter the video data has been animated to increase anonymity.

Consent was 'opt in' only, meaning that some children's data could not be analysed because their form had not been returned, rather than because of safeguarding or other concerns. This had a small impact on the quantity of data that could be used and necessitated additional editing and checking to ensure only appropriate data was used. It also created inequality within the group as children without parental consent often wanted to appear in the video and, despite identifying them with coloured sticky labels or wrist bands (an ethical conundrum in itself), there were a number of occasions when it became apparent that children had removed these. Since the video artist attended only the first and final sessions with the children, identifying those without consent was difficult alongside the demands of catching the action and framing a shot. In some circumstances it may be more practical to use an 'opt out' policy; however, this would need careful ethical consideration and was not appropriate in this case.

Ethical responsibility of the first researcher as practitioner

My interpretations often made unconscious feelings and motivations explicit, potentially exposing the inner world of participants and risking emotional vulnerability. Where movement reveals something that the participant themselves is unaware of or they perceive an interpretation as incorrect or critical, this risk is acute. As a helping relationship (Rogers, 1995), I adopted an attitude of compassionate positive regard for participants and data at all stages to help build trust and safety. This involved careful choices about what thoughts I shared and how, and emphasising the collective humanity within individual experience. As an example, in the video clip there is potential for shame when a child suggests that the girl looks awkward. I am aware that this could be received as critical by her, while at the same time, the child making the suggestion could feel shame at risking a 'wrong' answer. Feeling compassion for both children I comment that *I* might make a shape like that if I felt awkward, while maintaining interested curiosity regarding the emotion she is expressing. Compassionate positive regard was encouraged among staff, who were supported to accommodate children's perspectives constructively, for example by recognising that trust and empathy take time to develop and realising that this has behavioural implications. There is also an ethical responsibility in ensuring that interpretations of embodied experience function as communicative resources to open up intersubjective awareness, rather than as a mode of fixing imposed understandings.

Conclusion

This method could be used to elicit interpretations of movement when lack of shared language or verbal ability is a barrier to understanding embodied experience. For example, in work with young children, refugees, or people with learning disabilities, it offers a many-pronged approach for understanding non-verbal communication that can then be reflected upon. Where possible, as was the case in this project, participants can engage in elements of the analytical process to deepen awareness of their own and other's embodied behaviours and improve interoceptive skills and emotional literacy.

It is important to preserve the integrity of this as an embodied method. The translation of movement into words, as we have seen, is inevitably second-order and language brings in another interpretative layer. Further, while linguistic free association often conveys subconscious meaning, language largely provides access to the conscious level, whereas a focus on the body and emotions, along with a freely associating, creative method, can guide us to layers of meaning that remain inaccessible through language alone. This chapter shows that this composite method provides tools to support an embodied response to the observation and analysis of movement so that our analytical exploration remains grounded in experience.

Notes

1 5Rhythms® demonstration video link: https://photos.app.goo.gl/oq51CpZnqM8JZKiCA
2 The full video clip can be viewed at: https://photos.app.goo.gl/AkVHaheLjTAkAudP7
3 https://openfloor.org/
4 https://openfloor.org/

References

Brown, S.D., Cromby, J., Harper, D.J., Johnson, K., and Reavey, P. (2011) 'Researching "experience": embodiment, methodology, process', *Theory & Psychology*, 21(4): 493–515.

Davies, E. (2007) *Beyond Dance: Laban's Legacy of Movement Analysis*, Oxford: Routledge.

Dewey, J. (1958) *Experience and Nature*, New York: Dover Publications.

Garfinkel, S.N., Seth, A.K., Barrett, A.B., Suzuki, K., and Critchley, H.D. (2015) 'Knowing your own heart: distinguishing interoceptive accuracy from interoceptive awareness', *Biological Psychology*, 104: 65–74.

Horn, J. and Wilburn, D. (2005) 'The embodiment of learning', *Educational Philosophy and Theory*, 37(5): 745–60.

Lefebvre, H. (2004 [1992]) *Rhythmanalysis: Space, Time and Everyday Life*, trans G. Moore, London: Bloomsbury.

Lincoln, Y.S. and Guba, E.G. (1985) *Naturalistic Inquiry*, London: SAGE.

Payne, H. (2019) *Creative Dance and Movement in Groupwork*, Oxford: Routledge.

Rogers, C.R. (1995) *On Becoming a Person: A Therapist's View of Psychotherapy*, New York: Houghton Mifflin Harcourt.

Roth, G. (2011) *Maps to Ecstasy: The Healing Power of Movement*, Novato: New World Library.

Stanton-Jones, K. (1992) *An Introduction to Dance Movement Therapy in Psychiatry*, London: Routledge.

Thorburn, M. and Stolz, S.A. (2020) 'Understanding experience better in educational contexts: the phenomenology of embodied subjectivity', *Cambridge Journal of Education*, 50(1): 95–105.

Tortora, S. (2005) *The Dancing Dialogue: Using the Communicative Power of Movement with Young Children*, St. Paul: Redleaf Press.

Van Manen, M. (1997) *Researching Lived Experience: Human Science for an Action Sensitive Pedagogy*, London: Routledge.

Vanderstraeten, R. and Biesta, G. (1998) 'Constructivism, educational research, and John Dewey', *The Paideia Archive: Twentieth World Congress of Philosophy*, 2: 34–9.

Vargas-Gibson, M., Wolfaard, S., and Roberts, E. (2017) 'The 5Rhythms® movement practice: journey to wellbeing, empowerment, and transformation', in V. Karkou, S. Oliver, and S. Lycouris, *The Oxford Handbook of Dance and Wellbeing*, Oxford: Oxford University Press, pp. 717–30.

Section III

Creative performative analysis

9

Rewriting in role: inviting readers through imagination

Claire Coleman

Rewriting in role is a creative analysis tool that involves the deliberate and transparent process of rewriting existing data through fictional characters. This chapter details the theory, origins, practice, and potential of rewriting in role as an analysis method to invoke active questioning and ongoing thinking. The chapter begins by contextualising rewriting in role within my doctoral thesis, offering a detailed example of the theory and practice, positioning this alongside other creative methods, and finally offering a guide for those interested in adopting this creative analysis method.

Introducing rewriting in role

As a creative data analysis method, rewriting in role is a considered and deliberately partial reconsideration of initial data through which a richer story emanates. Rewriting in role aims to explore and interrogate key areas of the research through rewriting the data from the perspective of fictional characters. A drama-based method, it is heavily influenced by a range of creative approaches, including writing as a method of inquiry and narrative inquiry (Richardson and Pierre, 2005).

As an explicit theatrical device rewriting in role does not fictionalise original data sources or fabricate participant responses or impersonate the truth but employs imagination as an opportunity for liminality and resonance (Leavy, 2022). It provides a fictional space for destabilising ideological oppressions to play with 'elements of the familiar and defamiliarize them' (Turner, 1974, p 60). Rewriting energises the reader and generates alternative perspectives that deepen the research narrative. Writing through role offers the author a distanced perspective and safety within the fiction to express difficult ideas and ask tricky questions (Eriksson, 2007). The section 'Origin story' describes how the process of rewriting in role emerged through my doctoral work to become an integral part of my research.

Origin story

My doctoral thesis 'Dancing through the fourth wall' (Coleman, 2019) explored the potential for process drama as an enactment of critical pedagogy. Process drama employs drama conventions and techniques to create an immersive fictional drama experience created by and for its participants (Heathcote and O'Neill, 2015). It enables participants to explore questions or provocations and offers an opportunity to reflect on emerging understandings from both inside and outside the drama. In the drama participants question their worlds and play with identity. Process drama aligns with critical pedagogy as a relational and reflexive method that seeks to unveil social constructs. Critical pedagogy advocates for learning as a relational activity that values student knowledge and generates knowledges through problem posing (Freire, 2005). The two approaches share a commitment to social justice and creating spaces of agency and transformation.

Informed by a constructivist approach, this thesis employed arts-based qualitative methods and cumulative case study made up of two case studies. In the first case study, I applied a document analysis to materials generated from the process drama workshop 'The Seal Wife' conducted by Cecily O'Neill in Melbourne in the mid-1990s. The workshop begins with an Irish folktale about a mermaid captured by a fisherman, who seeks to return to the sea. This workshop utilised a range of drama conventions and is considered an excellent example of process drama. Notably documented in the monograph *Pretext and Storydrama: The Artistry of Cecily O'Neill and David Booth* (Taylor and NADIE, 1995), this event drew an international audience and remains a significant event in global drama education history. In the second case study, I observed at first hand a local school engaging with Mantle of the Expert (MOTE), a contemporary and popular form of process drama. Coincidentally also aquatically themed, this process drama concerned the conservation and breeding of the dumb gulper shark. As a participant observer, I took notes, conducted interviews, and anonymised responses in accordance with University of Auckland ethics approval. The next section, 'Rewriting role in "The Seal Wife"', briefly details how rewriting emerged during the Seal Wife drama and explains its further development within the Dumb Gulper Shark drama as a creative analysis method.

Rewriting in role in 'The Seal Wife'

Intended to exemplify process drama practice and facilitated by an expert drama practitioner, 'The Seal Wife' workshop is well documented (Fletcher, 1995; Taylor and NADIE, 1995). However, the late discovery of an unpublished and contradictory record of the workshop complicated the story. Seeking to create a coherent narrative out of multiple sources I wrote a diary account of the workshop from the perspective of fictional attendee, *Caitlin Anderson*. This rewriting operated as a field of play, brought the workshop to life, and incorporated the content from the original documents. Through this rewriting, I generated a singular narrative data source for analysis and discovered a creative method of engaging with research in a dialogic way. Creating a narrative beyond the 'official' dominant structure of academia aligned with the critical aims of the research to uphold mechanisms for challenging unequal power structures.

Rewriting in role allowed me to keep playing and questioning where it naturally evolved into a creative analysis method for the Dumb Gulper Shark case study.

Rewriting in role in the 'Dumb Gulper Shark'

The 'Dumb Gulper Shark' case study involved observing an MOTE process drama over a six-week period. MOTE is a dramatic approach that uses imaginary contexts to generate purposeful activities for learning and positions the learners as experts to create opportunities to study wide areas of curriculum. This involved a class of approximately 30 students in year 6 (aged 10–11), the classroom teacher, and the expert facilitator. Through the research, I interviewed the teacher and facilitator and employed drama-based research techniques to generate data from a group of students. These included students recreating significant moments in tableau, shifting into role as the facilitator, and generating playful scenes provoked by words recorded during the Mantle experience. The resulting data included interview notes and recordings, planning, resources, and photographs of the work. After rummaging within this data in relation to the emergent themes from 'The Seal Wife', of power, hope, ambiguity, and aesthetics, I struggled with the conflict between the seemingly requisite authoritative voice of the written PhD format and the dialogic foundations of critical pedagogy. Written as a monograph detailing a fictional conference panel between three scholars, rewriting in role within the 'Dumb Gulper Shark' study afforded an opportunity for critical analysis and playful creativity.

The fictional panel

The fictional conference panel invited three scholars into a facilitated discussion about the 'Dumb Gulper' MOTE drama. The three participants of the fictional discussion were *Bernard*, *Caitlin*, and me aka '*Claire*'. *Bernard Garcia Jameson* is a character wholly dedicated to a radical vision of critical pedagogy. A satirical character, he deliberately embodies the problematic aspects of critical pedagogy as predominantly male, abstract, pretentious, and lacking in self-awareness. The use of satire is deliberately employed to embrace the 'carnival spirit' and offer a temporary challenge to the dominant authoritative voice (Mikhail Bakhtin, cited in Sinclair, 2020, p 65).

The earlier character of *Caitlin Anderson* returns as an experienced drama educator some 30 years since 'The Seal Wife'. As a teacher-educator, she understands the theories and objectives of drama pedagogies and the realities of working in schools. Caitlin's return generated a tangible link between the two studies and signalled their relationship to one another within the cumulative case study. The final 'character' is *Claire*, a version of myself as an emergent researcher and recent doctoral student. As '*Claire*', I wrote from the perspective of shadow role as a version of myself albeit located within fictional social context. Staged as a conversation, the writing allowed space for the reader as a fourth participation to consider their own thoughts. This polyvocal writing enabled a dialogic method that honoured my critical stance, resisted the singular authoritative voice, and invited the reader to engage as a participant.

Bernard:

Once more, denying the visceral interests of students for failing to harmonise with the mythology of modern Western society as peaceful and benign. Primary school teachers are likely nervous about pursuing any interest in guns for fear of zealous, parental repercussions.

Claire:

Bernard, be fair. ... Sensing they were restless and needed to move, Lisa did allow the boys to act as security on occasion. But, as Aaron pointed out, asking these boys to check all the doors became something of a byword for being playful and wasn't really the business of the drama.

Bernard:

An interesting choice of words! How does swiftly rejecting the boys' interest in security equate with the student-centred rhetoric? Security guards typically possess physical power, which can be instrumental in enacting obvious and extreme oppression. Perhaps the boys are interested in exploring physical power? NO, this isn't the power that MOTE wants you to explore, sanitised power only please!!! Isn't experimenting with physical power and its responsibilities and consequences something they should explore?

This example demonstrates how rewriting in role appeared within the thesis as a transcript of a conference monograph to capture and evoke the energy of debate for the reader.

Rewriting in role sought to stimulate an affective response, to stoke feeling, and to draw the reader into their imagination. This created a dialogic space that incorporated academic arguments, reflected divergent perspectives, and invited the reader into a spontaneous and dynamic debate. Having described the origins of rewriting in role and its application within my thesis, I will relate it to creative data analysis and detail the methods and approaches that informed its development.

Rewriting in role as creative data analysis

Creative data analysis invites researchers to access new ideas and ways through engaging with artistic forms (Shagoury, 2011). In contrast to positivist research, which values fixed objectives and singular answers, creative research recognises and actively seeks complexity. Creative engagement gives us license to play, to spark new imaginaries, and to develop texts that resonate (van Amsterdam et al, 2023). It acknowledges the influence of sociocultural factors upon the research and values the position and influence of the researcher (Everley, 2021). I consider my research to dance on the border of critical theory and postmodernism; I reject the position of authoritative expert and uphold the role of the body as a site of knowing (Lather, 1991; Torres and Reyes, 2011). As creative analysis, rewriting in role aims for a participatory method that values embodied ways of knowing and is dedicated to social justice.

Engaging in creative data analysis requires creative thinking and the deliberate use of creative tools that facilitate different ways of knowing and generating knowledge. Rewriting in role re-engages with data to write the analysis from the perspective of fictional characters, providing an imaginary play space that liberates the analysis from positivist notions of truth. A form of de-norming, it disturbs the processes that typically govern academic writing and analysis and invites the reader to be an active participant in the dialogue (Schwartz, 2018). The next section, 'Unpacking rewriting in role', outlines the theoretical foundations for rewriting in role as a creative data analysis method.

Unpacking rewriting in role

An interpretivist method, rewriting in role is consistent with a constructivist epistemology that contends our individual concept of reality is dependent upon our perception (Kincheloe, 2012). Furthermore, social and contextual factors that shape our construction of reality are inextricably linked to power (Foucault, 1980), and how we make sense of the world influences the sense we make. This concept continues to apply when rewriting in role, as the author's concept of the world continues to colour their rewriting. My sense of the world continues to leak into my fictional writing because it is *my* writing. I cannot claim to write authentically or objectively for another, but through generating within a fiction I can reorientate myself beyond my typical understandings.

The research paradigm governing rewriting in role draws upon critical theory and its obligation to social justice through dynamic reflective praxis. It further seeks to incorporate a postmodern critique of critical theory by openly confronting the dominant ideologies underpinning the research (Torres and Reyes, 2011). Consequently, rewriting in role incorporates arts-based methods to oscillate between critical theory and postmodernism (Bagley and Castro-Salazar, 2019). Arts-based methods reflect a concept of knowledge as unstable, ambiguous, and multidimensional, and they apply artistic processes as the primary means of understanding and communicating an experience (Leavy, 2014). The arts align with a social justice agenda by resisting dominant paradigms, embracing ambiguity, accessing subjugated voices, and creating fresh insights (Leavy, 2014). As critical arts-based research, rewriting in role recognises the position of the researcher and re-engages with the world through the imagination, to disrupt commonplace attitudes and beliefs. As Gerber affirms 'Using imagination and the arts in research allows for the capacity to include unconscious sensory-embodied ways of knowing, explore multiple ambiguous and paradoxical dimensions of knowledge, integrate diverse perspectives and voices, and reach beyond the obvious and the known into the realm of the unknown, concealed, or soon-to-be-known' (Gerber, 2022, pp 72–3). The imagination allows us to touch, understand, and illuminate beyond traditional forms, aligning with the aims of critical pedagogy to challenge and explore through dialectical and deconstructive methods.

Rewriting in role reflects a concept of knowledge as unstable, ambiguous, and multidimensional (Richardson and Pierre, 2005). It emulates the goals of dialogical performance, bringing the self into dialogue with the other to question, challenge, and debate. Rewriting in role permits a renegotiation of existing data to problematise and accept hybridity. Rewriting flattens hierarchies of power to cultivate authentic dialogue, invite the reader as a critical agent, and embrace a pluralist community of inquiry (Bradbury-Huang and Reason, 2007).

Transparent in its use of the fictional frame, this method fractures the authority of the sole academic through multiple voices to generate a fictional duo/multi-ethnography that enables the writer to dialogue internally and externally upon the page (Sawyer and Norris, 2013). This dialogue communicates the characters through their way of speaking so the reader quickly has a sense of them (Leavy, 2022). Grounded upon an extensive process of collating, reading, and sifting through original data, rewriting does not fabricate new material but offers a reflexive re-engagement with existing writing to create 'déjà vu prose that makes the familiar hum with newness' (Adams and Holman Jones, 2011, p 108).

Rewriting addresses the collaborative aims of qualitative research by presenting multiple (albeit fictional) voices in the text and exposing the shaping and reshaping of the researcher's questions to the reader. Rewriting is an explicitly creative act of construction and draws upon several related literary and drama arts-based methods to contribute a network of ideas, depicted in Figure 9.1.

Contributing techniques and theories

Writing in role is a process drama technique in which the participants write to or from the perspective of a fictional character to reflect, communicate, or further the action of the drama (Neelands and Goode, 1990). The writing acts as both a distancing and narrative technique as participants consider the viewpoint of a fictional character. Similarly, rewriting in role offers the researcher a distance from the research site and the opportunity to explore alternative perspectives.

Rewriting in role operates as writing as inquiry because it both encompasses the crystallisation of numerous accounts (Richardson and Pierre, 2005) and offers an analysis capable of reflecting a world of uncertainty (Leavy, 2014). Rewriting in role relies upon the strategic use of fiction to illuminate, complexify, and create

Figure 9.1: Contributing techniques and theories to rewriting in role

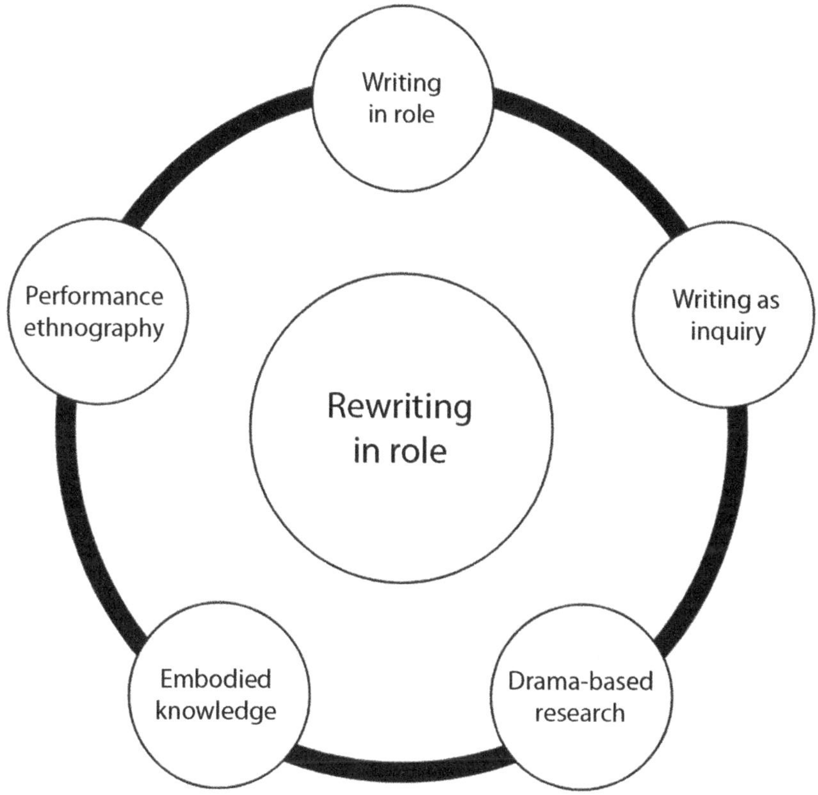

identifiable spaces that invite a range of participants into dialogue (Frank, 2016) and operates as a piece of theatrical social fiction (Leavy, 2022).

Rewriting in role, as a form of drama-based research (Norris, 2000), uses drama to generate meaning through the act of imagination for both the researcher and the reader. While rewriting in role is not a fully realised piece of theatre, it generates an imagined 'play' world to provoke and invite the reader as participant. Its reliance upon embodied knowledge and inherent theatricality further aligns rewriting in role with play building as inquiry (Harris and Sinclair, 2014). This performative element and commitment to creating a sensory understanding beyond words also resonates with performance ethnography (Denzin, 2018).

Rewriting is a pedagogical and ideological act of doing to evoke curiosity, criticism, and reflection from reader as participant. A deliberately dialogic form, rewriting privileges a polyphonic hearing that understands that utterances work simultaneously at multiple levels (Bakhtin, 1981, cited in Fecho and Meachu, 2007). I cannot 'be' another but, through rewriting, I can disrupt my comfortable position to engage with something new and different. This capacity to playfully shift the researcher's perspective may reignite the writing, illuminate new insights, and break from conventional ideas. The next section, 'How to rewrite in role', provides a guide on how to plan for and engage with rewriting in role.

How to rewrite in role

Rewriting in role may be employed in a variety of circumstances. While it may appear simple, rewriting in role requires forethought, planning, and the flexibility to deal with a messy, reflexive, and artistic process. You will need to write, rewrite, reread, rewrite, reread, and so on. It moves beyond simple reworking, through the use of the imagined other to engage in a dialogue with and between the author's emergent and ongoing thinking. It is a creative process and not a silver bullet and you never know what might happen. I cannot guarantee that the rewriting in role will appear in your final work, but, if you engage thoroughly, it can make a vital contribution to your analysis process. The following section, 'Resolving to rewrite', provides a schema for rewriting in role and reasons why you might employ this creative method.

Resolving to rewrite

The first crucial element to identify is why you want to rewrite in role and how it complements your existing methodology. This is not about what you hope to achieve through rewriting in role but establishing that your research aims and methodologies align with this method. You will need a fascinating complex question or inquiry to motivate the writing and keep you going when your creative powers might be lagging. This provides the tension – an essential ingredient for drama – that can arrest attention, motivate, advance the narrative, and sustain the energy required to propel the fiction (Heathcote and O'Neill, 2015).

Digging into data

Prior to the analysis you need to have spent time in and among the original data. You need to know the material thoroughly, so that when discussing it from the perspective of a fictional other, you can be confident in the viewpoint they hold and maintain its integrity. A creative method of research relies upon the repeated revisiting of data and inductive analysis. To ensure the analysis remains grounded in the rewriting as interpretation and analysis, it needs to be continually reflective and reflexive (Everley, 2021). You will need to take pauses to recognise and unpack the shifting social contexts and power dynamics within the research and between you and the research (Fletcher, 2013). This critical reflexivity is a continuous task that correlates with creative analysis methods as ambiguous and unsettling. In qualitative analysis, the analysis phase frequently begins during data collection, so you may find many of your character's thoughts beginning to form as you progress. This does not mean you have already done the analysis, so spend time digging through the data, doodling questions in the margins, and reading, rereading, and pondering. This preparation and ongoing engagement with the data is essential to rewriting but hopefully should not be painful.

Inviting imagination

Be joyful and engage thoroughly in the imaginative space, create the world of the character or characters, and flesh out the detail prior to writing. This will take time and a willingness to play around with ideas, change things, abandon things, rework, and reimagine. If you can commit to rewriting and engage with it, as an iterative, sometimes wonderful, and sometimes deeply frustrating artistic process, it will be worth it. Prepare like an actor for a role, forage for images, sounds, objects, phrases, or any other stimuli to help immerse yourself in the fictional space. When writing as *Caitlin*, I listened to mid-1990s pop music to stimulate my imagination and further build belief in the fictional world she inhabited.

Creating character/s

Rewriting in role is a purposeful act and a theatrical event and the characters are part of that theatricality. The most important aspect of rewriting in role is the selection and development of the role/s to ensure they make a valuable contribution to the writing. You need to consider the alternative voices, perspectives, or aspects of self you wish to bring into dialogue with the research. Once you know the type of voice, expertise, and perspective required, you need to 'create' that character so that you have a solid reference point and clear guide when writing in the role. Although this character will likely evolve during the process, you need to start somewhere. If your character is vague, it will be difficult to invest in them or write in 'their' voice. If they are indistinct and uninteresting, the voices risk becoming muddled and confusing rather than clarifying your thinking. The drama convention role-on-the-wall offers a useful place to begin creating your character/s.

Role-on-the-wall convention

As illustrated in Figure 9.2, draw a figure on a sheet of paper to represent a body with no bias towards age, class, gender, or ethnic origin. On the inside of the image add what the character thinks about themselves – the private aspects of the character – and on the outside of the image their external characteristics and what others think or know about them.

Gather any other materials that will help you flesh out the character; these might be images, quotes, and film clips. The better developed your character is, the stronger their voice will be and the easier it will be to write. Give them a fully realised background, family, occupation, hobbies, favourite films and so on. This is your opportunity to be creative and inventive. This is your character; they

Figure 9.2: Role-on-the-wall template

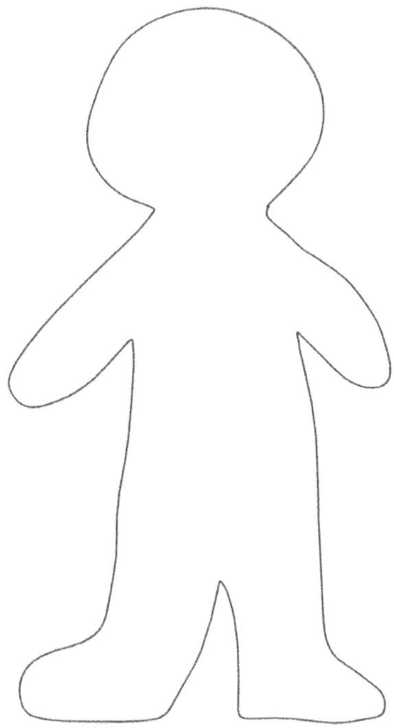

need to have an internal coherence for you, they need to be purposeful but also playful and not simply replicate stereotypes. You may, however, make deliberate decisions about the nature of that character; *Bernard* was deliberately designed in part to lampoon critical pedagogy scholars because I needed a character capable of hyperbole and controversy. Here are some of the details around the character of Bernard Garcia Jameson discussed earlier.

Bernard Garcia Jameson is an eminent scholar in critical pedagogy. He is a frequent guest lecturer at La Fundación McLaren de Pedagogía Critica and co-director of the Centre for Freirean Education (see also Figure 9.3).

Name: Bernard Garcia Jameson
Age: 57
Nationality: Irish/English
Hometown: Kent
Current Residence: London
Occupation: Professor
Income: $170,000
Talents/Skills: Critical pedagogy
Favourite Film: Akira Kurosawa's *Seven Samurai*
Favourite Music: Bob Dylan

Figure 9.3: Bernard Garcia Jameson (Sean Coleman)

Writing/reading/writing/re-acting

Once you have created the character, write a short excerpt in role about one of your research themes or questions as an opportunity to find their voice. This is likely not the finished 'voice' of the character but will give you material to generate and inspire ideas. As with any writing, let it sit and then reread it, in relation to the character description and the wider context of the 'imagined space' the character is operating in. Consider the social and contextual setting for this 'writing' and ask yourself the following questions as you read.

- How does this rewriting relate to the original source material?
- Does the rewriting capture the important ideas of the research?
- Does the rewriting leave space for the reader to wonder?
- Can you imagine someone responding to the writing? What might they say?
- Can you write a response to the character's writing?

If it does not work then go back and try again to see if you can identify what is missing: Do you need to develop the character more or saturate yourself with more images/texts/material that may inform the character? While this is not a performance piece, establishing a good foundation for the text will help to identify whether it really works as dialogue or the voice of another. Even an informal reading will enhance your understanding of the rhythms of the dialogue and signal where the writing does not sound like a genuine voice. This parallels with the creative process of the dramaturg who will workshop the initial script with actors (Wiltshire, 2020). While writing the '*Dumb Gulper Shark*', I recruited several colleagues at various points to stage excerpts of the fictional conference panel via Zoom and workshop the writing through performance. An example of this is available on YouTube at https://youtu.be/jg8HndO_fa0.

Rewriting in other spaces

Since developing the method of rewriting in role, I have shared this with colleagues as a method for exploring research through an alternative lens, refreshing their connection to the research and crafting their writing for different audiences. Many have recognised the value of rewriting in role, sometimes as an analysis tool or to process and reconsider data. As a creative analysis tool, it can be used to reframe and reconsider existing data from an alternative and seemingly extraneous viewpoint to gain new insights. These responses have encouraged me to apply rewriting in role as a mechanism to reconsider my multiple roles within teacher education, as teacher, teacher-educator, parent, and researcher.

Conclusion

Reflective of process drama, rewriting in role draws the research to the reader and the reader into the research. It exposes the research process and allows space for the reader to engage as a fourth voice, as research for provocation, not proof. Meaningful characters with specific viewpoints, knowledges, and questions are developed to ensure that both the reader and writer are invested in and affected by the fiction.

Rewriting rejects the dominant narrative available in the academic discourse and invites voices to question and counter the authority of privileged perspectives. This restaging allows for a performance of possibilities, not superior or definitive but exploratory. It blurs the lines between the giver and receiver of knowledge to facilitate a collaborative generation of knowing. Rewriting in role presents and engages with the data to reflect meaning (Everley, 2021) and combines inductive and deductive thinking to develop and explore theories in collaboration with the reader (Konecki, 2019).

I have found rewriting in role, as a creative analysis method, to be enjoyable, surprising, and academically vigorous. It requires a thoughtful and extensive engagement with the data that honours the research while generating resonances that go beyond a restatement of dry text.

Through this exploratory form, I can playfully dialogue with ideas and resist habitual thinking. I believe rewriting in role has value for analysis, because it aligns with the nature of research as a process of curiosity and wonder. Research is an act of imagination; it relies on asking new questions of the world and imagining different answers. Rewriting in role is an opportunity to keep asking questions, to interrogate thinking, to value diversity, and to engage our creativity for an unknown future.

References

Adams, T.E. and Holman Jones, S. (2011) 'Telling stories: reflexivity, queer theory, and autoethnography', *Cultural Studies ↔ Critical Methodologies*, 11(2): 108–16.

Bagley, C. and Castro-Salazar, R. (2019) 'Critical arts–based research: a performance of provocation', *Qualitative Inquiry*, 25(9–10): 945–55.

Bradbury-Huang, H. and Reason, P. (eds) (2007) *The SAGE Handbook of Action Research Participative Inquiry and Practice* (2nd edn), London: SAGE Publications.

Coleman, C. (2019) 'Dancing through the fourth wall: process drama as an enactment of critical pedagogy', PhD thesis, University of Auckland.

Denzin, N. (2018) *Performance Autoethnography: Critical Pedagogy and the Politics of Culture* (2nd edn), London: Routledge.

Eriksson, S. (2007) 'Distance and awareness of fiction: exploring the concepts', *NJ*, 31(1), 5–22.

Everley, S. (2021) *Using Creative Arts-Based Research Methods in School Settings: Understanding and Empowering Children and Young People*, London: Taylor & Francis.

Fecho, B. and Meacham, S. (2007) 'Learning to play and playing to learn: research sites as transactional spaces', in C. Lewis, P. Enciso, and E. Moje (eds), *Reframing Sociocultural Research on Literacy: Identity, Agency, and Power*, New York: Routledge, pp 163–87.

Fletcher, G. (2013) 'Of baby ducklings and clay pots: method and metaphor in HIV prevention', *Qualitative Health Research*, 23(11), 1551–62.

Fletcher, H. (1995) 'Retrieving the mother/other from the myths and margins of O'Neill's "Seal Wife" drama', *NJ Drama Australia Journal*, 19(2), 25–38.

Foucault, M. (1980) 'The eye of power', in C. Gordon (ed), *Power/Knowledge: Selected Interviews and Other Writings: 1972–1977*, New York: Random House, pp 146–65.

Frank, K. (2016) 'The management of hunger: using fiction in writing anthropology', *Qualitative Inquiry*, 6(4), 474–88.

Freire, P. (2005) *Pedagogy of the Oppressed*, New York: Continuum International.

Gerber, N. (2022) *Imagination and Arts-Based Practices for Integration in Research*, Abingdon: Routledge.

Harris, A. and Sinclair, C. (2014) *Critical Plays: Embodied Research for Social Change*, vol 11 (1st edn), [online]. Rotterdam: Birkhäuser Boston. Available from: https://www.researchgate.net/publication/301219725_Critical_Plays_Embodied_research_for_social_change

Heathcote, D. and O'Neill, C. (2015) *Dorothy Heathcote on Education and Drama: Essential Writings*, London: Routledge.

Kincheloe, J. (2012) *Teachers as Researchers: Qualitative Inquiry as a Path to Empowerment*, London: Routledge.

Konecki, K. (2019) 'Creative thinking in qualitative research and analysis', *Qualitative Sociology Review*, 15, 6–25.

Lather, P. (1991) *Getting Smart: Feminist Research and Pedagogy within the Postmodern*, Critical Social Thought, New York: Routledge.

Leavy, P. (2014) *Method Meets Art Arts-Based Research Practice*, New York: Guilford Press.

Leavy, P. (2022) *Re/Invention: Methods of Social Fiction*, New York: Guilford Publications.

Neelands, J. and Goode, T. (1990) *Structuring Drama Work: A Handbook of Available Forms in Theatre and Drama*, Cambridge: Cambridge University Press.

Norris, J. (2000) 'Drama as research: realizing the potential of drama in education as a research methodology', *Youth Theatre Journal*, 14(1), 40–51.

Richardson, L. and Pierre, E (2005) 'Writing: a method of inquiry', in N. Denzin and Y. Lincoln (eds), *The SAGE Handbook of Qualitative Research* (3rd edn), London: SAGE Publications, pp 959–78.

Sawyer, R. and Norris, J. (2013) *Duoethnography*, Understanding Qualitative Research, New York: Oxford University Press.

Schwartz, V. (2018) 'De-norming the syllabus', in R. Winkle-Wagner, J. Lee-Johnson, and A.N. Gaskew (eds), *Critical Theory and Qualitative Data Analysis in Education*, London: Routledge, pp 61–73.

Shagoury, R. (2011) 'Crystallization: teacher researchers making room for creative leaps in data analysis', *LEARNing Landscapes*, 4(2). Available from: https://doi.org/10.36510/learnland.v4i2.401.

Sinclair, C. (2020) 'Parody: fake news, regeneration and education', *Post Digital Science Education*, 2: 61–77.

Taylor, P. and National Association for Drama in Education (Australia) (1995) *Pre-Text and Storydrama: The Artistry of Cecily O'Neill and David Booth*, Brisbane: National Association for Drama in Education (Australia).

Torres, M. and Reyes, L. (2011) *Research as Praxis: Democratizing Education Epistemologies*, New York: Peter Lang.

Turner, V. (1974) 'Liminal to limonoid, in play, flow, and ritual: an essay in comparative symbology', *Rice Institute Pamphlet*, 60(3): 53–92.

Van Amsterdam, N., van Eck, D., Kjær, K., Leclair, M., Theunissen, A., Tremblay, M. et al (2023) 'Feeling clumsy and curious: a collective reflection on experimenting with poetry as an unconventional method', *Gender, Work & Organization*, 20(4): 1–20.

Wiltshire, K. (2020) 'Revising, dramaturgy and the theatre workshop', in G. Harper and J. Kroll (eds), *Creative Writing: Drafting, Revising and Editing*, London: Macmillan International, pp 96–111.

10

A mosaic of siblings of cystic fibrosis: a creative dramaturgical analysis

Amie Hodges[1]

Introduction

When I first started this project, the idea I had of creativity in my research initially lay within the use of visual and creative research methods to engage children in their narrative story. I can honestly say that I had not anticipated what would unfold as I moved out of my comfort zone into the field, towards the analysis phase of the data. I had not heard of dramaturgy or even considered a dramaturgical analysis. I think that in those early days my thoughts were in a systematic analysis mode. Anything other than that, in my 'imposter syndrome' moments, I thought would have been beyond my scope. I was undertaking my doctoral fieldwork and the thought of getting to the analysis phase scared me. My thoughts were: 'How do I do this?' 'What if I am analysing data incorrectly?' 'How do I get this right?' I managed to put the imposter to one side and gradually progressed in my journey. Keeping an open mind, I listened, read, networked, and reached a stage where I was not afraid to be creative. My approach, I was informed, was like Marmite, as one could love it or hate it.

I write these few words to begin my chapter in case these thoughts of imposter syndrome or fear or the 'how do I do this?' resonate with you as the reader, particularly if you are just starting out in considering ways of analysis.

This chapter will set the context of the project and articulate my analytical practices in the exploration of the experiences of child siblings within their family, in the context of cystic fibrosis (CF). It will examine my intertwined use of creative data analysis based on Goffman's (1959) dramaturgical framework, as well as the systematic use of my adapted version of Wolcott's (1994) framework in working with my data set. I use examples from data that include narrative scripts and dramaturgical scenes encountered during my field visits to demonstrate application. Sensory moments that aided the analytical process are highlighted, as well as discussion around imagination. This was used to bring the data together, through what I termed dramaturge fantasia. In using the imagination one can focus on an aspect of data to create an image that can bring a fantasy to life and provide a new meaning to the context of a situation (Edgar, 2004). This can serve to help in terms of problem solving, so researchers should not be afraid to visit what may feel like the extraordinary spaces of imagined worlds, as these can create new logic (Marsten et al, 2003; Edgar, 2004).

My roles and positioning, which are pertinent to the work, are also explored to provide greater insight into consideration of self when working creatively with data.

Context

The project

This project aimed to achieve specific insight into sibling worlds in the context of CF, with research conducted within the home (n = 10 children aged 7 to 11 living with a brother/sister with CF) (Hodges, 2016). There was a need to inform health-care professionals and families of the sibling experience, due to its limited recognition in the literature. Analysis of the findings informed family-centred care, an important aspect of children's nursing practice.

My plan was to conduct four one-hour visits with non-CF children in their family homes. This consisted of time for introductions, 'getting to know you' activities, and introducing the participants to the idea of independently creating pictures, photographs, collages to represent their own visual impression and interpretation of their world. This was in relation to the topics of 'me and my family' and 'future selves', prior to them having two narrative interviews on separate occasions, where they could talk about their focused creations.

In congruence with dramaturgy, the family home was the stage set and the participants and their families were the social actors.

In the field

During my visits, family context and social interaction was also observed. All the families were very welcoming and even though the focus was on the non-CF sibling other family members wanted to tell their stories (Hodges, 2016, 2018). As I got to know the families over the four visits, their constructed and co-constructed social interactive performances unfolded (Goffman, 1959). It was this performance and impression that added to the data set, along with the creativities that the children also brought to their narrative.

The children and families employed a range of creative methods to tell their stories, and this went beyond my initial proposal of pictures, collage, and photographs. It included songs, psalms, singing, dancing, role play, and using props and artefacts such as dolls and a doll's house. Some children communicated with voice and some in silence. The creations the children used elicited their verbal and silent narrative performance, which I combined with my researcher observations. The children referred to the participatory activity as 'homework' as I would not be present during the creation making. This added to my data set of over 200,000 words.

It is important to note I was working with a vulnerable participant group and ethical approval was gained for the project. All the participants received pseudonyms to maintain anonymity. Children negotiated time scales and activities throughout the work, to prevent them becoming burdened or pressured in any way. A parent was at home during visits and within close range in case a child did become upset for any reason. This was an important consideration as some children did require parental support.

The concept of dramaturgy

For those who are not familiar with the concept of dramaturgy, I have added further explanation here.

Looking through a dramaturgical lens, Goffman (1959) views the world as a stage, and views people as social actors. Goffman's framework is focused on performance, expression, and impression management, which include frontstage and backstage presentations. These presentations can be contextual to a setting or environment. For example, in the frontstage presentation, someone may want to provide a good or professional image of themselves and may not show their true self or feelings. Whereas backstage in a different contextual situation, a person can present how they really feel, being their authentic self. This can be linked to a situated identity where a person is playing a particular role at a given time and may have motive for acting in a particular way. Frontstage can be identified with public self and backstage can be identified with private self. One's costume is relevant to the presentation and performance, for example, what someone chooses to wear, hairstyle, make-up. This holds relevance in the frontstage in terms of impression giving.

Choosing dramaturgy as a framework

It was not until I entered the field, beginning the process of producing data with participants, that the notion of using dramaturgy as a creative analytical framework emerged. It was almost as if dramaturgy chose me as opposed to me choosing it, as events unfolded.

It was during my first couple of family visits, observing the siblings and their dynamic interactions with each other and with me, that I felt as if I was in the audience as well as a participant of a stage set. At this point the notion of using dramaturgy as an analytical framework emerged. It was useful making this decision early on in my fieldwork, as my senses of observation and my level of self-awareness heightened as I progressed through the visits. As a researcher I was present, but also removed.

Positioning

The researcher

An important consideration as a researcher engaging in creative data analysis was my positioning of self within the work. My axiological position was highlighted in outlining my values, beliefs, reflexivity, and transparency in my journey to becoming a researcher in an autobiographical account, whereby I consider my own family and sibling experience (Durant-Law, 2005; Hodges, 2016, 2018). Furthermore, my physical positioning as a researcher was not always static as I recognised that I was also a social actor and interactor within the work when engaging with the siblings and their families, as well as taking on roles of dramaturge, bricoleur, and reflector using dramaturge fantasia within this study. My interaction within these roles was as an adult, a researcher, a professional with a nurse/education background as opposed to being a child or family member. This all had relevance in stating my position for creative data analysis.

The dramaturge

A dramaturge can be referred to as the playwright, the writer of plays. Being a dramaturge is a duty-bound role with the need for 'critical sensibility' in preparing a text for performance and advising the actors, as well as editing, interpreting, and reordering the scenes to create a final production (Katz, 2014, p 13; Cardullo, 2014).

This was significant to note as the role of the dramaturge can be helpful in the meaning-making process for the actors and the audience.

This held relevance in terms of the preparations in the study proposal in addition to being in the field, interacting with children and families, generating, producing, and collecting data before the move to skilful creative analysis, interpretation, and editing of findings. Hence, throughout the research process I was mindful, in terms of critical sensibility, to represent what was meaningful to the actors in terms of their narrative performance. This included their creativity through voice and non-voice. Being a dramaturge, using my knowledge, skills, and experience enabled analysed data to be written as a three-act play.

The bricoleur

A bricoleur is a qualitative researcher or craftsperson who uses multiple methods, media, strategies, practices and resources, with tools to explore and solve contextual problems, questions, or jobs (Denzin and Lincoln, 2000). I advocated using multiple creative methods with participants to enable freedom of expression in impression management. Their use of singing, dancing, riding a scooter, and creating pictures, alongside other activities, provided aesthetic flair to the project, creating a portrait of sibling experience that added depth and meaning to the data for analysis because siblings were the narrators of their own creations.

Working with the data

I used an intertwined systematic and creative approach to work with my data as it felt complex. Using the duality was helpful in the analytical process as I was trying to bring together participant storied narratives and performances, as well as my own observations of sibling interactions and impressions. Additionally, I was trying to separate my own voice from that of the participants, through imaginative reflexivity, to create a separateness of self but a togetherness of siblings within their family. Using more than one approach positions the communication between the narrator, the characters, and the audience to capture the doing and telling of a story (Llangellier and Peterson, 2003). This was pertinent within the analytical process as the approach I used enabled the experience of the participants to be captured within the context of their family group, providing meaning to the work. This is synonymous with Goffman's performative analysis in terms of viewing the social action that is ensued in the coaxing, influencing, and transitioning of an audience, through an accomplished skilled performance that moves beyond words alone.

Systematic approach

In my systematic approach I adapted Wolcott's (1994) framework, which suggests a three-stage process for analysing qualitative data, to include description, analysis, and interpretation. My adapted version incorporated an organisational phase and referred to the descriptive phase as preliminary, these additional phases were used in my analysis. Table 10.1 outlines my systematic approach.

Creative approach

From a creative stance I applied Goffman's (1959) dramaturgical framework of social interaction to my work, with the focus on impression management of the performance given.

My key areas for consideration included participant roles and positioning, their narrative scripts, costumes, stage, and sets as well as their expression, impression, and their frontstage and backstage performances. Table 10.2 outlines key elements of the creative focus.

The creative analytical process included the exploration of the sibling narrative. This integrated narrative interviews, observation of performance, and family social interaction, creations, visual activity, and presentation of the sibling self within the context of the family environment. This provided insights into sibling position, space, roles encompassed in their overall self-presentation in their familial and cultural milieu, performing on the stage set of family life within their home. Through interactive performances non-CF siblings acted out scenes of multiple presentations of self to reveal the ideology and reality of their lives as a consequence of the opposing force of their siblings' CF.

It is important to note that CF was an opposing force due to it being a life-limiting condition that can have a burden on family life as a result of the complexity of the disease and the focus of attention being on the child with CF, due to daily treatment regimens and additional vigilance (Hodges, 2016, 2018).

Table 10.1: Systematic approach to data analysis and interpretation

Phase 1 Organisation Using Microsoft Word, individual files were created for each participant in their sibling/family group. Individual files were placed in order of visit (chronological). Creative artefacts, pictures, collages were labelled and placed in the sibling/family file. Sonic software program was used to upload audio tapes. Creative analytical framework was chosen (see Table 10.2 for creative process).
Phase 2 Preparation An iterative process was used. Narratives and observations were transcribed following each participant/family visit. At the moment of transcription, each narrative/performance was read. Each recording was listened to alongside reading each narrative. Each recording was listened to separately. Initially this cycle was completed with the first five participants, before engaging the same cycle with the remainder of the siblings.
Phase 3 Analysis An iterative process was used. I repeatedly reread, listened to, and visualised transcripts pertaining to narratives and performances. Each sibling/family group visit narrative and performance was discussed with my PhD supervisors. I attended masterclass workshop. Key themes were identified and collated, synonymous with the creative framework (see Table 10.2). Identified themes collated in relation to stage presentation, roles, and performances identified and described. Completion of data generation.
Phase 4 Interpretation (Links with meaning making and conclusion in Table 10.2.) Three key acts were identified, with thematic/individual scenes outlined to formulate a drama production and a discussion.

The intertwined process

The analysis process was iterative due to the cyclical and repetitive stance taken in sifting through data generated. Participants had presented their creativities, including pictures, songs, objects, and artefacts. They provided their own narrative interpretation, which was recorded and transcribed, and verbatim observer notes were also taken.

Following organisation of the data and choosing a framework for the analysis, my immersion commenced with repeated listening to audio transcripts alongside reading written narratives. Where a creative artefact was being spoken about, for example a collage or poem, I would have this alongside me, to provide context.

Where audio transcript was not used, only the field notes were repeatedly read. During this first phase of analysis, I was only halfway through my data generation with five participants and did not have my full data quotient. During this preliminary stage it was my aim to become familiar with the transcripts and stories that were emerging. My focused strategy was to gain an understanding

Table 10.2: My creative analysis approach

Participants and role players Identify key social actors and role players. Differentiate between the protagonist and antagonist.
Plot Link to question, study aims, and objectives. Consider if there is an opposing force with rationale for its identification.
Stage setting Identify each key stage set (eg the home was relevant to my study). Identify the substage set (eg specific rooms used within the home).
Costume What clothes are the social actors wearing? How are they presenting themselves?
Stage presence What is the stage presence of social actors? (eg frontstage, backstage). What social spaces are they occupying? And why? What is the situational context? (Overlaps with performance.)
Performance How are they presenting themselves? What is the situational context? Consider each scene (may be drawn on further beyond stage presence, ie in the act itself). What is the performance management/expression of the social actors and players? What impression are social actors giving to the audience? (The researcher and or other players, characters present.) Identify individual social interactions. What tools, props, artefacts are used within social actors' performances? What character and role are they playing? What is the emotional tone/stance of the characters? Also consider authenticity, contradictory performance.
Meaning Capture the meaning making of performance management. What meaning are the narrative stories and performances providing for the social actors and the researcher? Revisit the performance narratives acted out, spoken, or silent. What is the overall story trying to tell you as a researcher? What is the researcher's impression and interpretation of the performance and story?
Key scenes and themes Outline key scenes and themes (I did this using the drawing of the stage settings and headings). Conclude with key acts and production of the findings as a written play/drama.

Source: Table based on Goffman's (1959) dramaturgy.

of individual accounts of the participants. Therefore, I was getting to know the non-CF sibling to understand the context of their world, then made comparisons between each sibling/family group to look for common themes.

Initially I felt lost within the work and found it difficult to categorise and code the information but I highlighted and coded chunks of text in relation to key events of complexity in the disease trajectory, such as when a CF sibling was admitted to hospital or had an intervention, as well as highlighting the chaos of emotional responses of the non-CF sibling during our interactive dialogue in the transcripts.

Congruent with Goffman (1959), during this first phase I identified stage performances in terms of early presentation of self, space, and visual images during my initial visits. In trying to make sense of it all I wrote reams of draft notes in my attempt to capture themes. However, some early codes in the preliminary phase lay dormant and were not placed into Goffman's (1959) frontstage or backstage category until further data was generated.

This preliminary immersion provided early insight into sibling worlds and heightened my awareness of my presence within the research environment. As further data were collected from subsequent children, I read and reread the transcripts, listening to the recordings where available.

I used self-visualisation when reflecting on the visits, to relive the scenarios and stories encountered with the children and their families. In my visualisation I would place myself back in the participant's family home, to the stage set and the situational context of the narrative. Along with the transcript and the voices I could place myself back into situations, to hear the stories, see them, and feel them over again. While this process weighed me down emotionally, drawing on the senses (re-experiencing the emotional tone in the situational context along with my emotion) for the purpose of analysis was something I could not escape and it served to add to the analytical meaning-making journey (Hodges, 2018). This resulted in further codes being generated through Goffman's (1959) dramaturgic lens. During this process, support was gained through supervisory discussion meetings.

I generated extensive notes from this holistic sensory approach. Sibling roles and positions within the family became more readily identifiable in addition to my heightened awareness of my own position as a researcher who had subtly (but not consciously) shifted a family dynamic in placing focus on the non-CF sibling (Hodges, 2016, 2018). I also highlighted recurring codes across all the sibling family groups in relation to occupied spaces, well-being, and emotional status. At times I felt I was breaking things down too much, for example, only focusing on key words and short segments of information. Hence, I pieced the segments back together to further examine the fullness of narratives to see what each one was trying to say. Like a jigsaw, I was deconstructing and reconstructing the storied performances to provide researcher interpretation.

At each stage of the process, individual cases, identified codes, and extensive notes were discussed with my supervisors. I began to consider the question 'what is each story trying to tell me?' I reread each story with all codes and notes removed, and I finally saw a theatrical play with the actors, the stories, the frontstage and backstage performances. In the iterative process I had finally come full circle. During discussion with both of my supervisors I was able to begin to talk out and bring the key acts together to formulate and frame my data findings as a dramaturgic theatrical production.

Part of this process included the adaptation of creatively drawing an image of a curtained theatre stage set, representing a dramaturgical lens. It contained three stage settings with three headings and, under each of the headings, key aspects of each of the stories were captured.

1. *Presentation of character*
   Non-CF sibling performance themes were noted, including situated stage presence, as well as sibling self-expressions. Places and spaces were also listed as themes alongside character performance. A question mark was placed for sibling motivation for presentation of their character.
2. *Performance within the family*
   This included a list of the different roles that the non-CF siblings undertook. Authenticity, emotional stance, and expectation were also listed in terms of performance data.
3. *Risk of going out of character*
   This identified performance behaviour such as acting out, wisdom, disruption, reality. A subheading of stage fright was also captured within the drawing.

Drawing the stage set to discuss key performances, narratives, visual representations, and dialogic interactions enabled what I called 'segments of the sibling mosaic' to be foregrounded (Hodges, 2016). Using Goffman's (1959) dramaturgic lens I portrayed the findings of the parallel lives of siblings living with a child with CF. I presented the findings of different stage performances as a written three-act play. A play that was based on real people, places, and experiences, which were temporal to the moments of my research visits.

Creatively working with the data, the non-CF siblings were referred to as 'protagonists' as they were central to the work. The CF siblings were referred to as 'antagonists', regardless of whether their sibling relationship is that of friends or rivals, or viewed with ambivalence. Using the term antagonist is not meant to label CF siblings in a negative sense, but merely provides a dramaturgic view of character along with an experiential context of performance and impression.

The key family places and spaces that were visited in the siblings' homes were referred to as stage sets and were relevant to the context of the situation. Parents and other family members were classed as key players, as, through the dramaturgic lens, they are viewed differently to the antagonist and protagonist, because they are the main carers in the child's world. The protagonists were presented within the context of their family world to ensure greater understanding of their experience. They were not viewed in isolation, because the family is the mainstay of their life.

In presenting the data as a play, the disease of CF was the opposing force within family life, because the sibling narratives exposed that living in the context of the disease can present challenges for the protagonist. Having a sibling with CF who is acting as an opposing force (antagonist) to the protagonist in terms of their experience was identified in the data and presented within the acts' scenes and stage sets within the play.

Stage scene

Here I have outlined a short case scene example to represent an element of an interaction with David (age 10), a non-CF sibling, and his brother Henry, aged 5 (Hodges, 2016, 2018).

> 'Are you here to see me?'
>
> While I am sat with David on the sofa and talking to his mum in the lounge, David's brother Henry runs into the room semi-naked having been in the paddling pool and runs over to me. He is full of life and shouts, 'Do you want to see me?' 'Are you here to see me?' Henry runs up to me and stands directly in front of me while David is sat quietly to my left on the sofa and he does not say anything. David looks sullen with his head slightly bowed.
>
> When he [Henry] realised my focus was on David as a protagonist, he immediately jumped into his mum's arms.
>
> She gave him a momentary cuddle, much to the dismay of David who pulled a face to show that he did not want Henry to be getting his mum's attention and he asked his mother why she was holding Henry like a baby when he was no longer a baby.

In presenting the data here I am viewing the stage set of the lounge in the family home. The social actors are David, Henry, their mother, and me, and the roles in this scene are brothers, parent, and researcher.

In terms of costume, David is in his school uniform and Henry was in a swimsuit as he had been in the paddling pool. What was pertinent in the situational context of this scene was the social interaction, identification of positional space, and performance. Within this scene Henry's character is excited, lively, and vocal. He has not met me before, but he was keen to engage in conversation and he was upfront about wanting to be noticed as he ran around the living room with only his swimming shorts on. He then ran straight up to me to ask if I was there to see him. He was not used to people coming to see his brother as normally it was he who was being seen by someone. Henry had assumed and expected me to be there to specifically for him due to his CF.

When he realised my focus was on David as a protagonist, he immediately jumped into his mum's arms for a cuddle, much to the dismay of David, who pulled a face. With Henry's words 'Do you want to see me?' and his position of directly facing me, in theatrical terms, as an antagonist sibling with CF, he was forward in his approach, wanting to be the focus of attention and letting the dramaturge know that that was his normal place.

In this scene Henry gave the impression that he wanted to maintain a central focused position, which David did not like.

It was a challenge working through and analysing all the data and it was not possible to include all the information into the findings. However, in choosing the data represented in the scenes, it was of ethical importance to represent each child using the information that was meaningful to them.

Dramaturge fantasia as a form of reflexivity

In keeping with dramaturgy, I adopted what I referred to as 'dramaturge fantasia' as a form of imaginative reflexivity to present and express my emotional self and bring the participant data together (Hodges, 2016, 2018). Gabriel (2015) advocates for the use of imagination in research, as it helps to identify the creative opportunities that data can provide.

Using researcher reflexivity through imagination enabled the challenge of separating my voice, thoughts, and feelings from those of the participants. I considered my situated self in relation to the impression management and performance data collection and analysis. This was supported by writing my reflexive notes and engaging in discourse with my supervisors.

Dramaturge fantasia allowed my own metaphorical expression, which provided greater transparency, reflexivity, and understanding of self and participant performance, and impression management within the context of the data. This was also helpful in terms of dealing with the emotional labour of the research journey, which is also an important consideration for creative data analysis (Loughran and Mannay, 2018).

I used dramaturge fantasia (imagination) as a way of bringing the participants' data together. From my first visits I began to understand their worlds and family lives, and how they presented, performed, and interacted. I was conscious of the separateness of each of the families I worked with and how I could not bring them together in their real worlds due to the high risk of cross-infection that can impact in the context of CF (Hodges, 2016, 2018). Using the dramaturge fantasia also enabled me to capture shifting moments in terms of the data analysis.

What follows is an example of dramaturge fantasia reflexivity that captures a moment of recognition in the shift of data presented.

> The dramaturge sits facing the front of the stage, for some time she has been watching, waiting and interacting with all of the protagonists as she has visited each and every one of them in their homes and she has spent time in their key stage sets. She has gotten to know them, and they have begun to get used to her. She has a feeling that something is different today. All of the children are lined up on front of the stage no longer in their parallel isolated worlds where families with children with CF are unable to meet but they are together in mutual connection and something is changing, something is happening.
>
> The dramaturge watches as all of the children take one step forward and one by one they each join hands Charlotte takes Molly's hand and Molly

> reaches out for Josh taking his hand as he reaches out for Elinor and she smiles at Nancy as she clenches their hands together tight, Raff doesn't speak he just gives a nod then David, Erin, Hattie and Adam all move together. Charlotte takes the lead as she walks down the steps at the centre of the front stage, all the children follow still holding hands as they move towards the dramaturge and then they beckon her to follow. They walk around the edge of the stage and up some steps to the right where a black curtain is draped; it is partially concealing a locked door that Charlotte reveals. She looks to the children, and they all give a mutual knowing look. 'It's time' they all say as Molly moves forward to unbolt the lock on the dark door. As the door opens, they all move to enter a dark space, but it's okay because they are not alone today and they feel safe. As they go inside their hands separate and they look behind, they are looking for the dramaturge, as they want her to be there. When they see that she is with them, they all say, 'It is time to move backstage. Come on in, we are ready to show you.'

I chose this example as it represents a noticeable shift in the data that is pertinent to me as a researcher. It identifies the moment in data analysis when the protagonists were ready to move from what Goffman (1959) would see as presenting a frontstage stance in the data set towards a readiness to providing data that represents letting me into their backstage worlds, as they had come to know and trust me as a researcher and visitor in their lives. They had perhaps felt a readiness to let their guard down to take me backstage where they talked about the realities of living with a brother and / or sister with CF. This shift was a key defining moment within the data and demonstrated the benefits of four family visits, as I may not have had access to such privileged information with one research visit per child. In some ways my analysis started before I even realised; I was analysing, but the analysis became more formal with my research trajectory.

Conclusion

Preparing for dramaturgical analysis requires time, commitment, motivation, and determination to provide the scaffolding to work creatively with the data. Initially, actors and audiences (child, family, researcher) need to be prepared for their participatory roles, which can be enhanced using the appropriate creative resources, and safe spaces and places need to be provided to amplify their voices. Through persuasive, impressionistic, metaphorical, and creative storied participant performances data was generated for analysis and interpretation using Goffman's (1959) framework and adapting Wolcott's framework (1994). Using a dualistic creative and systematic approach, and using the reflexive imagination through dramaturge fantasia, provided robust and meaningful findings that enabled presentation of the data as three staged acts, presented as a written narrative play called *Segments of the Sibling Mosaic.*

Note

1 I would like to thank the participants who made this study possible, in addition to my doctoral supervisors Professor Daniel Kelly and Professor Katie Featherstone.

References

Cardullo, B. (2014) *What is Dramaturgy?* New York: Peter Lang AG.

Denzin, N.K. and Lincoln, Y.S. (eds) (2000) *Handbook of Qualitative Research* (2nd edn), London: SAGE.

Durant-Law, G. (2005) 'Soft systems methodology and grounded theory combined: a knowledge management research approach', *Online Journal of Knowledge Management*, 2(1): 13–23.

Edgar, I.R. (2004) *A Guide to Image Work: Imagination-Based Research Methods*, London: Taylor & Francis.

Gabriel, Y. (2015) 'Reflexivity and beyond: a plea for imagination in qualitative research methodology', *Qualitative Research in Organizations and Management*, 10(4): 332–6.

Goffman, E. (1959) *The Presentation of Self in Everyday Life*, Garden City: Doubleday.

Hodges, A.S. (2016) 'The family-centred experiences of siblings in the context of cystic fibrosis: a dramaturgical exploration', PhD thesis, Cardiff University.

Hodges, A.S. (2018) 'The positional self and researcher emotion: destabilising sibling equilibrium in the context of cystic fibrosis', in T. Loughran and D. Mannay (eds), *Emotion and the Researcher: Sites, Subjectivities, and Relationships, Studies in Qualitative Methodology*, vol 16, Bingley: Emerald Publishing, pp 49–64.

Katz, L. (2014) 'The compleat dramaturg', in B. Cardullo (ed), *What is Dramaturgy?*, New York: Peter Lang, pp 13–16.

Langellier, K.M. and Peterson, E.E. (2003) *Performing Narrative: The Communicative Practice of Storytelling*, Philadelphia: Temple University Press.

Loughran, T. and Mannay, D. (eds) (2018) *Emotion and the Researcher: Sites, Subjectivities, and Relationships, Studies in Qualitative Methodology*, vol 16, Bingley: Emerald Publishing.

Marsten, D., Epston, D. and Markham, L. (2003) *Narrative Therapy in Wonderland: Connecting with Children's Imaginative Know-How*, London: W.W. Norton & Company.

Wolcott, H. (1994) *Transforming Qualitative Data: Description, Analysis and Interpretation*, Thousand Oaks: SAGE Publications.

11

Theatrical research-based performance: an analytic method using theatre in an educational context

David Duncan

Introduction

Conventionally, research was conducted *on* children as objects to be observed or interviewed. My research, on child bereavement, did not use such methods. It would seem very unnatural and concerning to sit a child down, question them, and evoke an emotional response about someone who has died, where my only means of comfort would be to hand them a tissue. This unpleasant feeling acted as a catalyst for my interest in arts-based methods. My quest began to find methods for generating data that were appropriate for working with children on a sensitive issue. There was no checklist, outline, or flowchart that existed to tell me what would work best. I was guided by my supervisors and reflected on my own experience as a primary school teacher. With pupil-led learning high on the agenda in Scottish education, I wanted to maximise participation and have children involved at every step of the research process. My decision to use arts-based methods involving art, drama, and music came very early on in the process as a child-centred approach designed to encourage expression and generate multiple, creative voices other than oral. I wanted to facilitate opportunities for children to explore the topic of death and dying in creative ways to generate data. This was, essentially, my original plan. Soon, I started to question why I, as the lead researcher, should analyse the data without the children. It seemed I was stopping the creative process after data production and falling back into academic norms. It was at this point I deviated from my plan and designed an analytic method using theatre to involve participants in analysing and disseminating the data.

This chapter outlines how I facilitated the use of theatre to analyse data with a group of children. I reflect on my experience of generating plenty of creative qualitative data and subsequently using this in rehearsals to write a script and create a performance. Ownership of this performance belonged to everyone in the room: it was for the participants to write. Working with the data in this way added an extra layer of analysis and, in this chapter, I detail the process involved, including gaining ethical approval, key considerations related to performing data, and how to maximise children's participation in analysing data. Theatre can be a very powerful tool and there is no expectation when using it in research that the end result is an award-winning theatre production with moving scenery and dazzling costumes. I hope the process I undertook can be used by others, in whole or in part, to promote participants' involvement in data analysis. After all, participants are best placed to explore their own words and the words of their peers.

Practitioner enquiry stance

> The word theatre comes from the Greeks. It means the seeing place. It is the place people come to see the truth about life and the social situation. (Kissel, 2000, pp 29–30)

My first experience of using theatre in research was during my doctoral study at the University of Strathclyde. This study involved 29 children aged 9 to 10 and focused on the use of creative space to engage children with the topic of death and dying. Informed by the United Nations Convention on the Rights of the Child, the aim of the study was to promote children's voices in creative ways. The emergence of the children's rights discourse and a new perspective on the sociology of childhood repositioned children as social agents (James and Prout, 1997), and methodological shifts towards innovative research methods and techniques of analysis have responded to this reconceptualisation of childhood (Carnevale, 2020; Cuevas-Parra, 2020). More recently, (re)thinking about childhood studies has repositioned relationality as the core focus and individual agency can now be viewed as a barrier (Spyrou et al, 2018). Terms such as 'exercising agency' define children and childhood in singular, limited concepts instead of by way of the connections they have with other humans and with non-human aspects of the world (Spyrou et al, 2018).

The notion that research is restricted to academe has been long-standing (Gutierez, 2019). However, several lines of evidence suggest that teachers can be effective researchers (see, eg, Westbroek et al, 2022). The literature on research in education is well established and practitioner enquiry highlights several ways it can close the gap between theory and practice (Cordingley, 2015). Practitioner enquiry dates to Dewey (1916) and Stenhouse (1975). As defined by Menter et al (2011), practitioner enquiry is a way to approach an investigation with a clear rationale and opportunity to share findings more widely. I would like to see more teachers engage with enquiry as stance in schools; this chapter explores creative ways to bring about educational transformation.

My belief is that practitioner enquiry is doable, and I attempt to show how it can be completed as, first and foremost, a teacher. Through practitioner enquiry, I established my teacher-researcher identity. I had worked with some of the children who participated in this study in a teaching capacity and therefore established relationships existed. When taking on the researcher role, I was very clear that I was still a teacher. Moreover, I made it apparent that I too was human (children can often think teachers are instructing robots) who had also lived through bereavement. As my experience and understanding of death was very different from the participants', the fluidity and flexibility of my position as a 'transient insider' was important (Roberts, 2018, p 114). I cannot claim to know my participants' world completely and therefore I am not entirely an 'insider'. Many educators will find themselves in a similar position to me, and the creative approaches I used in my practitioner enquiry can be adopted and facilitated in other research contexts. By starting with learners' needs, theatrical research-based performance can be woven into the participatory process during a 'test of change'. At the time of completing my enquiry, I had been teaching for over five years and had extensive experience as a theatre director. Drawing on my performance expertise, I felt confident when working in innovative and creative ways throughout the research process. While this was beneficial, it is not a necessity for using this approach. No prior experience is required; however, to support the devising process, I suggest involving a drama practitioner from a theatre company. Their involvement can be as much or as little as is necessary to provide the research team with the confidence, knowledge, and skills to analyse creative data by devising theatre.

Study context

I found very few studies that involved bereaved children in the research process, and this is highlighted in my initial research (see Duncan, 2020a, 2020b). Evidence suggests that creative methods can help researchers to become trauma-informed by encouraging them to consider factors such as the environment, ongoing consent-checking and choice, emotional validation, and grounding techniques (Isobel, 2021; Petrone and Stanton, 2021). Despite this, creative approaches to using theatre and performance in research have not been used in this field with children. Notably, the method has been used in other fields and I used this to inform my own approach (see Gembus, 2018). Following data generation using arts-based methods, I facilitated the use of thematic analysis, incorporating theatre to analyse data with a group of children. I followed Braun and Clarke's (2021) six-phase process to thematic analysis and then, with the children, used the themes generated from this to write a script and create a performance. In working with the data in this way, it added an extra layer of analysis to the thematic analysis and increased children's engagement in research analysis.

After qualitative data was generated through a creative workshop in December 2021, a devising group was created to analysis the data. Thematic analysis was used initially, involving the children to some extent, to guide this process, and the group met in February 2022 to finalise the themes and structure of the performance. From May to June 2022, scenes were improvised and the play was formed. The script was written up by the children, with my input, and, at the end of June 2022, the performance took place. The audience comprised of those invited by the participants, including friends, family members, and teachers. Much of the work had happened behind closed doors and sharing the performance publicly was invaluable. Roy et al (2021) identify three different types of audience (known as 'publics') to describe that performances may have an impact on these different publics. I, along with the participants, acted as the first public. The second public consisted of audience members not involved in the process, but who had a relationship to one or more of the participants. The children chose not to share their work with the third public, which would have encompassed a wider audience completely independent to the participants. Others may choose to use the third audience to disseminate their findings widely.

Ethical considerations

The consideration of ethical issues was an ongoing, reflective process at the heart of my research study. I cautiously balanced children's rights to participation with rights to protection. The nature of this study could have created vulnerability in the populations being researched (Borgstrom and Ellis, 2017). Thus, specific ethical procedures were put in place. The topic attended to and assisted in coping with grief, communicating with gatekeepers, recruiting carefully, and negotiating particular challenges due to the (perceived and actual) sensitivity around death and dying. This research was granted ethical approval by the University of Strathclyde Ethics Committee. In various studies, death and dying was highlighted by pupils and parents as a sensitive topic (despite it not being highlighted so by researchers) (Borgstrom and Ellis, 2017). Further to this, the creative approach is crucial as sensitivity in research is not just about the topic under investigation. Topics such as death and dying can be ethically and methodologically challenging at the best of times (Stroebe et al, 2003; Dyregrov, 2004). Therefore, the main ethical concerns, including informed consent, risk to participants, and privacy and confidentiality, were addressed in the design of the study.

I fully respected the rights of all children involved and, as all participants were under the age of 16, the role of the adult gatekeeper was also taken into consideration. It was made clear that participants could opt out of the study at any stage, regardless of any initial consent. Consent was sought from the children and parents/carers: both formally agreed the child's participation in the study. To support this process, there was a Participation Information Sheet for parents/carers and an opportunity for them to ask any questions about the study. Child-friendly consent forms and Participation Information Sheets were given to the children at an introductory session. When recruiting, children were told about: the aims of the research; what time and commitment was required; what they had to do during the study; who would know the results and view their work; how feedback would be given; and how confidentiality and trust would be fostered. After the introductory session, the children were given the option to 'go and have a think' in case they wished to opt out. This helped ensure they were not in a context where they felt obligated to participate in the research.

Furthermore, given the timeframe of the study, consent was kept as a live issue through regular check-ins where the children were reminded that they could withdraw at any given time. Children's choice to participate in creative approaches may promote or silence their voice. A methodological approach incorporating an element of practitioner research enabled me to tackle some of these barriers as an insider familiar to the children and their families. Additionally, researching a sensitive topic can be distressing for some children and this can increase risk of harm to the participants (Silverio et al, 2022). The creative methods used aimed to facilitate age-appropriate expression and conversations, and were used to build trust and forge bonds between the group and the researcher. Children were closely monitored and observed during the sessions. If it was necessary, I intervened to provide support. In any uncertain circumstance, the group was paused so the situation could be dealt with.

When children were working creatively involving art, music, and drama, clear boundaries were set to ensure distressing and inappropriate content was not included. The participants were encouraged to express their pain, as part of the healing process, without it being too overwhelming. Where specialised help was required, children and their families were signposted to support networks or counselling services. This support ensured any risk to the children was dealt with appropriately, as reflecting on experiences did arouse strong feelings and emotions. The study did promote the expression of emotion and how intense emotion could be managed. Children became emotional during the process and could ask to take a break at any point. To safeguard the children, participants were made aware of procedures that would be taken if concerns were raised about their own or other people's well-being. In such research, children should also be made aware that personal information may be passed on in the interest of their safety, and to protect them from harm or abuse.

Throughout the process, I also prioritised my own well-being. Taking regular time out to reflect ensured that I prioritised happiness and good health. Regularly, I sought support from my supervisors on an academic and well-being level. I was fully trained in working with children of this age and had experience of exploring sensitive topics. This was an important aspect of my role, and any similar study should involve a team of individuals with experience in working with children and in the field of theatre. This could be achieved by working with outside agencies and other partners.

In creative work it can be challenging to ensure participants are not identifiable in the investigation outputs. To protect the identity of the participants I anonymised all data that was generated and any names used were pseudonyms. Any group data, such as script excerpts and props, was anonymised for the purposes of respecting the confidentiality of others. The creators had access to their own research data as the first public. Participants assessed the overall quality of the theatre performance before it reached the second public. They agreed that the ideas could not be identified as an individual's story. Proudly, the participants had ownership of the generated data and I had overall authorship of the performance, ensuring that the participants were given credit. The final performance was a genuine and respectful representation of the children's experiences.

Creative data and thematic analysis

Firstly, thematic analysis was approached in an inductive way: coding and theme development were directed by the content of the data. I concluded that this approach to thematic analysis was most appropriate to the study as it was driven by the data (and therefore child-centred). Secondly, in thematic analysis, a common approach of sorting and sifting can be used to identify themes (patterns of meanings). Under time constraints this enabled me to delve into the data and fully understand it. Thirdly, this approach can be used with a range of data set sizes and with various kinds of data. Our study generated a large amount of qualitative data. The data was also very diverse, including artwork, recorded music, drama scripts, and written post-it notes.

The data was analysed in discrete groups depending on the method of data collection, then across all groups and codes. A participatory approach enabled me to check the findings and interpretations with the participants. In doing so, confidence can be placed in the truth of the research findings as participants believe they are a correct interpretation of their original views. After the data was generated, an in-depth analysis was carried out using Braun and Clarke's (2021) six-phase process. I created Figure 11.1 to illustrate the process involved at each of the stages.

Notably, children participated to varying extents throughout each phase. I adapted this approach by using theatre.

Initially, I set about familiarising myself with the data generated from a creative workshop through mind mapping. An example of a mind map is illustrated in Figure 11.2. I immersed myself in the data, looking at each area separately then weaving it together. I began to delve deeply into the data and attached meaning to the artwork, drama scripts, music tracks, post-it notes, and written reflections. At this point, all data was anonymised before being uploaded to NVivo. The data was organised into broad folders (art, drama, music, and reflections) so I could begin initial coding. At this point, I took a note of casual observations of initial trends to extract the core story. Although I used NVivo to code the entire data set and create nodes, the raw data was never out of my reach.

Coding was an important part of turning the raw data into a communicative and meaningful story. Once the data had been uploaded to NVivo, I began systematic coding. I identified important parts of the text, music, or artwork and summarised these with key phrases/words. Figure 11.3 shows how the data was coded. In this example, the node 'Curiosity' was coded to a participant's

Figure 11.1: Process of using thematic analysis

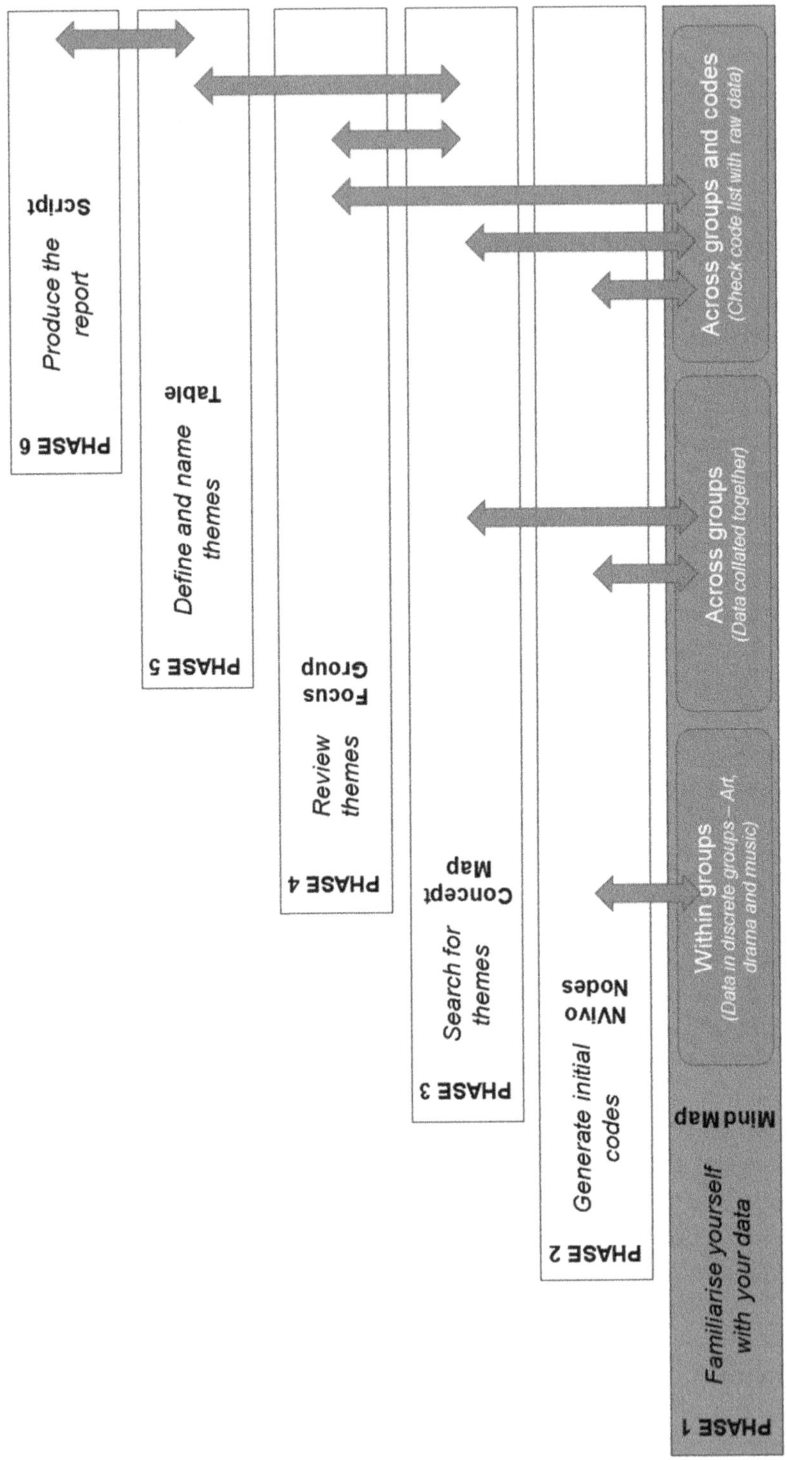

Source: Adapted from Braun and Clarke (2021).

Figure 11.2: Example of mind map created from generated data

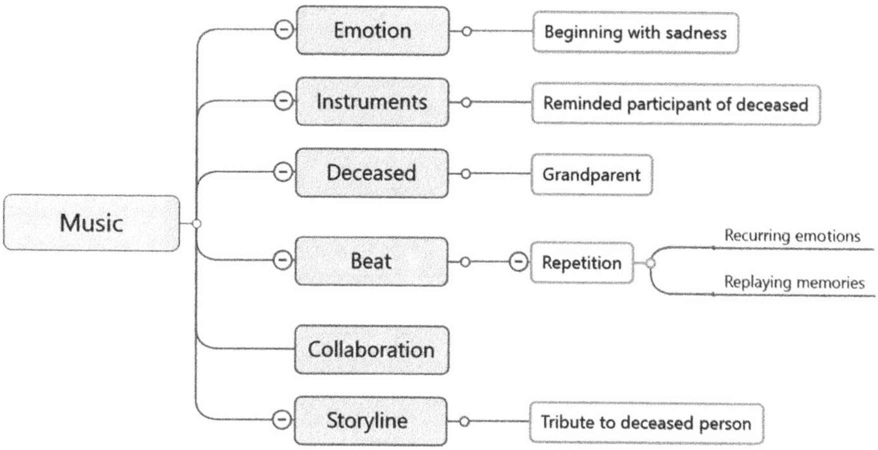

Note: Mind map created by David Duncan at a primary school in East Renfrewshire, 2021. Credit: Author.

Figure 11.3: Example of how nodes were coded and referenced in NVivo

Note: Created by David Duncan at a primary school in East Renfrewshire, 2021. Credit: Author.

Figure 11.4: Word cloud created to search for themes

Note: Word cloud created by David Duncan at a primary school in East Renfrewshire, 2021. Credit: Author.

artwork. The questions asked in the artwork highlighted a desire to know more information. Similarly, in the drama data set the characters in the script showed an interest in death and, through music, there were elements of suspension and tension. After I established the codes, I then identified significant broader patterns of meanings. At this stage, these became possible themes. During this phase, the eight children who had volunteered, and subsequently given consent, to be part of the analysis process became involved. All children from the original cohort of 29 participants were invited; eight volunteered to be involved in the theatrical research-based performance.

The group of eight children, herein known as the devising group, took part in one analysis workshop before rehearsals. I created word clouds from codes to share with the children at the workshop to gather their views of possible themes (see Figure 11.4). I used words from the mind maps to make the word clouds. This was a way of providing the participants with a child-friendly visual representation of the data with which they could engage.

At the analysis workshop, the eight children in the devising group shared their views on the word clouds and selected the most prominent words to become the themes. Refinement of the themes took place and appropriate changes were made; for example, as 'scared' and 'don't feel safe' were amalgamated to become 'fear'. The group then discussed themes and subthemes at length, and a decision was made on establishing these. I then checked these against the original data set, with a particular focus on the research questions, to assess for reliability and validity. At this stage, some themes were rejected. For example, 'difference' was not included as a theme after some discussion and debate. I encouraged the children to dismiss any themes that were not truly representative of the data. This also avoided having too many themes and the inclusion of themes that lacked coherence. I regularly spoke to the children about the relevance of themes to all creative methods.

At this point in the analysis process, we diverted from Braun and Clarke's (2021) thematic analysis process. Instead of duly producing a report, the children and I devised a piece of theatre using the themes. This enabled us to delve deeper into the data. The process involved in analysing the data using theatre is discussed in the sections following, including an outline of the opportunities and challenges.

Taking it further: devising theatre with children during the analysis stage

The eight children from the devising group met once a week (over eight weeks). All rehearsals took place after school hours, were carried out in the school's drama studio, and lasted two hours. The children took on various roles and this is outlined in Table 11.1. Importantly, we did not start the rehearsals with a pre-existing script. By devising theatre, the script was formed through improvising and experimenting with voice, movement, mime, characterisation, physical theatre, and monologues.

Games. At the start of the process, the devising group spent a lot of time playing drama games and taking part in activities to promote collaborative work, readiness, and discussion.

Ensemble work. Time was given for the devising group to come together and to bond. Ensemble exercises included choral speaking, synchronised movements, and singing together.

Improvisation. The devising group discussed possible scenarios and improvised these as scenes. There was a 'writer in the room' (one of the participants) who captured the dialogue. This dialogue became the first draft of the script.

Table 11.1: Participants involved in the data analysis process and their roles

Pseudonym	Age	Gender	Bereavement	Additional role
Anna	10	F	Uncle	Script supervisor
Ben	10	M	Gran	Props
Edward	10	M	Sister	Writer
Isla	10	F	Great Grandparent	Scenery
Leah	10	F	Uncle	Costume
Max	10	M	Papa	Stage manager
Orla	10	F	No	Director
Sophie	9	F	Gran	Producer

Emotion. In-depth discussion took place around the overall tone of the performance. The group felt the topic of death had connotations of darkness, crying, and pain; however, they did not want the play to have a sombre feel nor lack energy and excitement because that did not reflect the data. The group explored different emotions through voice, body language, and facial expressions.

Space. The group explored physical proximity and how the audience could be part of the space and action. The participants began to sketch out possible set designs.

Creating characters. Characters were created for the play and their descriptions were written up. Actors began to walk, talk, and move like these characters. Character sketches and visual references were added to a storyboard map to communicate their appearance, personality, and background story.

Movement and music. Visual stimuli from the data were used to create physical theatre. Small sequences of movement were repeated, and music was added to set the mood.

Working from text. Members of the group performed small pieces of text that had been written up from improvisations. This included scenes and monologues.

Creating the script

A play structure began to evolve and a scene-by-scene breakdown was agreed. A script was drafted during rehearsals and, ultimately, the children had to agree that it was true to the data set and was reflective of their own experiences. An annotated excerpt from the script is detailed below, showing links to the original data set and themes. The story is fictional and involves character names.

Scene 4 – Death Is Uncomfortable

Jack:	I could smell the bitter antiseptic and taste the cleanliness. All I could hear was a murmur of electronic devices. My eyes were fixed to my dad.
Nimbus:	You do remember it well.
Jack:	Why are you making me do this? This doesn't feel right!
Nimbus:	You don't want to remember?
Jack:	No! It's too soon. You are making it awkward. Stop talking about death.
Nimbus:	Why not?
Jack:	Because you shouldn't! People don't talk about it. They tell you not to. I feel uncomfortable.
Nimbus:	Uncomfortable you say? About something that everyone faces at some point in their lives.

Vivid images were recalled by the children.

Children expressed what was right for them. They questioned whether death should be spoken about based on adults' reactions.

The theme of oppression became apparent as children were silenced.

Death is inevitable and children wanted adults to be more realistic about it.

Jack: That's not the point. You don't understand.
Nimbus: Oh, I do. I know what it feels like.
Jack: You don't know ~~what it feels~~ like ~~for me.~~
Nimbus: I am here to listen. This is your story after all.
Jack: What story?
Nimbus: There once was a boy who lost his father.
Jack: He isn't lost. He is dead.
End.

An imagined character highlighting the need for children to experience empathetic healing.

Children wanted adults to use clear language so they can face reality and grieve healthily.

The following principles were crucial when creating the script.

Roles within the group. The participants confidently volunteered to take on different roles, including producing, researching, checking links to the data set, directing, and stage managing. Not all participants were cast in the performance as some wished to take on a backstage role. This ensured the group felt they had full ownership of the performance.

The creative environment. As the show began to emerge, we then created a busy, imaginative environment where creativity could flow. The space was filled with sketches, storyboards, images, props, written material, and visual material to help bring the vision alive. The script was followed precisely under the eye of a participant: the script supervisor.

Keeping notes. Throughout the rehearsal process I took notes on the original script to assist the group. Various photographs were taken for the purposes of reflecting on the process in relation to the overall aims of the research study.

Rehearsing. The cast regularly performed sections of the play to one another and asked for feedback. They had the chance to offer new ideas and insights. The rehearsal process promoted trial and error.

Production value. Towards the end of the process we began to select materials to include in the performance. This included music, props, costume, and scenery. The children began to weave all the elements together for added production value.

Perform. In June 2022, the show was performed to a selected audience.

Overall, by structuring data into a piece of theatre, and consequently analysing it further, new perspectives on the topic under investigation were found. I was able to facilitate children's engagement with the topic of death and dying through creative methods, including theatre. Schools are data rich and opportunities should be built in to generate creative qualitative data through practitioner enquiry. Subsequently, this can be used to write a script and create a performance. It could also be used to plan drama workshops for the children in which they explore the main themes. Schools should provide more opportunities for children to engage with inspirational and powerful theatre so that children can express their views, be heard, and influence decisions. I have outlined key factors to consider when using theatre with children to explore death and dying. However, it should be noted that this can be adapted and focus on other sensitive themes and vulnerabilities.

This approach to analysing data was very complex and careful consideration was given to the presentation of findings. My thesis did not capture the excitement and nerves leading up to the performance on the night. It is difficult for the reader to feel the humid temperature in the room as the lights shone down onto the stage and to hear the soft, hollow sounds that evoked feelings of sadness and loss. Theatre is a very 'in-the-moment' experience and describing the performance in words or showing photographs or a recording of the play would not achieve what we set out to do. Thus, in my thesis, I presented the findings as a script and supported this with discussion, in subsequent chapters, to highlight the opportunities and challenges of the approach.

While this was a very useful process, it came with challenges. The idea of performing data undoubtedly sent a shiver down the spine of my supervisors. They questioned its relevance, validity, and necessity. I had to argue my case and trust my instinct (hoping that I was not jeopardising my doctoral journey). While I was encouraged to think about originality, there was only one professor at the time from Strathclyde University who encouraged the use of creative methods, and her passion inspired me. There was no consensus among research staff, so breaking the 'how to write a thesis' norm was challenged. However, my ideas did excite my supervisory team, and I received full support and guidance.

Furthermore, using theatrical research-based performance does require time and resources. It can be very costly, so I kept my materials to a minimum. The ethical procedures were lengthy and rigorous, which restricted some activities and caused delays. Uncertainty was inevitable, particularly as I was a novice researcher working with an uncommon data analysis method. When conflicts and unpredictable situations did arise, they were addressed with solutions. I often took advice from others as I was by no means an expert at using this method. Although my theatre experience and teaching skillset were invaluable, these are not requirements for researchers wishing to use theatrical techniques. There are ample opportunities in communities for researchers to link up with theatre groups, drama workers, or directors/writers.

Conclusion

Research-based theatre can be a valuable and creative analysis tool. This method of analysis can be applied to a range of research fields with success in education, health, psychology, and humanities. Donmoyer and Yennie-Donmoyer (1995) claim that when data is structured into theatre this is a form of analysis. When the participants' data is used to create a script, for example, this is an analytical act as new meanings are created by the act of finding an appropriate way to convey them. It can provide a degree of flexibility, adaptability, and openness for participants, including children, when they are involved in data analysis.

Theatrical research-based performance can act as an excellent means of engaging children in non-traditional ways that are developmentally and age appropriate. This approach challenges how children are represented and shows their capabilities when taking ownership of research data. Participation in this analytical method can become messy and complex if a range of factors, such as time, resources available, and ethical procedures, are not thought out carefully. Facilitating opportunities for children to express their views, be heard, and to influence decisions that affect them is not a gift; it is their right. As a teacher, I was intrigued by creative ways to engage children with research. At the start of my doctoral journey, I considered using theatre then quickly dismissed the idea. As my research began to unfold, it felt unnatural to be promoting creativity through my method of data collection yet not analysing or presenting my work in this way. This chapter captures my own creative journey from data collection through to dissemination of my results. The explanation of how I analysed my data using theatre is guidance for others. I wanted to practise what I was expecting my participants to do: to be creative. I now challenge you to do the same.

References

Borgstrom, E. and Ellis, J. (2017) 'Introduction: researching death, dying and bereavement', *Mortality*, 22: 93–104.

Braun, V. and Clarke, V. (2021) 'Can I use TA? Should I use TA? Should I not use TA? Comparing reflexive thematic analysis and other pattern-based qualitative analytic approaches', *Counselling and Psychotherapy Research*, 21: 37–47.

Carnevale, F.A. (2020) 'A "thick" conception of children's voices: a hermeneutical framework for childhood research', *International Journal of Qualitative Methods*, 19: 1609406920933767.

Cordingley, P. (2015) 'The contribution of research to teachers' professional learning and development', *Oxford Review of Education*, 41: 234–52.

Cuevas-Parra, P. (2020) 'Co-researching with children in the time of COVID-19: shifting the narrative on methodologies to generate knowledge', *International Journal of Qualitative Methods*, 19: 1609406920982135.

Dewey, J. (1916) *Democracy and Education: An Introduction to the Philosophy of Education*, New York: Macmillan.

Donmoyer, R. and Yennie-Donmoyer, J. (1995) 'Data as drama: reflections on the use of readers theater as a mode of qualitative data display', *Qualitative Inquiry*, 1: 402–28.

Duncan, D.A. (2020a) 'Context and implications document for "Death and dying: a systematic review into approaches used to support bereaved children"', *Review of Education*, 8: 480–5.

Duncan, D.A. (2020b) 'Death and dying: a systematic review into approaches used to support bereaved children', *Review of Education*, 8: 452–79.

Dyregrov, A. (2004) 'Educational consequences of loss and trauma', *Educational and Child Psychology*, 21: 77–84.

Gembus, M.P. (2018) 'The safe spaces "in-between": plays, performance and identity among young "second generation" Somalis in London', *Children's Geographies*, 16: 432–43.

Gutierez, S.B. (2019) 'Teacher-practitioner research inquiry and sense making of their reflections on scaffolded collaborative lesson planning experience', *Asia-Pacific Science Education*, 5: 1–15.

Isobel, S. (2021) 'Trauma-informed qualitative research: some methodological and practical considerations', *International Journal of Mental Health Nursing*, 30: 1456–69.

James, A. and Prout, A. (1997) *Constructing and Reconstructing Childhood: Contemporary Issues in the Sociological Study of Childhood*, London: Falmer Press.

Kissel, H. (2000) *Stella Adler: The Art of Acting*, Applause.

Menter, I., Elliot, D., Hulme, M., Lewin, J., and Lowden, K. (2011) *A Guide to Practitioner Research in Education*, London: SAGE Publications.

Petrone, R. and Stanton, C.R. (2021) 'From producing to reducing trauma: a call for "trauma-informed" research(ers) to interrogate how schools harm students', *Educational Researcher*, 50: 537–45.

Roberts, E. (2018) 'The "transient insider": identity and intimacy in home community research', in T. Loughran and D. Mannay (eds), *Emotion and the Researcher: Sites, Subjectivities, and Relationships*, Bingley: Emerald Publishing.

Roy, A., Kennelly, J., Rowley, H., and Larkins, C. (2021) 'A critical discussion of the use of film in participatory research projects with homeless young people: an analysis based on case examples from England and Canada', *Qualitative Research*, 21: 957–74.

Silverio, S.A., Sheen, K.S., Bramante, A., Knighting, K., Koops, T.U., Montgomery, E. et al (2022) 'Sensitive, challenging, and difficult topics: experiences and practical considerations for qualitative researchers', *International Journal of Qualitative Methods*, 21: 1–16.

Spyrou, S., Rosen, R., and Cook, D.T. (2018) *Introduction: Reimagining Childhood Studies: Connectivities … Relationalities … Linkages …*, New York: Bloomsbury Publishing.

Stenhouse, L. (1975) *An Introduction to Curriculum Research and Development*, London: Heinemann.

Stroebe, M., Stroebe, W., and Schut, H. (2003) 'Bereavement research: methodological issues and ethical concerns', *Palliative Medicine*, 17: 235–40.

Westbroek, H., Janssen, F., Mathijsen, I., and Doyle, W. (2022) 'Teachers as researchers and the issue of practicality', *European Journal of Teacher Education*, 45: 60–76.

Section IV

Creative visual analysis

12

Co-creation of a sensory assemblage as data analysis

Mel Roberts and Anne Collis[1]

Introduction

This chapter describes the how and why of a co-created method of creative data analysis. It came about when a creative practitioner (Mel) and a neurodivergent researcher (Anne) used their different forms of expertise to co-create a sensory assemblage and a set of interpretation notes.

Anne was nearing the end of her doctoral studies and needed to synthesise a complex array of transdisciplinary theoretical explorations and empirical research conducted using a number of collaborative and co-design methods. She had a deep feeling that her accumulated evidence and fragments of findings formed a coherent whole. However, she could not put into words her sense of direction, organising principles, or overall shape for this coherent whole. Nor could she find words for the analysis and synthesis required. In everyday life Anne relies on sketching and mind mapping when words fail her. Given this, her university agreed to a collaboration with a North Wales creative practitioner. The aim was to see if the partnership could help Anne synthesise her work first visually and then in words. This collaboration indeed solved Anne's problem and in the process gave rise to a method of creative data analysis that both authors think has wider application.

Both the authors believe that knowledge is contingent, partial, and subjective. Anne uses the language of cognitive maps (Furnari, 2015) to describe the subjectivity of sense making. Mel prefers the metaphor of personal geography (Hall, 1994), believing that we all have our personal geography, a map helping us make sense of our journey through life. Hall wrote, 'Orientating ... reflects a need of the conscious self-aware organism for a kind of transcendent orientation that asks not just where I am but where do I fit in this landscape?' (Hall, 1994, p15). Through making a personal geography of a journey of exploration, be it a physical one or one of research, Mel believes we create a visual representation of the journey, which can then be explained to and explored by others.

What both approaches to knowledge and knowing share is the belief that we make sense of the world using maps of the world that exist within our minds and that can be made visible through creative art practices. The product of this method – the sensory assemblage – is a visible, tangible representation of the personal geography of the researcher's journey through the data analysis of their research.

Context

Anne's doctoral research (Collis, 2021) was funded through the Knowledge Economy Skills Scholarships (KESS2) programme.[2] She was exploring public involvement in policy making in order to understand and address challenges in communication and engagement between those making policies and those who would be impacted by the policies. This exploration led Anne into a blend of transdisciplinary theoretical research and collaborative empirical research. Ethical approval was granted in March 2015 by Bangor University's College of Business, Law, Education and Social Sciences Research Ethics Committee. Data were generated in three phases: interviews and focus groups with people involved in policy-making-related public consultation, collaborative autoethnographic work to develop a new approach to consultation, and analysis of how Barod Community Interest Company (CIC) implemented and iterated the new approach. The most notable empirical element was collaborative autoethnographic work with four North Wales women – Diane, Ellie, Helen, and Rachel – as part of which they co-created a new consultation method they called Coffee Shop Conversations. The research became centred on epistemic justice, noting the ways in which the technical language of policy making and academic research sustains existing epistemic injustices (Fricker, 2007).

The researcher and her then-supervisors, Dr Roger Slack and Dr Sara Louise Wheeler, had identified her research approach as sitting close to bricolage (Pratt et al, 2022), with thoughts that perhaps a mosaic was a better metaphor for drawing together the multiple elements of the research into a coherent whole. Accordingly, the researcher arranged to meet a local creative practitioner, Mel, to learn more about mosaics and mosaic making.

The concept of 'thesis as mosaic' was described to the thesis committee chair (and later doctoral supervisor), Dr Clair Doloriert. She enquired whether a physical mosaic was going to be made and submitted alongside the written thesis. It was not part of the plan, but her comment planted the seed.

Anne is neurodivergent (Stenning and Bertilsdotter Rosqvist, 2021). Her brain is wired in ways other than those expected of 21st-century Western thinkers. Her form of neurodivergence includes autism, ADHD, and aphantasia. She daily harnesses the power of the visual and kinaesthetic as a bridge between an inexpressible feeling and words. In common with many neurodivergent people, her creativity, thinking, and problem solving take place in webs and pictures, not linearity and words. This is not uncommon for neurodivergent qualitative researchers (Grant and Kara, 2021).

When it became clear that Anne could not find words to enable her to synthesise the elements of her research into a coherent story, the KESS programme commissioned Mel to work with Anne to see if they could co-create a visual representation of the thesis. They wanted to find out whether combining two areas of expertise – Mel's creative arts practice and Anne's sociological research practice – could unlock Anne's ability to articulate what she sensed but for which she had no words and thereby enable Anne to write her thesis.

Situating work methodologically

The term Mel and Anne use for their co-creation is 'sensory assemblage'. This term has meanings in creative arts practices and sociology.

An assemblage is an art form, recognised by the Tate as 'art that is made by assembling disparate elements – often everyday objects – scavenged by the artist or bought specially' (Tate, nd). The artefact Mel and Anne co-created is an assemblage rather than a mosaic because it incorporates a teacup, CDs, cotton wool, wire mesh, and a toy unicorn, in addition to more traditional mosaic materials such as tiles. It is sensory in that it is designed to be experienced through touch as well as sight. It is also multilayered, in keeping with an interpretation of the complex interactions between the elements of the research and the comparative visibility and hiddenness of different elements in any everyday explanation of the research.

The term 'assemblage' also has sociological meaning (Deleuze and Guattari, 1987; DeLanda, 2016). At its heart, assemblage theory understands that there is no fixed objective reality but rather a continuous forming and re-forming of assemblages of material and immaterial elements in the process of subjective, contingent sense making. In Mel and Anne's context, the assemblage is explicitly sensory because it is with the senses that the physiological, emotional, and cognitive 'world-within-the-skin' (as Mel and Anne describe it) interacts with the 'world-beyond-the-skin'. Masny (2016) describes 'reading' a data assemblage using rhizoanalysis. Rhizoanalysis is a form of Deleuzean data analysis where there is no final fixed analysis or interpretation. Rather there is a never-ending, always-becoming analytic process punctuated by resting places described as 'intermezzo'. In this context, we see the artwork sensory assemblage as a tangible, visual intermezzo in a never-ending meaning-making process of constituting and reconstituting the sociological assemblage. The artwork sensory assemblage is supported in this by a set of interpretation notes indicating the meaning of the artwork to the researcher. The rhizoanalysis is not solely the preserve of the creative practitioner and researcher because the sense making will continue to occur in the space between the sensory assemblage, the interpretation notes, and a person's own sense making.

Mel and Anne see their collaboration as akin to an act of collaborative witnessing. Collaborative witnessing involves two people bringing their different knowledges to co-create a narrative. For Ellis and Rawicki (2013) this involved co-creating a narrative of Holocaust survival. Rawicki brought lived experience of surviving the Holocaust. Ellis brought academic knowledge of co-creation of narratives. Unlike a collaboration of equals where participants bring all their knowledge to co-create something with meaning to both, in collaborative witnessing one brings their knowledge to support co-creation in line with the sense making of the other. In the method described in this chapter, the creative practitioner takes the role of using their knowledge to support the co-creation of an artwork that has meaning to the researcher. The researcher brings lived experience of creating and handling the research data. The creative practitioner brings the knowledge of how to make visible and accessible the knowledge within and of that lived experience.

Description of the method

At the heart of the method is rich co-production. This requires the creative practitioner to understand the thesis and the research methods carried out. An effective working relationship has to be established and clear communication is paramount. In this case, Mel and Anne had a pre-existing relationship. Even so, the initial discussions needed spacing over a period of months to allow for the rich and trusting co-productive relationship to mature sufficiently for the two to work intensively and effectively at the three-day workshop stage.

Initial discussions took place via Zoom calls and social media messaging. Anne outlined the different elements of her research while Mel made notes and asked questions to clarify points or interpret the information, sketching early thoughts about layout and ideas. As illustrated in Figure 12.1, Anne had previously sketched

Figure 12.1: All Roads Lead to Goffman

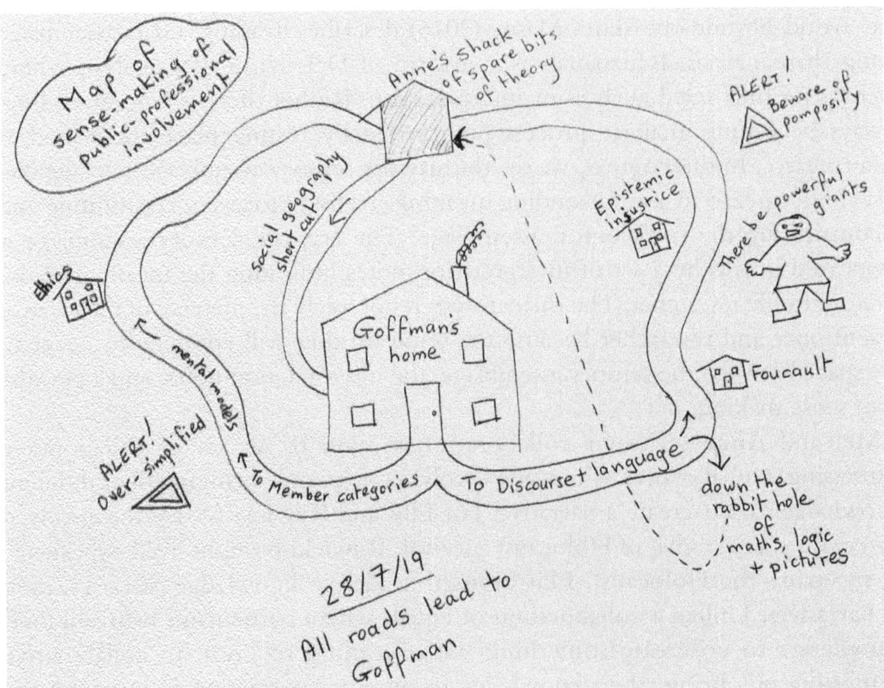

a map she called 'All Roads Lead to Goffman', and she shared this with Mel. Mel began imagining and sketching textures, colours, and designs as Anne described elements of the research.

For the creative practitioner, the ideas for the shape, colour, framing, and texture of the base of the sensory assemblage are literally as well as sociologically foundational. They set structural and artistic boundaries for what can be created. For the researcher, decisions about the base in these initial exploratory discussions can bring clarity to what they consider the solid foundation and external boundaries of their research. The conversations involve an iterative cycle of talking and questioning, with the creative practitioner proffering sketched and spoken ideas and the researcher explaining their visceral and intellectual reaction to these ideas. Through this process the creative practitioner is enabled to discern something of the researcher's underlying sense making and personal investment in their role as researcher. On occasion, it may also enable the researcher to find clarity and give voice to their understanding of these matters – as it did for Anne.

Mel used her knowledge of mosaic and materials to make 'best guesses' at interpreting what Anne was saying. Her questioning drew out and clarified for Anne that the foundation and unifying principle of her research was her ontology, not an element of the theoretical or empirical research. Anne's responses to options for the baseboard shape were visceral; it needed to be circular. Trying to rationalise her visceral reaction helped Anne articulate that epistemic justice in research was of primary importance to her. She told Mel the following story from the empirical research with a group of collaborators. Despite Anne's desire for an equal relationship, her role had become more dominant. As Anne told her collaborators her feelings of failure, explaining how she had meant them to be equal, like to five Olympic rings, one collaborator doodled five circles with one much larger than the others; another collaborator commented it looked like a diamond ring, which must make them the diamonds with Anne being the gold band holding everything together. Anne told this story to Mel, and this determined the shape of the assemblage. The circular baseboard was Anne; the four diamonds set within and extending beyond the baseboard were her four collaborators.

At this point, Mel began to see that they were creating a personal Mappa Mundi, a map of Anne's world. Mappae Mundi were mediaeval maps that indicated something of physical geography and much of the mapmaker's personal geography (Alington, 1996). Anne only recognised this once the assemblage was completed and she stepped back to look at it. Until this point, the sole guide was what 'felt right' for the positioning and representation of each element of the assemblage.

At this stage of the process, the creative practitioner needs to be open to possibilities rather than try to conform to a particular art form. It is important to understand that in the planning stage many options are necessary because things will change during the making process.

This iterative questioning, proffering, and discussing cycle meant Mel began to understand the importance to Anne of Goffman's dramaturgy (Goffman,

1990a) and Charles L. Dodgson's (Lewis Carroll's) mathematical logic of infinite regress creatively woven into *Alice Through the Looking Glass* (Carroll, 2014). Mel began imagining textures, colours, and designs for the layers of the assemblage. It was at this point she realised that the original concept of a simple single-layer mosaic in tiles was not going to be adequate to represent Anne's growing understanding of her research. It was only later, when Mel introduced the word 'assemblage', that Anne recognised her sociological work was also better described as assemblage than mosaic or bricolage. This reframing of the artwork led to a breakthrough and a reframing of how Anne approached drafting her written thesis.

Central ideas were decided in this planning stage, for example the need to represent the centrality of one element of the research, Coffee Shop Conversations. Although that was fixed, Mel made and brought a number of halved cups so the decision as to which one 'felt right' could be made at the workshop stage. Seeing a vintage teacup triggered Anne's realisation of how much the Alice stories, including the Mad Hatter's tea party, had influenced her thinking. The intended coffee shop type of mug was rejected in favour of the vintage teacup.

The hands-on workshop took three days, continuing the iterative cycle of dialogue about the research, Mel proffering materials and Anne explaining her reaction to them. It would be too easy at this stage for the creative practitioner to focus on the creation of an artefact that supports their own understanding of the researcher's research. While this has its role in creative data analysis, the goal of this method is making visible and tangible the researcher's understanding and in so doing to give words to the researcher in an act of collaborative witnessing (Ellis and Rawicki, 2013). Accordingly, the creative practitioner's role is to use their knowledge to support the researcher's understanding of their research.

Often, Mel would open her treasure chest of materials and offer a suggestion. Anne would then stand, eyes closed, sensing whether the texture of a material 'felt right' for the intended purpose. They played with materials and shapes, which included spirals, shards, and reflective tiles. This need for something to 'feel right' was a compulsion for Anne because, unless it 'felt right', any proffered addition to the sensory assemblage was experienced as psychologically and physically jarring and the idea of keeping that addition was distressing. This phenomenon is not yet well understood but seems related to neurodivergence (Beljaars, 2022) where it is possible to know viscerally but without words when something is 'right' or 'not right'. Mel and Anne followed the flows of what 'felt right', rather than treating this as problematic or irrelevant. As a piece of cloth was dismissed with a shudder, Mel would gently probe to identify what felt wrong and – if possible – why it felt wrong. She would then proffer alternatives or simply open her box of materials to see what Anne chose. Such playing with materials, shapes, placement, and layering is essential to the method. How the creative practitioner and researcher enact this will depend on their individual ways of working and the communication style they develop within their collaboration. It would be interesting to know whether this part of the method is experienced differently where the researcher is not neurodivergent.

Making a sensory assemblage also enables the researcher to analyse key principles and concepts in theory without verbosity. The analytic process is about considering the theories and concepts as physical objects and responding to the materials on multiple levels – cognitive, emotional, sensory, and symbolic. The section 'You, Me and the Other' (see Figure 12.2) illustrates this, referencing Goffman (1990) and Garfinkel (1967) and the need to categorise people, thereby creating social hierarchy and social inequality. Simple decisions like using the same pen for each word and the same leather for a word's encapsulation with a leather noose were deliberated and discussed until the researcher felt clear in their presentation. The words were tacitly understood to be of equal status and the lasso, despite having negative connotations of a noose and the Other, reflected the strong shared vision.

Anne's belief in the power of stories and science meeting and the resulting interaction is scattered across the assemblage. The saucer exemplifies this, being made of CDs, as its reflective surface mirrors the process of analysis and it encodes words and pictures but they can only be accessed using a compatible decoding device. The CDs work metaphorically for the cultural construction of language and the contingency of meanings and therefore communication (Crompton et al, 2020).

Figure 12.2: You, Me and the Other

Figure 12.3: Thesis Made Flesh

As illustrated in Figure 12.3, while Mel and Anne worked together, they made Anne's knowing-without-words visible and tangible. This involved developing an understanding of how elements interrelated through the act of deciding spatial relationship, position, and inter-reaction. This enabled clarity of thought and, for Anne, 'the physicality provided something external to my head, not bounded by words. So for the first time I could explain my thesis to myself and others' (taken from reflexive notes, 19 October 2021). Anne went on to produce everyday language 'interpretation notes', replicating how she had walked a colleague with learning disabilities round the assemblage to explain her research. The assemblage and interpretation notes are co-dependent; meaning occurs somewhere in the space between them and a person's own sense making.

What was learned

This method of creative data analysis is dependent on an effective working relationship between the researcher and the creative practitioner. The creative practitioner acts as collaborative witness in the process, never taking over but merely questioning decisions on how a theory or research is to be conveyed and giving alternatives. It provides a method of sense making beyond working solely with data.

For Anne, creating the sensory assemblage (in its art meaning) enabled her to create a sensory assemblage (in its sociological meaning). Once created, it provided a representation of her thesis for the wider public. Mel and Anne would argue this reduces the disadvantage and exclusion of anyone without the vocabulary to articulate or read knowledge using conventional academic language. Their argument is grounded in the ease with which Anne and Alan Armstrong, a colleague with learning disabilities, could discuss her research once there was a visible, tangible assemblage to act as a map. Alan had been excluded from the knowledge about her research when it was only available in words – whether written or spoken. By contrast, he was able to understand, discuss, and question her research when she treated the sensory assemblage as a visual and tactile map of her research and narrated her way around it. As an aside, if Alan had not died in 2021, he would have been the third author of this chapter, bringing his own insights to a discussion of the contribution of creative data analysis to epistemic justice.

Three potential uses of this method

These three potential uses are:

1. creative data analysis as a use by itself;
2. for neurodivergent researchers who need a method to get them from 'sensing without words' to 'knowing with words';
3. epistemic justice and knowledge mobilisation.

This third potential use is closest to Mel and Anne's hearts. They believe this approach to creative data analysis can make it possible for people within and outside the academic world to interact with the research and co-create knowledge as both researchers and audience without privileging those with academic knowledge. They also argue that the movement of knowledge from academic circles to being useful in everyday life relies on people being able to make sense of the knowledge using their own personal geography. It is possible that the method facilitates this by combining the researcher's interpretation (interpretation notes) with the opportunity to interpret and reinterpret the research within their own personal geography (the artwork).

This method can involve practical, epistemic, and institutional barriers. Practically, working relationships like Mel and Anne's are reliant on serendipity and personal connectedness. There is no easily found directory of artists and academics aware of the possibility of working together. Once people find each other, it is a challenge to build a creative relationship of trust and mutual respect that allows the co-creators to comment on each other's areas of expertise. For Mel and Anne the relationship grew into a collaborative, participatory, and reflective process of equals, where one (Mel) offered her knowledge for the benefit of the other (Anne). With such a relationship in place, this method can bring together empirical inquiry, theoretical exploration, and artistic, analytic interpretation in the form of the co-creation of a sensory assemblage and set of interpretation notes.

It can be a challenge to find a suitable working space that meets the artist's practical needs while working cognitively and emotionally for the academic. In this example, the co-creation was akin to an artistic residence where the artist entered the academic's space and the academic had suitable physical space for the artist.

This chapter offers an epistemic framing of the method, but Mel and Anne are transdisciplinarians who lack the depth of knowledge associated with disciplinarity. There is still much epistemic work to be done to ground and develop this method.

The most significant barriers to anyone wishing to adopt and adapt this method are probably institutional. It requires supervisors and managers to be aware of the possibility of approaches such as this and to be willing to broaden their thinking. Obtaining funding and approvals require academic institutions to see the value and significance of creative data analysis, for example valuing analysis beyond that involving numerical data and words. It requires a freedom to fail in the attempt for creative leaps in knowledge rather than an institutional insistence on incremental addition to existing knowledges.

Finally, Mel and Anne believe that the drive towards managerialism has reduced space in the academic world for neurodivergent and visionary thinkers. Anne was fortunate to have supervisors who supported and encouraged her unconventional thinking. Their support enabled the co-creation of this method. In turn, this method allowed a neurodivergent researcher to make her tacit knowledge available to her in words through the creation of a visible and tangible representation of her world – her Mappa Mundi. The method's potential for knowledge mobilisation is hinted at in Anne's ability to explain her research to a non-academic friend using the art assemblage. Mel and Anne believe the method they used has potential for any situation where academics are failing to communicate with those beyond the ivory towers.

Notes

1 The PhD, including the creative partnership described in this chapter, was funded by KESS 2, a pan-Wales higher level skills initiative led by Bangor University on behalf of the HE sector in Wales. It is part-funded by the Welsh Government's European Social Fund convergence programme for West Wales and the Valleys. The PhD also involved a company partner, Barod CIC.

2 KESS2 was a pan-Wales higher-level skills initiative led by Bangor University on behalf of the higher education sector in Wales. Each project involved a university, postgraduate student, and a company partner – in this case Barod CIC.

References

Alington, G. (1996) *The Hereford Mappa Mundi*, Leominster: Gracewing Publishing.

Beljaars, D. (2022) *Compulsive Body Spaces*, London: Routledge.

Carroll, L. (2014 [1871]) *Through the Looking Glass*, London: Macmillan Children's Books.

Collis, A. (2021) 'Not just the usual suspects', PhD thesis, Bangor University. Available from: https://research.bangor.ac.uk/portal/en/theses/not-just-the-usual-suspects(eae97035-4f93-4c5f-8e82-fe79eddfc817).html

Crompton, C.J., Ropar, D., Evans-Williams, C.V., Flynn, E.G., and Fletcher-Watson, S. (2020) 'Autistic peer-to-peer information transfer is highly effective', *Autism*, 24(7): 1704–12.

DeLanda, M. (2016) *Assemblage Theory*, Edinburgh: Edinburgh University Press.

Deleuze, G. and Guattari, F. (1987) *A Thousand Plateaus: Capitalism and Schizophrenia*, trans. Brian Massumi, London: Continuum.

Ellis, C. and Rawicki, J. (2013) 'Collaborative witnessing of survival during the Holocaust: an exemplar of relational autoethnography', *Qualitative Inquiry*, 19(5): 366–80.

Fricker, M. (2007) *Epistemic Injustice: Power and the Ethics of Knowing*, Oxford: Oxford University Press.

Furnari, S. (2015) 'A cognitive mapping approach to business models: representing causal structures and mechanisms', in C. Bade-Fuller and V. Mangematin (eds), *Business Models and Modelling*, vol. 33, Advances in Strategic Management, Bingley: Emerald Group Publishing, pp 207–39.

Garfinkel, H. (1967) *Studies in Ethnomethodology*, Englewood Cliffs: Prentice-Hall.

Goffman, E. (1990 [1956]) *Presentation of Self in Everyday Life*, London: Penguin.

Grant, A. and Kara, H. (2021) 'Considering the autistic advantage in qualitative research: the strengths of autistic researchers', *Contemporary Social Science*, 16(5): 589–603.

Hall, S. (1994) 'I, Mercator', *Orion Magazine*,Spring: 8–16.

Masny, D. (2016) 'Problematizing qualitative research: reading a data assemblage with rhizoanalysis', *Qualitative Inquiry*, 22(8): 666–75.

Pratt, M.G., Sonenshein, S., and Feldman, M.S. (2022) 'Moving beyond templates: abricolage approach to conducting trustworthy qualitative research', *Organizational Research Methods*, 25(2): 211–38.

Stenning, A. and Bertilsdotter Rosqvist, H. (2021) 'Neurodiversity studies: mapping out possibilities of a new critical paradigm', *Disability & Society*, 36(9), 1532–7.

Tate (nd) 'Assemblage'. Available from: https://www.tate.org.uk/art/art-terms/a/assemblage

13

Creating artworks from data

Charlotte Barratt

Introduction

New ways of thinking can occur by accident while planning for something more conventional. This chapter describes how I used drawings as data in my project working with people from different heritages, who were learning English because they were new to the UK or English was not their first language, in a local social history museum. To support this, I will provide some alternative uses to drawing in research, how it can benefit the researcher and participants, touching on how drawing can be a tool for the writer, and finally show how it can be used to mitigate research bias and Westernised thinking. Analysing drawings, whether they are self-created or another's, is not confined to the analytical step of the research process. This chapter will show how we analyse images, in particular drawings, during the data collection, processing, and analysis stages of research.

My project introduced me to different ways of thinking and showed how experiences have a multitude of influencing factors that are not easily translated from one language, or person, to another. I realised that by bypassing linguistic structures in creating images of experiences, I was also bypassing my Westernised semiotic interpretation of the moment. In presenting data this way, it enables each viewer to make a considered analysis of the image and removes the hurdle of attempting to verbalise an intangible experience, potentially with a different semiotic lexicon. This process also allows for interpretation of experiences where people involved in the research have declined, or are unable, to have images of themselves included, either for religious or personal reasons. As I will describe in more detail, some of the people I worked with were asylum seekers, refugees, or held religious beliefs that did not permit them to have their photograph taken.

I would stress that using drawing does not require any training or talent. The images of experiences do not have to be realistic or show faces to convey an emotion or moment. This chapter will explore using drawing as a thinking process and how it can decolonise one's analysis.

The people and the project

In my project, I wanted to explore the experiences of people of diverse heritage in my local social history museum. Leicester is one of the most super diverse cities in the UK (if not *the* most diverse). As of the Census 2021, the White (English, Welsh Scottish, Northern Irish, or British) population in the city stood at 33.2 per cent, with the largest community being Asian, Asian British, or Asian Welsh: Indian at 34.3 per cent (ONS, 2022). To be a truly super diverse community more than one demographic characteristic needs to show no majority (Vertovec, 2007); Leicester has no religious or socioeconomic majority. I wanted to see how the local social history museum represented the story of migration to the city and how some of the people from different heritages experienced this representation of migration. I volunteered at two English for Speakers of Other Languages (ESOL) centres over two academic years and, as part of the classes, we visited the museum. The first group of 11 people learning English came from nine different countries and spoke nine different languages. They were new to the UK, most having been in Leicester for six to twelve months, and I met them once a week or fortnight for two hours over one academic year. This group were from countries such as Poland, Venezuela, Eritrea, India, Iran, and Kurdistan. The second academic year I met three groups who were on six-week courses that put language learning to use in real situations. The museum visit was part of their course. The people in those groups were from more settled communities in Leicester, such as the Indian, Bangladeshi, and Somalian communities, and some had been in Leicester for as long as 20 years.

To enable my analysis of diverse visitor experiences in museums, I took photographs and films, and made field notes. I had intended to conduct photographic elicitation interviews; however, the timing and opportunities led instead to informal discussions following the visits, so my data collection changed into data generation.

I began my data analysis using NVivo to code the data I had gathered on the visits, pulling out themes around the senses, emotions, and memories. As I was analysing my notes and images, I felt inspired to draw some of the photographs I had taken. The process of drawing my photographs opened up new ways of thinking about their meanings above and beyond the NVivo analysis. As I was working on describing other people's experiences and emotions, I found that this method gave me the opportunity to spend time concentrating on the captured moment and interpreting that time. This method opened my mind to the emotions of others through subsequent informal conversations during our time together.

Drawing then became an integral part of my research process. I tried to include an opportunity to draw with one of the groups and create artwork from both my own thoughts and their collaborations around visiting the museum. I had hoped we could have co-created data in this way but there was a reluctance by participants to add to my original artwork. I found myself drawing from the photographs that had been taken during my fieldwork almost as an analytical 'palate cleanser'. These were not illustrations of research data to accompany the text, but part of a thought process around exploring the emotions and experiences of the people involved. This not only gave me thinking time to consider the emotions and sensory engagements of the encounter, but also produced an additional data set with which to work. This additional data set provided an alternative analysis of the new data for myself and any viewers. My data collection and analysis developed organically from the project rather than being pre-planned.

Other approaches that use drawing

My project plan had not included drawing as a methodology, but drawing has increasingly been used to gain insight into the experiences of a wider range of people. Using the definition of drawing as 'intentional mark making on a receptive surface' (Lyon, 2020, p 298), since the mid-1980s images and visual methods have been growing in areas such as anthropology, sociology, organisational studies, and education (Kearney and Hyle, 2004). For a more detailed background, Lyon (2020) outlines the breadth and depth that have developed by using drawing within experiential and participatory projects. I will focus on a few examples that are relevant to my project.

There are a number of approaches that involve using drawing as part of the data collection process – such as Ward and Shortt's (2012) use of 'participant-produced drawing' that allowed participants to 'articulate thoughts, feelings and impressions' (p 439). This method works similarly to photographic elicitation interviewing in which participants were able to take their own images. Ward and Shortt chose this method to more accurately reflect 'customer satisfaction' through collecting rich detailed data that included emotional reflection and the temporal aspect of learning (pp 436–7). In this case participants were encouraged to produce a drawing in 15 minutes, using it as a prompt to explain their emotional experiences of the teaching on their university module. One of the considerations of this approach was around any anxiety caused by unconventional methods such as drawing. Ward and Shortt (2012) make the point that the interviews were essential to the process as the participants were not required to be good at drawing:

> From one objective, reading drawings may appear void of emotion, but to the illustrator, a different set of meanings may exist and to them the image may be imbued with feeling and emotion only decipherable to them. (Ward and Shortt, 2012, p 443)

While they credit the creator with all the knowledge of the emotion or experience behind the image, Ward and Shortt dismiss the interpretation of others in reading the images.

Kearney and Hyle (2004) also used drawing in this way in an educational setting and confirmed the use of drawings as a 'powerful confirmation of the participants' verbal reports and observations' (p 380). In projects where school-aged children have been asked to express their feelings around change or experience through drawing, the benefits have been cited as allowing the participants to bypass the usual confinements of research such as answering interview questions or filling in surveys. This allows the children to contribute their voices to the findings and, in some cases, is a more culturally appropriate expression than more conventional methods (Bland, 2018, p 343). Bland (2018) insists that any image created must be accompanied by text or explanation from the creator to avoid any misinterpretation. This method of using drawing to elicit conversation with children is also used by Literat (2013) and Tay Lim and Lim (2013).

There is a benefit to the participant and the associated contemplative process in using drawing as a prompt. There is a focus on the outcomes and not the journey around articulating emotional progress through drawing in comparison to answering focus group questions or ticking boxes on a survey. In the cases described the participants are asked to realise the intangible and then to verbalise the intangible as an explanation.

The analysis of many of these examples sits within the generated text or words rather than the images. Although the images provide the questions to the resulting conversations, any themes that are not raised are overlooked. In the same way Samuels (2007) found that dictating the themes of images to be collected using photographic elicitation interviewing restricted the resulting conversation and actual experiences of the novice monks with whom he was working. His approach began with listing tasks and objects he thought would be important to each monk and assumed they would be important to his participants. The conversations from these guided images did not really allow for true experiences to be explored. By not taking advantage of analysing the drawing, we are restricting our findings to only our assumed themes. This links with our own cultural assumptions around analysing images, which is discussed in the section 'Using drawing to bypass Western interpretive influences'.

Drawing as a thinking process

During the writing-up stage of my research I found myself drawn (pun intended) to visualising my thoughts. Figure 13.1 illustrates the process of considering how the literature and my thinking were intertwined. The use of the Celtic knot structure tied the themes together where they appeared to intersect. I kept coming back to this drawing as I was untangling my writing and thinking, and yet I never finished it. The visualising of my thinking process is not a new concept; Arnheim (1969) advocates using drawing to work through problems and find solutions.

> [D]rawings, paintings, and other similar devices serve not simply to translate finished thoughts into visible models but are also an aid in the process of working out solutions of problems. (Arnheim, 1969, p 129)

Later in my project, one of the participants drew a story they wanted to share with the group. They had drawn a childhood memory of playing marbles (Figure 13.2)

Figure 13.1: Thinking through my literature review (unfinished)

as part of a sharing activity where everyone was asked to talk about an object that meant something to them. This opened up the possibility for me to include others' drawings in my project.

I thought that showing and contributing to each other's drawings might generate further insights during the process of sharing our experiences. There was, however, general approval of the picture I had drawn and a firm reluctance to 'ruin it', even when it was presented as copies of the original. This could have opened up a new line of participation and I started to draw one of the moments we had all shared in the museum. I had hoped that the group would contribute to building on a shared artwork based on this picture, but they did not engage. However, as we were in a language class so there was constant encouragement to speak and write in English this could have been part of the reasoning behind not wanting to draw more. Nevertheless, the act of drawing our moment enabled me to develop a better understanding of what was taking place in the shared moment. Realising the benefits of my own engagement with drawing, I found myself drawing more of these moments during the writing-up process as they

Figure 13.2: Playing marbles as a child

Note: Participant-created drawing (included with permission).

gave me mental space to contemplate what I was seeing and how someone was feeling and bring these to the paper.

I have to admit that the process of drawing did not feel very 'academic', regardless of the benefits it was providing to my analysis. The following quote boosted my confidence in my process:

> They may have said they were drawing; they may have said they were researching; they might even have said, as Leonardo did, that they were searching through drawing (Brizio, 1981), but the term 'drawing research' is relatively new. (Garner, 2008, p 16)

My thoughts developed as time passed and I decided that my drawings should be snapshots of the times we spent together as a group (Figure 13.3); as Jephcott (1964, p 9) explained, a drawing can provide a greater understanding of a moment or emotion in that moment compared to field notes. In her work she employed artists to illustrate some of the situations she was writing about. It was during the process of drawing the moments that my thinking around these images changed.

There has been debate around the use of images as illustrations rather than data in their own right and this is where I found myself. An individual may draw further knowledge from the image than words could ever convey. Rosenberg (2008) describes the process of creating the artworks as *Denkraum*, which means

Figure 13.3: My first drawing of our museum visits

using the subjective space as a thinking space (p 123). This term is more used in design rather than other areas of art; however, it fits the situation of using art as a thinking process. The term *Denkraum* was coined by Aby Warburg (1866–1929) by combining the German words for thought, thinking, or reflection (*denken*) and space (*kraum*). His use of the thinking space allowed him to create an exhibition of moveable artworks that ended in what was called 'a *Denkinstrument*, a tool to think with: not the work itself, but the condition for its emergence' (Murphy, 2021). In many uses of drawing in research the *Denkraum* is applied to the participants, they are given the space to articulate their thoughts before or during answering questions. However, this *Denkinstrument* is as valuable to the researcher as it is to the participant in engaging with the participant's and researcher's thoughts. Warburg's end creation was (unintentionally) a collection of images rather than a body of written work.

In a similar way to *Denkraum*, *enstatic* drawing gives the artistic thinking process a more creative and reflective direction, working towards an end product.

> In this respect, rather than being a form of communication, enstatic drawing may be used to stimulate thinking, describing a state and process used for developing creative thoughts. (Ings, 2013, p 125)

From Ings' (2013) description of this method, art is used to support the process of developing creative thoughts rather than an end artistic product. However, enstatic art describes the process I used to stimulate my thoughts and engage with the emotional and sensory experiences of the people with whom I was working. As many projects require a written aspect, the process allowed deeper, Geertzian-style *thicker*, thinking around the experiences of the people with whom I was working and a greater connection to those moments. The written explorations of experience were easier to articulate following this process, while having visual descriptions of the moments linked the process more closely with Warburg's *Denkraum*. For example in Figure 13.3, I was pulled into a moment where people were sharing a memory brought on by a nursery rhyme. I felt that the act of sharing was important in this particular moment. I also toyed with adding a visual indication so it was clear that they could hear the sounds.

With this chapter I found myself choosing subject matter from photographs I had taken and beginning to draw. At points, when I was lost for words, I would pick up the drawing and make some more progress. This drawing not linked to the subject matter as I did not have any images that would evoke a specific memory or moment that was entirely relatable to the words I was writing. I chose a subject that had a neutral association, although now I associate this image with these words (Figure 13.4).

Drawing can also be used as an in-field data collection method. Forde (2022) used gesture drawings as a process of remembering and communicating embodied experiences. Their fluid working sketches of people playing football reminded them of the moment and the movement and gave them time to compose comic strips around the links between football and the fractious local social history of the South African town of Makhanda. The use of comic strips is becoming an increasingly accessible way of communicating findings in research or as teaching tools (Salvador-Cisneros and Wonsang, 2022).

There are two opposing arguments around illustrations: one, that providing them as an accompaniment to academic writing is redundant (Banks, 2001), and the other that illustrations alongside text give further depth to research findings (Jephcott, 1964). Academia is a text-orientated field and often it seems that there is little space for alternative interpretation methods. Nonetheless, I would argue that in the humanities and social sciences in particular there is space for co-interpretation in exploring experience. As experiences are specific to the individual – even shared experiences – interpretation can also be specific to the interpreter. Visual outcomes allow for creative thinking and should not need to be accompanied by text (Barone and Eisner, 2012, p xi; Garner, 2008). Providing images as evidence or results in their own right allows for 'otherness … in the way that poetry does and uses this otherness of thinking (poetic thought) as a process of poiesis or invention' (Rosenberg, 2008, p 111).

The theory around using drawing in research almost solely lies within the teaching of art and art theory. The artist Michael O'Toole (1994) drew on sociolinguist Halliday's (1973) insights into labelling types of art and their uses, this was then gathered by Kress and Van Leeuwen (1996), who decried the lack of uses for visual representation in research. In continuing this thinking, Riley

Figure 13.4: Water lily

(2008) proposed a theoretical framework to describe different types of drawing, dependent on the depth of viewer's engagement and the proposed function of the drawing. Interpersonal and experiential are two of the proposed functions in this theory (p 159). The interpersonal function relates to how the creator of the image has portrayed the content and what that content is for. The experiential function is how the drawing communicates the experience of that moment itself, whether that is a physical or emotional representation (p 158). This moves away from the strict belief that viewers of art require some training in how to interpret and view images to gain understanding from them, and it puts the onus onto the artist as to what they are trying to convey.

In moving away from the heavily text-based world of academia, drawing is by no means a less rigorous or valid method of analysing or presenting data. Tracy (2010) provides eight criteria by which we can measure the quality of qualitative data and qualitative analysis without bending to quantitative requirements. Two of the criteria are sincerity and credibility, which both include taking into account researcher bias and values and also multivocality (p 840). In particular, *resonance* – 'research's ability to meaningfully reverberate and affect an audience' – applies when using drawing as a method to share findings (p 844). By removing the requirement for words this opens up the act of analysis to a wider audience, which allows for more nuanced and culturally wider results.

This manner of drawing also bridges language gaps due to age or inequities in shared languages. Scott (2021) advocates that drawing is accessible to everyone, having developed before the written word (p 4); an example of this in practice is the method Draw and Write. This method was developed and tested by Wetton and McWhirter (1998) as a tool to explore children's experiences and understanding of health education. The emphasis in this method was around the analysis of the images with a little help from written words from the children and a few short conversations. This method is commonly used in education to gauge understanding in early years age groups and can be easily put into practice by teachers and educational practitioners.

Drawing has now become part of my thinking process as well as an alternative method of presenting my data. The act of picking up an image and drawing it, regardless of whether it is from a linked source or not, is an effort to clarify my thinking and has great benefits to my thought process. Often the tricky part is finding a subject to draw! Some images look like they want to become drawings rather than photographs as they may have a graphic theme that appeals to my style, or I have a small sketchbook of drawing prompts that can be dipped into for those quicker fixes. Scott (2021) outlines the many health benefits of drawing for researchers and participants, from formulating thoughts generally or for studying how to communicate concepts and experiences. As researchers we should be utilising all available methods of disseminating our findings or helping to clear our minds.

Using drawing to bypass Western interpretive influences

As researchers we impose our own interpretation onto our data based on our heritage and place of cultural upbringing, for example. While there is no foolproof way of being an entirely unbiased researcher, there are methods of mitigating this bias and imposition of understanding. Western researchers often look to fit sensory encounters neatly into their own understanding of the senses – usually sight, taste, sound, touch, and smell. This can also apply to some emotional encounters. Interpreting others' emotional responses to a situation is complex and can be intensely personal. For example, outpourings of grief can be anything from silent contemplation to loud outbursts. Using a drawing to convey the emotion of a moment allows the viewer to interpret the situation as per their own cultural heritage and understanding.

Supporting this method of interpretation, Kress and Van Leeuwen (2001, 2005) advocate for thick description. This multifaceted method of interpreting social situations draws on as many verbal, physical, emotional, and sensory modes as possible to create a detailed or *thick* record of the encounter. There is also an understanding that the social and cultural capital of both researched and researcher have an influence on how the encounter is then communicated, interpreted, and understood.

> Meanings belong to culture, rather than to specific semiotic modes. And the way meanings are mapped across different semiotic modes, the way some things can for instance, be 'said' either visually or verbally, others only visually, again others only verbally, is also culturally and historically specific. (Kress and Van Leeuwen, 2005, p 3)

With most uses of drawing in research, the drawing is then interpreted by an 'authority'. Any interpretation has the bias of the interpreter added, even with a written account of the images' meaning from the creator. This also applies to the use of comic strips to communicate research findings, as the text adds the creator's bias to the image. Removing this level of interpretation allows the viewer to have agency in how they interpret the image.

Research seeks to standardise findings, make them reliable and rigorous in order that someone else may repeat the experiment and find the same or comparable outcomes. These outcomes are usually gathered together in a Westernised format. In an increasingly globalised world, culture and experience is not as easily packaged into a Western method of semiotics. As mentioned in the section 'Drawing as a thinking process', making space in the analysis process for participants from differing heritages and culture to interpret the results allows for a more rigorous and less Westernised process. I would have liked to have had the shared time to explore this further in my research by gathering the group who participated together to draw and share their thoughts and memories. This would have created an additional set of artworks to be analysed.

Practical uses in projects

In much the same way that using photographs in research allows a greater exploration of experiences and emotions, using drawing can produce similar outcomes. The start of this chapter has some examples of how drawings are used as methodology, in eliciting further conversation that is then transferred into text and analysed. By not imposing a strict analysis of the artworks each new viewer is invited to make their own interpretation. In the same way as having multiple people code an interview, multiple viewers will find some of the same interpretations and also new nuances in the artworks from their own reading of them.

My personal use of this method came about when exploring others' experiences from a wide range of backgrounds and heritage. I would suggest that in any instance where experience and emotion are considered, using drawing as part of the analysis process and dissemination of the data broadens others' understandings of the moment portrayed. If you are able to capture the moment, rather than interpreting it in words, it allows others to interpret using their own semiotic understanding. Where there are language or cultural differences among the researcher-artist, participant, and reader, the use of drawing to explain the moment gives each person involved the opportunity to translate that moment into an emotional vocabulary that resonates with them.

For some participants, drawing is an alternative to photography, which may not be appropriate for cultural or personal reasons. Unless the artist is depicting a lifelike portrait, a certain amount of artistic licence can be added to the image to make the person unrecognisable while still maintaining a level of emotion or the experience. When working with children or vulnerable adults this can be a matter of safeguarding, and for some people this allows their religious beliefs or even personal preferences to be respected. Some of the people in the groups I worked with did not want their photograph taken; following discussions around this, in some cases people were happy to be included in group shots from behind so their faces were not showing but their stance, involvement, and presence could be included. This translated into some of the drawn images later on.

Using drawing can make your findings accessible to a wider audience. Using the image bypasses language competency, whether that is due to age or differences in language. It allows for multiple readings of the data, and for multiple, multifaceted understanding based on each reader's personal interpretation. In addition to uses where experience and emotion are considered during the data collection process, using drawing to disseminate the findings to a broad range of audiences would be appropriate. More often than not, any imagery used to show outcomes is accompanied by text. In museum settings where exhibits show art or images to explain a moment, quite often the text is too comprehensive, the language used technical, and so the words will go unread. The image should be enough to explain the moment.

Drawing may not be appropriate in all projects. However, where there is space for exploring cultural, experiential, or emotional moments, drawing provides a platform for the researcher to think more deeply through the data. It also opens up the option for others to apply their knowledge to the analysis process.

Conclusion

This chapter shows how drawing can be used in research for more than facilitating conversation. The act of drawing is not limited to the participant and can aid the researcher with analysis and connecting with experiential and emotional moments. Particular moments can be revisited and drawing can be used as a contemplative tool that allows for focusing the mind and formulating a visual thick description.

Drawing is a good way of eliciting information from participants but researcher interpretation will always have some influence over the findings, which are more often than not expressed in writing. By using images as data that stand alone the influences of the researcher are mitigated and each viewer can interpret the moment within their own frame of understanding. Drawing has as much rigor as a piece of written interpretation by allowing others to contribute to the analysis in a similar way to multiple codings of written data.

Data around emotional experience is always subject to an interpretation and drawing allows the culture of the reader to be included in the interpretation. In many arenas where experience and emotion are being explored, drawing could be considered an option for the researcher to open up new ways of thinking about their own practice and work with others.

While I stumbled onto this method in my own research analysis, it could be used to explore any number of experiences in a multitude of research arenas. It could also be used for collaborative artwork projects describing experiences. I would envisage it being a useful tool in working with children, in health-care environments, with groups that do not share a common language, or to gain a better understanding of others' emotional experiences. The resulting artworks do not need to be professional, and the creator does not require any formal training in order to create an artwork that captures a situation. As it is a creative method, each creator would be able to use their own interpretation of the resulting image to remember and analyse the moment, or to visually describe an experience. This, however, does present a challenge in encouraging readers of the results to accept a creative piece as part of the results rather than purely just part of the data.

References

Arnheim, R. (1969) *Visual Thinking*, Berkeley: University of California Press.

Banks, M. (2001) *Visual Methods in Social Research*, London: SAGE.

Barone, T. and Eisner, D.W. (2012) *Arts Based Research*, London: SAGE.

Bland, D. (2018) 'Using drawing in research with children: lessons from practice', *International Journal of Research & Method in Education*, 41(3): 342–52. Available from: https://doi.org/10.1080/1743727X.2017.1307957

Brizio, A.M. (1981) *The Words of Leonardo* in *Leonardo the Scientist*, London: Hutchinson.

Forde, S.D. (2022) 'Drawing your way into ethnographic research: comics and drawing as arts-based methodology *Qualitative Research in Sport*', *Exercise and Health*, 14(4): 648–67. Available from: https://doi.org/10.1080/2159676X.2021.1974929

Garner, S. (2008) 'Towards a critical discourse in drawing research', in S. Garner (ed), *Writing on Drawing: Essays on Drawing Practice and Research*, Bristol: Intellect Books, pp 15–26.

Halliday, M. (1973) *Explorations in the Functions of Language*, London: Edward Arnold.

Ings, W. (2013) 'Enstasy: immersive drawings as a design process', in E. Brandt, P. Ehn, T.D. Johansson, M. Hellström, T. Markussen, and A. Vallgårda (eds), *Nordic Design Research Conference: Experiments in Design Research*, pp 119–26. Available from: http://www.nordes.org/nordes2013/pictures/Nordes2013Proceedings.pdf

Jephcott, P. (1964) *A Troubled Area: Notes on Notting Hill by Pearl Jephcott*, London: Faber and Faber.

Kearney, K.S. and Hyle, A.E. (2004) 'Drawing out emotions: the use of participant-produced drawing in qualitative enquiry', *Qualitative Research*, 4(3): 361–82.

Kress, G. and Van Leeuwen, T. (1996) *Reading Images: The Grammar of Visual Design*, London: Routledge.

Kress, G. and Van Leeuwen, T. (2001) *Multimodal Discourse: The Modes and Media of Contemporary Communication*, London: Arnold.

Kress, G. and Van Leeuwen, T. (2005) *Reading Images: The Grammar of Visual Design* (2nd edn), London: Routledge.

Literat, I. (2013) '"A pencil for your thoughts": participatory drawing as a visual research method with children and youth', *International Journal of Qualitative Methods*, 12: 84–98.

Lyon, P. (2020) 'Using drawing in visual research: materializing the invisible', in L. Pauwels and D. Mannay (eds), *The SAGE Handbook of Visual Research Methods* (2nd edn), London: SAGE, pp 297–308.

Murphy, K. (2021) 'With his cryptic clusters of images, Aby Warburg remapped the art of the past', *Apollo: The International Art Magazine*, [online] 13 February. Available from: https://www.apollo-magazine.com/aby-warburg-bilderatlas-mnemosyne-review/

ONS (Office of National Statistics) (2022) 'Ethnic group, England and Wales: Census 2021' [online]. Available from: https://www.ons.gov.uk/peoplepopulationandcommunity/culturalidentity/ethnicity/bulletins/ethnicgroupenglandandwales/census2021

O'Toole, L.M. (1994) *The Language of Displayed Art*, London: Pinter Press.

Riley, H. (2008) 'Drawing: towards an intelligence of seeing', in S. Garner (ed), *Writing on Drawing: Essays on Drawing Practice and Research*, Bristol: Intellect Books, pp 153–67.

Rosenberg, T.E. (2008) 'New beginnings and monstrous births: notes towards and appreciation of ideational drawing', in S. Garner (ed), *Writing on Drawing: Essays on Drawing Practice and Research*, Bristol: Intellect Books, pp 109–24.

Salvador-Cisneros, K. and Wonsang, W. (2022) 'Innovative ways to communicate research in the humanities: comic strips', in *EDULEARN22 Proceedings*, IATED, pp 7862–71. Available from: https://doi.org/10.21125/edulearn.2022.1839

Samuels, J. (2007) 'When words are not enough: eliciting children's experiences of Buddhist monastic life through photographs', in G. Stanczak (ed), *Visual Research Methods: Image, Society and Representation*, Thousand Oaks, CA: SAGE, pp 197–224.

Scott, C. (2021) *Drawing (Arts for Health)*, Bingley: Emerald Publishing.

Tay-Lim, J. and Lim, S. (2013) 'Privileging younger children's voices in research: use of drawings and a co-construction process', *International Journal of Qualitative Methods*, 12: 65–83.

Tracy, S. (2010) 'Qualitative quality: eight "Big-Tent" criteria for excellent qualitative research', *Qualitative Inquiry*, 16(10): 837–51.

Vertovec, S. (2007) 'Super-diversity and its implications', *Ethnic and Racial Studies*, 30(6): 1024–54. Available from: https://doi.org/10.1080/01419870701599465

Ward, J. and Shortt, H. (2012) 'Evaluation in management education: a visual approach to drawing out emotion in student learning', *Management Learning*, 44(5): 435–52. Available from: https://doi.org/10.1177/1350507612459169

Wetton, N.M. and McWhirter, J. (1998) 'Images and curriculum development in health education', in J. Prosser (ed), *Image-Based Research: A Sourcebook for Qualitative Researchers*, London: Routledge Falmer, pp 263–83.

14

Using emoji as a creative tool for data analysis

Anuja Cabraal and Lauren Gawne

Introduction

This chapter outlines the use of emoji as a tool for creative data analysis that can provide a new perspective on data by channelling new ways of thinking about and approaching coding and analysis. Our aim in this project was to explore the ways emoji could be used for coding and analytical purposes, as well as exploring how using emoji as coding labels could influence our perspective on the data. We wanted to explore the process of using emoji for analysis instead of words and see if this made any difference to the lexical analytical processes, and potentially the outcomes of the analysis. One of the areas we seek to explore more broadly in our work is what influences the way we analyse data and how things – including the different tools that we use – influence our analysis.

This project brought together two researchers from two different disciplines. Anuja is a qualitative researcher who uses, and trains people in, a range of qualitative analytical approaches and tools, including reflexive practice and research journaling. Lauren is a linguist who is interested in how people use language in social interaction, including emoji use in online communication. In this project, Anuja created, implemented, and documented the creative qualitative methods, and Lauren provided the data to code, the expertise in the technical and social features of emoji, and was a supportive conversationalist for talking through implementation challenges. Collaboration was key to this project, but Anuja was responsible for the majority of the reflexive practice. Therefore, throughout this chapter, the use of the 'I' voice reflects the experience of Anuja as the lead author, and the use of the 'we' voice reflects shared decision making.

We report on our novel use of emoji for analysis, outlining how this influenced or impacted the analytical process. The data used for the emoji analysis was also analysed in another project by us, and, therefore, our emphasis here is on understanding the role of emoji for analysis. As researchers, it is important that we are aware of the reality we are exploring, and also recognise that we are not external to that reality (Richards, 2021). It is also important to try and understand and reflect on the reality of the participants in the research, as it can differ from our own. For example, see Linares, López and Rada (2022). The ontological view for this research was primarily what Hesse-Biber (2017) terms the *interpretive position*, where we believe the social world is being constructed (and reconstructed) through social interactions, and, thus, people are engaging in meaning making through their interactions (Schwandt, 2000). The epistemological stance in this work was most closely aligned with a constructionist perspective; however, using emoji to interpret, understand, code, and analyse added a more subjective perspective to the research. In order to work cohesively, meaning and understanding was applied by the researcher through the visual representation of the emoji.

Data and ethics

Our data for this project used a set of five published interviews, as part of an interview series for the blog *Superlinguo* that was conducted via email by Lauren between 2015 and 2020. Interviewees were either approached (or opted in) to be interviewed by email as part of the public interview series showcased on the *Superlinguo* blog. The full set of 51 interviews were used in a larger qualitative research project on the relationship between linguistic education and subsequent employment (Gawne and Cabraal, 2023). The interviews were published under a Creative Commons licence that allows for reuse, and we therefore did not use any formal institutional ethics process for this project. However, ethics is more than a tick box. We felt it important to inform participants of the research. They were provided with an opportunity to withdraw their interview, as well as clarify or contest any aspect of the research. This was done for both research projects.

Ethics is not something to be considered at the beginning or the end of a research project, nor should it only be a consideration at the data-gathering stage; it is something that needs constant attention throughout all stages of the research, and this is particularly important for creative research (Kara, 2018). This research primarily involved choosing data to analyse and the process of analysis itself. The work was focused on the process of analysis, rather than the findings or outcomes of what people said in response to questions, so the reporting of the findings did not feature as heavily in terms of ethics as they would have otherwise. For these reasons, the key ethical consideration in this process was transparency and accurate and ethical understanding of the data. It was important to keep detailed notes throughout the decision-making process, as well as choices and meaning assigned to emoji, including all analytical interpretations. This chapter continues our commitment to transparency by sharing the process.

We recognise that the process of data analysis is often poorly explained or even understood (Kara, 2020), but is an essential part of the rigour of the qualitative process. To maintain transparency, we provide a detailed description and reflection on our analytical approach using emoji in this chapter. For this reason, Anuja kept and maintained a research journal for this project. Anuja has experience in this form of reflexive research and kept a journal for the duration of the project; as Lauren was not directly engaged in the data-coding process she did not maintain a research journal. This journal detailed the purpose and aims of the project, the challenges faced (as presented in this chapter), and also the reasons behind selecting the emoji.

We believe that participant consent and engagement is extremely important. Even though we were using published, publicly available data under a licence that allowed for reuse, we ensured we contacted the participants and informed them that their interviews were being used for this work, and we gave them the opportunity to remove their responses from this research project. No participants requested their interview be removed. The participants who engaged with the research process were glad to hear that their initial contributions were continuing to be of use in linguistic and qualitative research beyond the original public education function. Given that this creative exploration of the data in this chapter is about the authors' own reflections as much, if not more, than the qualitative content of the five selected interviews, we have not provided any detailed metadata on the interviews or participants.

Process of using emoji for analysis

We opted to use emoji to explore the creative and subjective choices that are often present in qualitative research. Emoji are small images that are encoded as text. Because they are part of the set of symbols in text, they are very easy to use for social media and communication via messaging. Unicode is a consortium of technology companies that agrees on the standard text format for all of the world's written scripts, symbols, and also emoji. This means that the standard Unicode set of emoji can be used and shared across devices. At the time of working on this project, Unicode 13.0 had 1,809 emoji, not including variation for skin tone or gender.[1] Because of their inclusion in Unicode, emoji are one of the most easily accessible ways to share images online, but just because they are widely accessible does not mean that they are understood the same way universally; many are recognisable as images but those images can have varying meanings in different cultures and across generations (see, for example, Barbieri, 2016; Marston, 2019, 2020; Sampietro et al, 2022). For example, the Rose (🌹) can be used to symbolise romantic love, general affection, or communism, depending on culture and context.[2] The emoji that we see are part of the font set for a computer, such as Apple or Windows, or for a program, for example X (formerly Twitter) or WhatsApp. These formatting distinctions mean that the same emoji might render slightly differently depending on the operating system. This is similar to the way that one computer might open a document using Arial and another use Calibri font; the symbols still show the same characters with slight variation. These slight variations are usually not consequential, but it is worth being aware of this feature as it might influence your interpretation and use of, for example, a particular facial expression in your coding. As we discovered in our research, even very slight variations can influence emoji choice during analysis.

When starting the project, I spent time considering where the coding was going to take place, and which software we were going to use for the analysis, which had implications for the choice of emoji set we were going to use. When I first started the research, I was only aware of one qualitative research software program that had emoji capability for coding purposes (MAXQDA). We decided against using MAXQDA for two reasons; one, MAXQDA uses its own emoji forms, which are a limited subset of the full Unicode set and, two, we did not have access to the software and, taking cost into consideration, decided it was better to go with something we could both access. At the time of writing this chapter, Quirkos also supports emoji use. We return to a discussion of Quirkos later in the chapter.

I decided to do the analysis using Google Docs. This was for three reasons. The first was because it was easily accessible to both of us and is flexible for collaborative work – necessary as we were both in COVID-19 pandemic lockdowns in Melbourne, Australia, during most of this project. Secondly, it is a method that is easy for anyone to replicate for their own work, as Google Docs is easier to access than specialist qualitative research software. Thirdly, a black-and-white version of the Google Noto emoji set is available as an Open Source font, allowing us to easily include the emoji in publications.[3]

The actual mechanics of coding using emoji in Google Docs was more challenging than initially anticipated, as Google Docs does not have the same flexibility or range of tools as Quirkos. Firstly, it was not possible to add emoji using the 'comment' feature in Google Docs at the time of the analysis; this has subsequently been updated but would still be difficult to export or compile. I then trialled three other coding input methods. The first was to add emoji at the end of the section of text being analysed. This made it difficult to know where the section started and the scope of relevant content for the coding. The second method was to create a table, with the interview segmented into sentences in rows and a second column for adding emoji. I found this process removed the words from the context of the larger interview while reading through. Even though the data was very structured and in the form of long-answer responses to questions via email, I found that context was important for analysis of this data.

I opted for a third option for the coding. I printed out the transcripts, printed out the emoji by screenshotting, pasted these into a document, and then worked by hand. I used the visual emoji set given in Google Docs and coded sections using pen and paper. I placed the emoji in a grid formation and assigned an identifier for each emoji (a number and letter) in order to reference which emoji we were using. This was because a description of the emoji could be later misunderstood and had the potential for us to default back to thinking of this as a word-labelling rather than emoji-labelling exercise. This method of the emoji 'grid' worked effectively. Taking an 'old school' analogue approach to what is usually a digital tool also helped us think about what the emoji themselves were adding to the coding process, independent of any digital method.

I printed the full set of emoji and stuck them on my wall in front of me. During analysis, I read the transcripts and scanned the emoji to see which one seemed to represent that section of the text. I then noted that next to the transcript and, in my research journal, listed the emoji and provided a detailed explanation of my emoji choice, along with analytical notes. Choosing the emoji was a creative task that forced me out of using predetermined words in my mind. Sometimes, the choice was between two identical items, albeit with a colour variation. I also wrote down the emoji choice and the meaning I had assigned that emoji. In this way, I started to develop a 'codebook' for emoji that I referred back to as I continued with the coding and while writing this chapter.

The selection of the emoji was based on the researcher and their choice, rather than a focus on how the emoji is used in social contexts or how it would be interpreted or understood by others, including participants or even when disseminating the findings. In this way, the emoji choice was made with the focus and attention on understanding and analysing the data based on what that emoji represented and meant for the researcher, rather than on dissemination of the results or broader societal and cultural interpretations of emoji. It is important that we make this distinction for several reasons.

The first is, as we highlighted earlier, emoji are understood by different people differently. In addition, even with one coder, when using word-based codes there can be ambiguity among the codes (Saldaña, 2021). We mentioned this when we introduced emoji, but it is worth repeating again, as it is important to remember that your coding (including lexical) will not necessarily be understood by others, or even by yourself after some time, and this is also true of different coding techniques (Saldaña, 2021).

Rather than a limitation, this freedom in selection of emoji is a strength of the process, as it gives the researcher more freedom to choose emoji based on their understanding, rather than being concerned with how others will understand it or what it means in a social context. It does, however, require that extensive notes be maintained to explain and justify emoji choice, not only to other readers but our future selves. The questions that arise from this are: How will I (the researcher) explain this? How will I share the findings of the research and justify or defend the choices made? This comes with the help of the research and reflexive journal, which details the selection of the emoji and the reasons behind the selection process.

Findings from the creative process

The process we used for coding and analysing the data with emoji presented a number of challenges but also enabled strengths that lexical forms of analysis do not provide.

One of the biggest challenges was the logistical challenge of remembering the 1,809 emoji, and then trying to find one that I thought best represented the data. Developing a familiarity with the symbols takes time. When starting the emoji coding, there was a sense that I knew emoji, having used them in messages and on social media. However, when faced with using emoji for analysis, there was a difference, and this difference was in the creative process. When using emoji for messaging or in social media, they are often used alongside words, to add the kind of information that is usually conveyed by intonation (Wagner, 2016) or gestures when we speak (Gawne and McCulloch, 2019). Using emoji for analysis instead of words meant that emoji were used to depict and understand meaning in the data. In the process, I found there were three main ways that emoji shifted my thinking.

Moving beyond go-to emoji

The first observation, which became apparent almost immediately, was that I needed to look at, and be aware of, many more emoji than just the handful that I come across in everyday communication. A simple example of this is when coding sections where people were talking about graduating or being excited about what they had learned through their course. My immediate reaction was to find and select the 'person raising both hands in celebration' emoji (🙌); however, when I was unable to immediately locate it on my coding sheets, I looked for something else that could capture the excitement, but also the meaning behind learning and transitions. In the end, I selected the fireworks 🎆 emoji, because it conveys celebration and, often, big and meaningful events. Even after I located the waving hands emoji, I kept the fireworks as I felt they conveyed the meaning better than the hands.

The creative process was really in examining and selecting the emoji that best fitted with not only the words people were using, but also the context in and import with which they were saying it.

Thinking visually

Secondly, the move away from wanting to use words into shifting towards coding visually was also a creative challenge to habit. This difficulty was one of the key strengths of this method. The deep thinking had to be done upfront, in conjunction with the coding, in order to make sense of, and understand, the data. Finding a way to represent data using an emoji required deep reflection on what the data meant, as well as finding the most appropriate form for interpreting and representing the data. I discovered that there were extra steps in thinking (analysing) that were required when using emoji, in that not only did we have a word for the theme (for example, skills), but we also had to consider the context of that idea and how to best represent it within the emoji options.

Using the example of skills, if doing lexical coding (which I did using the same data set), I created a code called 'skills', and from there categorised the type of skill it was, what it was used for, and where it was gained; in particular, if it was gained from a participant's university education and, specifically, which part of their education. When coding using emoji, the context was immediately relevant. I found that finding a code for 'skills' and then subcategories later insufficient. The context of the skills, the type of skills, and even where they came from were all important when deciding on the choice of emoji. For example, for technical skills, I chose to combine two emoji – a hammer and book. The hammer represented the technical side while the book showed that the technical skills were academically specific to linguistics. For more universal skills that were broadly applicable, I chose an emoji that had the letters 'abc' on it.

The use of emoji forced a deeper consideration around the interpretation of the data in a non-linear form. I was able to use emoji to convey complex meaning, as well as storytelling, where words would have been difficult to use as codes or categories. One of the other valuable things with using emoji was that I could use the emoji codes to provide analytical details and descriptions within the coding itself; that is, to code using sentences or phrases where words may have been too much. For example, participants spoke of the skills they gained during their linguistics course helping them more broadly in their careers. For this, the emoji depicting linguistics, foundation, and career boosting were used 📃💪🧱🏹💲.

The choice of emoji here was not straightforward and needs explaining. The 📃 depicts education; I thought it looked a little like a graduation document, so this was used to represent the degree or education. The 💪 depicts strength, or in this case, I used it to depict skills gained, which may also be termed strengths of a person, setting their foundation 🧱, which then 🏹 helps boost their career. For career, I used 💲; however, I was not happy with this choice. I used it because there were such a variety of careers; in the end I decided the one thing most careers had in common was that they meant one earned money, so could also be termed employment.

Embracing creative expression

Emoji coding allows for, and encourages, creativity and expression during analysis, in a way that has meaning for the researcher (and even participants). We can use the example just given to discuss this. While, for skills gained, someone may have used different emoji to depict all the different skills, I felt it was better to go with a broad theme for 'skills' in the story, and skills, for me, also represented strengths – so I used the muscle emoji. Even for foundation, I chose the bricks because, to me, I thought of building something; however, a seed or seedling 🌱 could as easily have been used. Using these examples, hopefully it is evident that even for a researcher different emoji could be chosen to represent the same concepts/words when coding the data.

It quickly became apparent that it was not necessarily the choice of emoji that was important but the reason and meaning given for this choice. In some cases, the choice of emoji led to a much more accurate description, and thus better representation, of the data. Working with these visual images also encourages a deeper and more context- and meaning-driven understanding behind coding, which cannot always be captured or understood when using only words.

For all three of these observations about the creative process, the research journal writing that needed to accompany this meant that the analysis happened simultaneously and any doubts or confusions were clearly evident upfront. This required clarity of meaning and interpretation at the time of coding itself, rather than as a 'next step', as can sometimes happen during the coding process. This is one of the key strengths found when using emoji for coding and analysis.

Strengths and weaknesses

Having discussed the main findings on the creative process, we draw them together to discuss where we see some of the key strengths and weaknesses of this process.

A major strength of emoji analysis is that when you become stuck in a rut with your usual analysis habits and need a new way of examining the data, emoji can force you to slow down and reconsider choice making. The storytelling nature of combining visual elements in sequence also provides new ways to consider the relationships between themes: Do they need teasing out, or should they be grouped together? Importantly, it also forces a deeper reflection on the choice of labels, themes, and concepts to make you think about the context, meaning, and relationships around them at the time of coding, such that the coding, analysis, and reflexivity happen simultaneously. Silverman (2017) emphasises that good data analysis is not only about the methods and techniques, but also about the theorising that takes place during analysis. This is something that the process of using emoji for analysis, as a method, effectively invokes from the researcher.

One weakness of emoji coding is that it would be difficult to use across an entire research project as the only form of analysis, given the time it takes to code and analyse the data. It would be difficult to use initially, when concepts are new, and there is a lack of clarity around the themes and concepts presented in the data. As you can see, this weakness is really the other side of the coin of the strength of this method – forcing you out of your comfortable habits. Slowing down, thinking visually, and allowing yourself space for creativity are all valuable, but come at a cost in terms of the scope and scale of where this method could be feasibly deployed.

Further development of this method

For the purposes of this research, emoji coding was used as a creative tool to analyse and understand the data. It was used to create codes and interpret data. Given this is the very beginning of using emoji as an analytical tool, as far as the authors are aware, there is a lot of scope for this method to be developed and also refined. We look at two potential avenues for development of coding and analysis with emoji: one is the use of different methodological approaches, and the second is the use of different technologies.

Methodological ways to develop this method

For this research we started out by using emoji for analytical coding, in parallel to an existing project on the same data that used a phenomenological analytical approach. We believe it is worthwhile exploring the different stages in the analytical process where emoji coding could be useful. We think that using it in conjunction with word-based coding and analysis (inclusive of a variety of analytical approaches) would be beneficial, as would using it as a supplement when finding it difficult to move forward with more common analytical approaches. Feeling stuck is a normal part of the analytical process and a variety of coding approaches may be used within one project (Silver and Lewins, 2014; Saldaña, 2021) with many ways to move forward. We suggest it may be worth considering, or exploring, the option of using emoji as one of them.

An additional way to develop this method would be to explore its use in a team setting, with multiple people analysing the data. We found that one person doing the emoji coding and then discussing the process and findings worked well. However, given the amount of explanation on emoji choice that was needed, it would be interesting to explore the way that emoji coding could work in a team-based environment including the process of discussion and negotiation. Having used secondary data that was not initially intended for research purposes, we suggest exploring emoji coding using primary data. It would be interesting to see the differences and similarities in the processes and outcomes of the coding and analysis compared with the use of secondary data.

Finally, it would be interesting to undertake participatory analysis, where participants of the research were part of the coding and analytical process. Participation is not something that is static or absolute, but can fluctuate throughout the research process (Kara, 2020). Using emoji rather than words can provide a playful or alternative way for participants to approach and reflect upon their own contributions, potentially providing an accessible and fun way into data analysis. However, the usual caveats regarding emoji not being universal in their meaning apply and discussions about emoji choice would still be needed.

Technological ways to develop emoji for analysis

With qualitative software now beginning to include emoji in their toolkits, there are a number of ways that technology can drive development of the process of coding and analysis with emoji.

The tools used for coding and analysis influence the analytical process. We have explained how we found the use of Google Docs restricting, and then had to resort to coding by hand using a grid system. It would be interesting to see the effects that the functionality of being able to code directly using emoji in a software program such as MAXQDA and Quirkos would have on the analysis, particularly if it allowed for an integration of group projects, providing a practical way to implement teamwork, as discussed with regard to methodological developments. One of the biggest changes we would anticipate is in the choice of emoji. In using the analogue method of printing out emoji and hand coding, all emoji were on equal footing, whereas using a computer to input them brings up a set of favourites, potentially reinforcing a choice from a limited set or encouraging use of those that are higher up in the categories of emoji.

I took the opportunity to use Quirkos for a small mock project using textual data from short-form fictional interviews. Using the digital interface to interact with the data and emoji led to different creative choices than the way I initially analysed the data using emoji with Google Docs. Quirkos treats emoji the same as text, so images will appear in your own computer's preferred emoji font, making these feel more natural than the type of emoji MAXQDA employed in their software in 2020. Quirkos allows you to code and group data easily into themes and stories, including themes with emoji labels, and to maintain research notes (incorporating emoji) alongside the coding. Quirkos currently allows for emoji placement in a number of different spots in the software, from the research journal or notes section to the notes attached to the document itself, as well as the codes or 'Quirks' as they are known in Quirkos. This means that you can directly link, and find your reflections, to the emoji coding and analysis, and link this to the notes, the journaling, and even the codes or quirks as they get created. Having the layout as a map rather than linear as it is in NVivo and MAXQDA means that there is so much more scope for reflexivity and creativity, both in the placement and in the selection of the emoji. There is scope to tell a story using emoji – where words would often be too long for a code, the story or detailed analysis can be neatly described in anywhere between two and six emoji, which is not too many to display as codes. Of course, it is still important to document what a description of this means, making use of Quirkos' journaling function.

Conclusion

In this chapter we have introduced you to a novel coding and analytical method using emoji instead of words as code labels. Sidestepping lexical coding requires the coder to rethink the meaning they are ascribing to written text or transcripts, changing their approach to analysing data. This process was slower than lexical coding precisely because it required deeper thinking around the meaning, context, and stories being told. The emoji-coding method can be easily adapted and has possible uses, alongside other coding techniques, to find creative ways to approach your data that sidesteps the limitations of words or to add a depth of analytical thinking that helps move beyond the initial analytical stages. We have outlined some potential future directions for the use of emoji in qualitative research, and we would encourage others to try this approach to creatively disrupt existing analytical processes.

Notes

1 http://unicode.org/emoji/charts-13.0/full-emoji-list.html.

2 http://emojipedia.org/rose/.

3 For the purposes of publication, all emoji depicted in this chapter are Noto Emoji, not the original coloured version used during analysis. http://fonts.google.com/noto/specimen/Noto+Emoji.

References

Barbieri, F., Kruszewski, G., Ronzano, F., and Saggion, H. (2016). 'How cosmopolitan are emojis? Exploring emojis' usage and meaning over different languages with distributional semantics', in *Proceedings of the 24th ACM International Conference on Multimedia*, Amsterdam: ACM, pp 531–5. https://doi.org/10.1145/2964284.2967278

Gawne, L. and Cabraal, A. (2023) 'Linguistics education and its application in the workplace: an analysis of interviews with linguistics graduates', *Language*, 99(1): e35–e57. https://doi.org/10.1353/lan.2023.0003

Gawne, L. and McCulloch, G. (2019) 'Emoji as digital gestures', *Language@ Internet*, 17(2). www.languageatinternet.org/articles/2019/gawne

Hesse-Biber, S.N. (2017) *The Practice of Qualitative Research: Engaging Students in the Research Process* (3rd edn), London: SAGE.

Kara, H. (2018) *Research Ethics in the Real World: Euro-Western and Indigenous Perspectives*, Bristol: Policy Press.

Kara, H. (2020) *Creative Research Methods: A Practical Guide* (2nd edn), Bristol: Policy Press.

Marston, K. (2019) 'Researching LGBT+ youth intimacies and social media: the strengths and limitations of participant-led visual methods', *Qualitative Inquiry*, 25(3): 278–88.

Marston, K. (2020) 'Visual and affective analysis of social media', in L. Pauwels and D. Mannay (eds), *The SAGE Handbook of Visual Research Methods*, London: SAGE, pp 604–14.

Richards, L. (2021) *Handling Qualitative Data: A Practical Guide* (4th edn), London: SAGE.

Saiz Linares, Á., Ceballos López, N., and Susinos Rada, T. (2022) '#MakeYour YouthRing: participatory qualitative research with young people at risk of exclusion', *The Qualitative Report*, 27(9): 1847–64. https://doi.org/10.46743/2160-3715/2022.5525

Saldaña, J. (2021) *The Coding Manual for Qualitative Researchers* (4th edn), London: SAGE.

Sampietro, A., Felder, S., and Siebenhaar, B. (2022) 'Do you kiss when you text? Cross-cultural differences in the use of the kissing emojis in three WhatsApp corpora', *Intercultural Pragmatics*, 19(2): 183–208. https://doi.org/10.1515/ip-2022-2002

Schwandt, T. (2000) 'Three epistemological stances for qualitative inquiry: interpretivism, hermeneutics, and social constructionism', in N.K. Denzin and Y.S. Lincoln (eds), *The SAGE Handbook of Qualitative Research* (2nd edn), London: SAGE, pp. 189–213.

Silver, C. and Lewins, A. (2014) *Using Software in Qualitative Research: A Step-by-Step Guide* (2nd edn), London: SAGE.

Silverman, D. (2017) *Doing Qualitative Research* (5th edn), London: SAGE.

Wagner, M. (2016) 'How to be kind with prosody', in J. Barnes, A. Brugos, S. Shattuck-Hufnagel, and N. Veilleux (eds), *Speech Prosody 2016*, Urbana: Speech Prosody Special Interest Group (SProSIG), pp 250–3.

Section V

Creative written analysis

15

I poems and polyvocality: experiences of using a combined qualitative creative analysis technique to strengthen the voices of research participants and aid reflexivity

Jacqueline Dodding and Hazel Partington

Introduction

We are colleagues at the University of Central Lancashire who studied together in the same cohort for our doctorates in education. While we conducted studies in quite different areas, and used different primary data analysis techniques, we both independently decided to incorporate an additional phase of data analysis using techniques drawn from feminist ethicist and psychologist Carol Gilligan's Listening Guide method (1993) to construct I poems to strengthen the voices of the people who took part in our research. We have since used this combined approach in several other projects and feel that it has enhanced our approach to data analysis.

In this chapter we present our experiences of taking ownership of our research and allowing ourselves to be creative through use of this combined qualitative data analysis approach. We will illustrate our creative analytic process with examples of how we construct I poems to preserve individual experiences and balance them with a thematic presentation of the collective experience. We believe that a combination of different data analysis approaches can complement each other and provide additional dimensions to analysis and findings. We will also demonstrate how the use of I poems can deepen researcher reflexivity.

I poems and the Listening Guide

An I poem is a type of 'research poem', that is to say a poem that has been constructed from and is grounded in qualitative data (Furman et al, 2006). The construction of I poems forms the second stage of Gilligan's four stages of data analysis set out in her Listening Guide. As qualitative researchers, we do not try to reduce the contributions of people who take part in our studies to a single coherent voice. Rather, we try to present the polyvocal, 'naturalistic, complex, varied, expansive and [sometimes] cacophonous' world of the several or many voices that have spoken to us (Thody, 2006, p 2).

The analytical steps described in the Listening Guide offer a 'qualitative, relational, voice-centered feminist methodology' (Woodcock, 2005, p 49), which we believe can provide further insights into participants' worlds through the construction of I poems from selected verbatim participant quotes and thus convey a relatable essence of participants' experiences and preserve the polyvocality inherent in the data (Thody, 2006).

Using the Listening Guide to analyse qualitative data involves four stages,[1] described by Gilligan (2015) as beginning with three 'listenings' or passes through the data to:

- Identify the psychological features of the data landscape. This process involves paying attention to the story being told, while also noting the main features or landmarks of the interaction such as 'who is there, who or what is missing, are there repeated words, salient themes, striking metaphors or symbols, emotional hot-spots, gaps, or ruptures' (Gilligan, 2015, p 71).
- Consider how the 'I' or first-person voice moves across the landscape (this is the stage at which 'I' phrases are extracted in order to construct an I poem in the same sequence in which the phrases appear in the transcript).
- Identify and reflect on any contrapuntal voices in the interviewee's contribution (these are inconsistencies or variations in the voice used – the 'they say statements') (Gilligan, 2015).

The fourth stage of the process is to review and reflect on the resulting I poem and compose a final analysis.

As this method can also be applied to textual data, the researcher may sometimes be reading rather than listening, or reading an interview transcript while listening to a recording of the interview (Woodcock, 2016). Whether listening or reading, the stages outlined by Gilligan (2015, p 69) provide a 'trail of evidence' from which the researcher can construct an analysis or interpretation of the participant's experience. For a more in-depth explanation of the Listening Guide method, we recommend reading Gilligan's 2015 paper, her interview with Mechthild Kiegelmann (2009), and Woodcock's (2016) 'How-to' article.

Epistemology and ontology

Initially the Listening Guide approach was developed in the light of Gilligan's feminist epistemology and the notion that 'female' voices were not being heard in research evidence. Much of the available literature on I poems adopts this stance (Mauthner and Doucet, 1998; Balan, 2005; Petrovic et al, 2015) and most often the paradigm associated with I poems is the interpretivist paradigm (Gilligan, 1993; Balan, 2005; Zambo and Zambo, 2013). The interpretivist framework is essentially concerned with meaning and the quest to understand a shared definition of a situation (Schwandt, 1994). Ontology is assumed to be relative because knowledge and meaning are perceived as acts of interpretation. Therefore, objective knowledge, which is independent of thinking, reasoning humans, cannot exist. Within the interpretivist paradigm, epistemology is assumed to be subjective and constructivist, building a second-order theory or interpretation of participants' theories (Schutz, 1973). However, depending on the aims of each research project and the characteristics of the participants, the approach could lend itself to other epistemological frameworks. For example, in Jacqueline's thesis,[2] where widening participation in education was the focus, a critical theory lens was employed, and in Hazel's doctoral research on the impact of online learning on professional identity, a combined constructivist/constructionist lens was applied.[3] We believe that this combined data analysis approach fits well within a broad interpretivist framework, but can be adjusted depending on the aims of the research and the characteristics of the people taking part.

Why we use this approach

During our doctoral studies we were introduced to Gilligan's Listening Guide methods to which we both later independently turned to address our concerns that the voices of participants in our studies were being fragmented by the nature of the data analysis process. We used Listening Guide techniques to construct I poems from focus group and interview data to preserve polyvocality and strengthen the voices of participants in our studies. Reflecting on this, we can see that we were both strongly influenced by Ely et al's (1991) polyvocal text *Doing Qualitative Research: Circles within Circles*, and the experience of writing a collaborative journal article with our doctoral cohort and the course lead (Satchwell et al, 2015). We find the quote that follows a heartening reminder for qualitative researchers to take ownership of their data analysis process and not be afraid to adopt a creative approach:

> Whatever your approach to data analysis, it seems fair to say that you, the researcher, are in charge of making meaning, of making sense of your data. No-one else can do that job since you are so intimately bound up with shaping your study and with understanding what you studied. In this sense, all qualitative data analysis is idiosyncratic. To accept that you are at the helm is often frightening at first and comforting at last. It can be any of several shades of overwhelming, as well as liberating and powerful. (Ely et al, 1991, p 143)

The Listening Guide as an analysis method used concurrently alongside other methods of analysis is not a new concept. For example, Edwards and Weller (2012) also analysed data using thematic analysis and I poems; Balan (2005) used I poems concurrently with the analytic hierarchy method (Spencer et al, 2003);

and Corby et al (2018) used I poems together with five-phase hermeneutic phenomenology analysis. While we do not claim that our use of I poems with other methods of data analysis is especially new, we are offering our reflections to other researchers to encourage them to embrace creative data analysis approaches.

Benefits of the combined technique

We adopted the combined technique as a response to the challenges of faithfully representing the participants in our studies as 'whole people'. We consider that our decision to use I poems arose from an ethical imperative to mitigate issues of fragmentation by analysis, honouring the polyvocality inherent in our research and protecting the coherence of participants' voices.

By its very nature qualitative research is subjective and selective when analysing texts; however, the use of multiple data analysis techniques in this way strengthens the presence of the participants' individual stories. One of the strengths of this creative and systematic data analysis technique is the flexibility of application to various types of data collection and methodological approaches. During and since our doctoral studies, we have used this approach to create individual I poems and collective We poems with both oral and textual data collection, including interviews, focus groups, and qualitative free-writing submissions. We find that using Listening Guide techniques in addition to other methods such as thematic analysis can help to move our analytic thinking along.

I poems can be incorporated when using methodologies such as Action Research (Zambo and Zambo, 2013), hermeneutic phenomenology (Corby et al, 2018), and phenomenography (Dodding, 2018), and where such methodologies may have prescriptive analysis methods that may not 'fit' with the vision for the research. The Listening Guide and the I poems developed from such data may provide an extra dimension to the analysis, enabling, as previously suggested, multiple participant voices to be foregrounded in the reporting of the research.

Process

In this chapter, we are using thematic analysis as an example technique that we find works well in combination with the use of I poems (similar to Edwards and Weller, 2012); however, as discussed, I poem analysis can be used equally well with other qualitative analysis techniques such as interpretative phenomenological analysis and analytic hierarchy (Spencer et al, 2003).

Braun and Clarke (2022, p 4) describe thematic analysis as an 'accessible and robust' technique for the analysis of qualitative research data involving the systematic coding of data to develop themes. The recommended six stages of thematic analysis have been well documented by Braun and Clarke and others (see Terry and Hayfield, 2021; Braun and Clarke, 2022), hence in this chapter we are concentrating on describing the process of creating and analysing an I poem and how we use it in combination with thematic analysis. It is beyond the scope of this chapter to give a detailed description of how to use thematic analysis and we can do no better than recommend Braun and Clarke's (2021, 2022) recent publications for those who wish to explore this further.

This chapter reflects where we are thus far when combining the Listening Guide with other methods of data analysis, but we are certain that the technique will evolve with time and, just as we learnt from reading the work of Mauthner and Doucet (1998), Balan (2005), and Edwards and Weller (2012), readers of this chapter will be able to develop and adapt their own versions of combining I poems with qualitative data analysis techniques of their choice.

We tend to start our analysis of data using thematic analysis following the six stages laid out by Braun and Clarke (2022) to develop themes to answer our research questions. By its nature this process involves a fragmentation of the data set before reassembling it into themes that represent the collective experience of the participants. Once we have the themes, we return to the original transcripts or texts to re-engage with participants' individual voices.

How to create an I poem

As with many research approaches, researchers often make adaptations in their application of methods and the Listening Guide is no exception. Gilligan's initial development of the Listening Guide (1993) involved extracting phrases consisting of the I plus the verb, for example, 'I love', 'I want', and 'I don't know'. However, there have been several studies in which an extended version of the I poem has been used, wherein longer sections of text are used in order to make the poems more accessible and to give a meaningful insight into the participants' situation (see Mauthner and Doucet, 1998; Balan, 2005; Edwards and Weller, 2012).

As our primary intention in the use of I poems was to allow individual participants' voices to be heard in the reporting of the data, and because we were conducting educational rather than psychological research, we chose to use the extended method. As illustrated in Table 15.1, this meant we incorporated some contextual text to complete the anchoring 'I' statements to create extended I poems, in order to serve our purposes of telling our participants' stories and giving a sense of them as whole people with their own story to tell.

We look at the written transcript of the interview and highlight the 'I' phrases or where 'I' turns to 'you', 'we', or 'they' (this might be implied rather than overt). We then copy and paste these phrases into a word document in the order in which they are in the transcript. Anything that does not contribute to the story of the participant is removed from the draft I poem. Then we extract these to produce the final I poem and, from this, compose an analysis.

The following example is taken from a project we worked on with our colleagues Dawne Gurbutt and Liz Dinse at the University of Central Lancashire to explore how staff had experienced the sudden transition to working from home during the March 2020 lockdown. We used Microsoft Forms to invite free-writing contributions from academic and services staff at the university. We

Table 15.1: How to develop the extended I poem

Stage 1	Read the transcript and highlight every occurrence of the first person 'I' and associated verb. Include further text that may contextualise the participant's story. Most important is for the sense of 'I' to stand out in the I poem.
Stage 2	Remove the chosen original text using the cut-and-paste function and place in a new word document in the exact sequence it appears in the transcript. Each 'I' phrase should be placed on a new line.
Stage 3	Look for the 'contrapuntal' voices in the text. The presence of contrapuntal voices might be signalled by the voice switching to 'we' or 'they' for example. Or it might be where you attribute a word or phrase to each of the 'voices' in the I poem, for example: 'confident voice'; 'confused voice'; 'unsure voice'.
Stage 4	Compose the final analysis based on your review and reflection on the I poem, and on the story it is telling.

initially divided the contributions between us and analysed them thematically. We then compared the themes that we had each come up with to refine and finalise the themes from the data set. In the second phase of data analysis, I poems were constructed for each individual participant. To aid reflexivity and awareness of positionality, each member of the research team also contributed their own piece of free writing, which was analysed by another member of the team. Broad themes agreed on by the group included the challenges and benefits of working from home; feelings of isolation; impact on well-being; adapting your environment and technology; dealing with emotions; and being online. The resulting report encompassed both the thematic analysis and selected I poems.

Example of stage 1 of analysis

> I already worked from home a lot anyway, I like working at home. But this is different – there seem to be a lot more meetings that I get invited to now that everyone else knows how to have an online meeting.
>
> I have noticed that I am looking after myself better than I used to when working from home. I am making sure that I take time away from my desk and that I make time to do some yoga and take my daily walk. I am grateful for the wonders of the internet which allow me to keep in touch with friends and family. Thank goodness for broadband – what would we do without it? I am lucky that I was already 'techy' and have all these resources at my fingertips.
>
> I have noticed that I don't always feel like I am working as effectively as usual, and I think this is because of the background of this terrifying global pandemic. I hear this from other people too – students, colleagues, friends and family. We are hearing sadness every time we switch the news on. But moment to moment, day to day, my life is ok – it's quieter and I

miss being able to hug my grandchildren but there are some nice aspects to this sabbatical that we have all been put on.

Example of stage 2 of analysis

I already worked from home a lot anyway
I like working at home
~~I get invited to now~~
~~I have noticed~~
I am looking after myself better
~~I used to when working from home~~
~~I am making sure~~
~~that~~ I take time away from my desk
I make time to do some yoga and take my daily walk
I am grateful for the wonders of the internet
I am lucky that I was already 'techy'
~~I have noticed~~
I don't always feel like I am working as effectively as usual
I think this is because of the background of this terrifying global pandemic
I hear this from other people too
We are hearing sadness every time we switch the news on
I miss being able to hug my grandchildren

Final I poem

I already worked from home a lot anyway
I like working at home
I am looking after myself better
I take time away from my desk
I make time to do some yoga and take my daily walk
I am grateful for the wonders of the internet
I am lucky that I was already 'techy'
I don't always feel like I am working as effectively as usual
I think this is because of the background of this terrifying global pandemic
I hear this from other people too
We are hearing sadness every time we switch the news on
I miss being able to hug my grandchildren

Final analysis

This I poem tells the story of someone adjusting to the rapid and unexpected changes that many people experienced in the early stages of the UK's lockdown in response to the COVID-19 pandemic. It has not been such a significant change for this participant as for some other participants as they were accustomed to working from home. The poem conveys a sense of needing to pay attention to self-care and an awareness of being in a relatively fortunate position because they were already used to working with technology. However, it is clear that the participant was also feeling adversely affected by the awfulness of the pandemic and how it was impacting on people. They tapped into the collective grief with a switch to a contrapuntal 'We' when they wrote about hearing sadness in the news, before switching back to their more personal sadness of not being able to hug their grandchildren.

When compared to the thematic analysis of the whole data set, this participant's experience contrasted sharply with those who were having difficulty adjusting to the sudden shift to working from home but aligned with the wider group in terms of the background of sadness and fear.

Listening for contrapuntal voices

As mentioned, one of the stages of working with I poems is listening for contrapuntal voices, described by Gilligan (2015, p 72) as a way to pick up 'the tensions, the harmonies and dissonances between different voices … and … modulations and silences (such as where "I" turns to "you" or drops out completely) …', which helps researchers to 'resist binary categories, and to hear complexity rather than flatten the data'.

This example, taken from Hazel's doctoral research, gives another example of how I poem analysis can help to identify contrapuntal voices.

> I am a scientist
> I might not be a scientist in the conventional sense but
> I am a scientist
> I don't like to be boxed in
> 'You've done your degree, now get on with being a mum and a little bit of
>   a complementary therapist'
> I want to be more than that, I need to break free from that
> Here I am working for a doctorate

In this I poem extracted from an interview, the participant's voice clearly expresses a need to stand her ground – she sees herself as a scientist but, as a complementary therapist, she is aware that not everyone will see her that way. She is resisting being boxed in by those who are sceptical about her chosen profession, but also has to resist well-meaning friends and family, whose patronising voices we hear telling her to be content now that she has her master's degree and to rein in her aspirations. This example helped Hazel to move her analytic thinking along, as in the participant cohort of complementary therapists graduating with MScs there were several who talked about their professional identity as perceived by the wider public and sceptical voices in the media; but this participant revealed another layer of influence from her closer social circles.

How to use in combination with other data analysis techniques

We tend to begin by analysing data thematically to gain a perspective on the whole data set, before then turning to the Listening Guide techniques to construct I poems. However, sometimes when immersing ourselves in the data in the initial stages of thematic analysis we find an I poem might 'jump out' at us from the transcript, in which case we will work with that and then return to complete the stages of thematic analysis. Depending on the project, our presentation of findings might be predominantly thematic, with one or two I poems used to highlight particular stories, or we might feature the I poems more strongly and present the themes as an equally weighted or secondary strand.

In her doctoral thesis, Hazel used I poems to produce introductory vignettes for each participant and presented her main findings thematically. Jacqueline, however, undertook a phenomenological analysis to provide an overview of themes within the data and then, in order to foreground the student participants' voices, developed I poems for each individual, thus highlighting their individual educational journey. Other researchers have developed their own approaches when combining the Listening Guide with other methods of analysis. Balan (2005) constructed I poems from data collected during individual interviews, which she then invited participants to member-check, giving them the option to modify or delete words, phrases, or stanzas as they saw fit. Following this, she held a focus group with the same participants, which she analysed using the analytic hierarchy method. Edwards and Weller (2012) began their data analysis process with thematic analysis before moving on to construct I poems, while Mauthner and Doucet (1998) began with the Listening Guide approach to produce I poems before moving on to use thematic analysis.

Using I poems to enhance researcher reflexivity

As well as being invaluable to offer a different analytic angle on participants' contributions, we have both found I poem analysis to be an excellent tool for researcher reflexivity. This fits with Edwards and Weller's (2012, p 203) notion that I poem analysis 'not only draws attention to the research subject's subjectivity – how they understand and speak about themselves, it also produces a particular analytic ontology – how the researcher is placed in relation to the subject and their social reality'. We certainly found this to be the case when we constructed I poems from reflective journals (Jacqueline) and reflective sections in the doctoral thesis (Hazel). This was used by Hazel to locate herself in relation to the participants in her research, while Jacqueline found the approach useful to trace her own journey as a mature student with a marginalised qualification. This is a practice we have since taken into other projects to enhance our reflexive processes, using journals, text messages, free writing, or participation in baseline interviews. We find that the focus on the 'I' phrases and the stripping away of extraneous words helps us to focus on that which we need to pay attention to.

We now present extracts from the I poems we composed from our reflective writing during our doctorates.

Jacqueline's reflexive I poem

I was a 'non-traditional' student
I had been discouraged at school
People labelled me without understanding who I was and what I wanted
I don't like labels in any form
I realise that my identity is shifting
I now describe myself as a lecturer in research
I love research – maybe that's the direction I should take?
I want to do something that challenges me – in a good way

Jacqueline's doctoral research investigated the educational experiences of mature female students and why they had not engaged with higher education (HE) at age 18 years. Her reflexive I poem describes her own educational journey, her feelings about being labelled, how her identity shifted during her doctoral journey, and recognition of her 'new' professional identity.

Hazel's reflexive I poem

I considered myself neither a novice nor an expert
I had a vague idea I wanted to research
Is it doctoral enough?
I can see parallels, I was amazed
I reflected, I hold
I have reflected, I have reflected
I am acutely aware

Hazel's doctoral research looked at the impact of online master's-level study upon professional identity; her use of a piece of reflective writing enabled the construction of an I poem that illuminated the parallels between her own shift in identity through her doctoral journey and that experienced by the participants in her study.

In both cases, the text that we used to construct the I poems had been written without any intention to use it in this way. While one can write reflectively with intent to produce an I poem, our feeling is that the most authentic expression comes when there is no premeditation attached to the writing from which an I poem is constructed.

Conclusion

It may be that our initial use of I poems during our doctoral research was born out of our not yet fully developed skills in using the primary analysis methods that we had chosen, which left us feeling there was a gap to be filled. Other qualitative data analysis techniques, such as thematic analysis, are, of course, complete in and of themselves. However, having used this combined approach now in several projects, working individually, with other researchers, or on the same team, we do feel that it provides a useful creative spark.

We would encourage researchers who are considering a combined approach to data analysis to feel able to experiment and trust their instincts as to how to do this, while also being able to articulate a clear rationale for the choices they make. Making decisions about how to analyse data is an intrinsic part of the journey to becoming an autonomous and creative researcher who is happy to 'take the helm' (Ely et al, 1991, p 143) and thereby take ownership of their own research.

Notes

1 The number of listenings involved in the Listening Guide varies according to different authors, Gilligan (1993, 2015) writes about three listenings followed by the composition of a final analysis, Woodcock (2016) recommends four listenings before composing the final analysis.

2 Thesis title: 'Looking beyond the label: what are the educational experiences of a cross-section of four cohorts of students labelled as "non-traditional"?' Jacqueline's research explored the past and present educational experiences of mature female students who were labelled as 'non-traditional' in a college-based HE establishment in the 13th most deprived borough of the United Kingdom. Ethical approval was granted by the Business, Arts, Humanities and Social Sciences (BAHSS) Ethics Committee at the University of Central Lancashire (UCLan).

3 Thesis title: 'Post-qualification education and professional identity in the contested landscape of complementary and alternative medicine: a case of transformative learning in an online setting'. Hazel's research explored how complementary and alternative medicine practitioners' professional lives were influenced by their MSc studies, and achievement of a master's degree from three international, online courses delivered by a UK university. Participants were from Central, Southern, and Northern Europe, the UK, South-East Asia, and North America. Ethical approval was granted by the BAHSS Ethics Committee at UCLan.

References

Balan, N. (2005) 'Multiple voices and methods: listening to women who are in workplace transition', *International Journal of Qualitative Methods*, 4(4): 1–21.

Braun, V. and Clarke, V. (2021) 'Can I use TA? Should I use TA? Should I not use TA? Comparing reflexive TA and other pattern-based qualitative analytic approaches,' *Counselling and Psychotherapy Research*, 21(2): 37–47.

Braun, V. and Clarke, V. (2022) *Thematic Analysis: A Practical Guide*, London: SAGE.

Corby, D., Taggart, L., and Cousins, W. (2018) 'Meeting the challenge of interpretation: hearing the voices of people with intellectual and developmental disability through I-poems', *Journal of Intellectual Disabilities*, 22(2): 197–205. https://doi.org/10.1177/1744629517690043

Dodding, J. (2018) 'Looking beyond the label: what are the educational experiences of a cross-section of four cohorts of students labelled as "non-traditional"?' EdD thesis, University of Central Lancashire. https://clok.uclan.ac.uk/23819/

Edwards, R. and Weller, S. (2012) 'Shifting analytic ontology: using I-poems in qualitative longitudinal research', *Qualitative Research*, 12(2): 202–17.

Ely, M., Anzul, M., Friedman, T., Garner, D., and McCormack Steinmnetz, A. (1991) *Doing Qualitative Research: Circles within Circles*, London: Falmer Press Teachers' Library.

Furman, R., Lietz, C., and Langer, C.L. (2006) 'The research poem in international social work: innovations in qualitative methodology', *International Journal of Qualitative Methods*, 5(3): 24–34. https://doi.org/10.1177/160940690600500305

Gilligan, C. (1993) *In a Different Voice: Psychological Theory and Women's Development*, Cambridge, MA: Harvard University Press.

Gilligan, C. (2015) 'The Listening Guide method of psychological inquiry', *Qualitative Psychology*, 2(1): 69–77.

Kiegelmann, M. (2009) 'Making oneself vulnerable to discovery: Carol Gilligan in conversation with Mechthild Kiegelmann', *Forum Qualitative Sozialforschung/Forum: Qualitative Social Research*, 10(2). https://doi.org/10.17169/fqs-10.2.1178

Mauthner, N.S. and Doucet, A. (1998) 'Reflections on a voice-centred relational method of data analysis: analysing maternal and domestic voices', in J. Ribbens and R. Edwards (eds), *Feminist Dilemmas in Qualitative Research: Private Lives and Public Texts*, London: SAGE, pp 119–44.

Petrovic, S., Lordly, D., Brigham, S., and Delaney, M. (2015) 'Learning to listen: an analysis of applying the Listening Guide to reflection papers', *International Journal of Qualitative Methods*, 14(5). https://doi.org/10.1177/1609406915621402

Satchwell, C., Partington, H., Barnes, L., Dodding, J., Gurjee, R., Ramsdale, S., et al. (2015) '"Our breadcrumb trail through the woods": reflections on the use of a secret Facebook group as a strategy for surviving and thriving on the doctoral journey', *International Journal of Doctoral Studies*, 10: 465–82.

Schutz, A. (1973) *Collected Papers 1: The Problem of Social Reality*, The Hague: Martinus Nijhoff.

Schwandt, T. (1994) 'Constructivist, interpretivist approaches in human inquiry', in N.K. Denzin and Y.S. Lincoln (eds), *Handbook of Qualitative Research*, London: SAGE, pp 118–37.

Spencer, L., Ritchie, J., and O'Connor, W. (2003) 'Analysis: practices, principles and processes', in J. Ritchie and I. Lewis (eds), *Qualitative Research Practice: A Guide for Social Science Students and Researchers*, Thousand Oaks: SAGE, pp 200–18.

Terry, G. and Hayfield, N. (2021) *Essentials of Thematic Analysis*, Washington: American Psychological Association.

Thody, A. (2006). *Writing and Presenting Research*, London: SAGE.

Woodcock, C. (2005) 'The silenced voice in literacy: listening beyond words to a "struggling" adolescent girl', *Journal of Authentic Learning*, 2(1): 47–60.

Woodcock, C. (2016) 'The Listening Guide: a how-to approach on ways to promote educational democracy', *International Journal of Qualitative Methods*, 15(1). Available from: https://doi.org/10.1177/1609406916677594

Zambo, R. and Zambo, D. (2013) 'Using I poems to hear the voices and understand the actions of EdD students conducting Action Research', *The Qualitative Report*, 18(42): 1–17.

16

Composite narratives, developing characters: a method of creative data analysis in developing public engagement artefacts

Lauren White, Adam Carter, and Katherine Davies[1]

Introduction

In this chapter, we reflect upon our experiences of translating research findings for public audiences through the production of two artefacts. In both cases, we created composite 'narratives' – a collection of participant accounts combined and reshaped to produce central messages and stories (Willis, 2018), and 'characters', fictional participants combined from several participant characteristics whose lives the narratives described. We did this primarily because our data pertained to personal and sensitive issues which, when presented in the form of a fuller narrative, risked making our participants recognisable. However, in the process of creating our composite characters, we realised that this work was more than a process of anonymisation – it involved rich and creative analytical thinking. In producing both artefacts we found ourselves moving iteratively between cases and themes and thinking critically about diversity and how this can be represented. We also engaged in productive processes of generalisation in our attempts to harness the 'spirit' of our data, striving to identify the nub of how relationships, practices, and the rhythms of everyday life worked for all of our participants.

Katherine Davies and Adam Carter, researching how Brexit was experienced in the everyday lives of UK families, worked with a web designer to create an online exhibition representing a year in the life of three 'families' in 'Brexit Britain'. The timeline came out of an Economic and Social Research Council (ESRC)-funded qualitative study conducted between July 2019 and July 2021, that sought to explore how Brexit was being lived and experienced in everyday family life in the UK.[2] Working with 15 families, the project explored how people's views, behaviours, and attitudes towards Brexit were influenced by the relationships in which they were embedded, and vice versa. The project took a multimethod qualitative approach and produced a variety of data forms, including interviews, diaries, video observations of participants watching Brexit-related television (or 'Goggleboxing'[3]), field diaries, notes on 'hanging out' with the families, and texts, social media, and email communications. Katherine and Adam worked with each family for at least a year, through the various parliamentary votes, elections, intense media coverage, and rancorous, relentless public debates.

Lauren White, whose UK-based research sought to understand the everyday experiences of people living with the common health condition IBS (irritable bowel syndrome), created an animation depicting everyday life with IBS through the eyes of three characters: 'Anna', 'Jim', and 'Sheila'.[4] These three characters were based upon accounts from 25 people who self-identified as living with IBS and who completed diaries and follow-up interviews detailing how IBS affected the places they could go, the routines of their day, and the people and things that came to matter. Lauren worked in close collaboration with the university public engagement team, a scriptwriter, and an animation team, recognising the collective endeavour of creative processes and the dialogue involved in creation.[5] While participants were not directly involved in the production of the animation, a more collaborative approach to analysis (Sou and Hall, 2021) is desired in future work.

Both research projects are rooted in an ontological orientation that recognises the importance of everyday life, and our research methods were designed to access the fine-grained, mundane, and routine aspects of participants' lives. As such, our analysis has been driven by an epistemological orientation towards noticing, valuing, and understanding often taken-for-granted and banal relationalities, routines, and practices. The process of transposing this analysis into tales of everyday lives capable of capturing the nuances and complexities of our data pushed our analysis into more iterative, critical, and conceptual territory. In this chapter, we reflect on the process of creating composite narratives and assembling characters *as analytical work*. In so doing we emphasise the analytical potential of research dissemination, troubling the assumption that analysis is a 'stage' in the research process that can be completed by demonstrating how research communication beyond the academy is a fundamental part of the creative and analytical process.

Why composite narratives or characters?

Though there are numerous reasons why researchers opt to collectively assemble accounts in communicating their research (see, eg, Arjomand, 2022; McElhinney and Kennedy, 2022), composite narratives and associated characters are most often utilised to ensure confidentiality and anonymity. Drawing on Wertz et al's use of 'composite first person narratives' (2011, p. 5882), Piper and Sikes' (2010) use of fictionalised stories created from ethnographic and interview data, and Markham's (2012) adoption of composite accounts for online contexts, Willis (2018, p 471) defines composite narratives as 'a number of interviews ... combined and presented as a story from a single individual'. Most often the creation of composite characters similarly combines characteristics from a number of lives to create a single non-identifiable character. Wolseth (2019), for example, used composite characters as well as composite events as writing strategies when faced with reporting difficult ethnographic relationships. Edmunds (2010) used the term 'composite characters' to describe characters he used to engage readers with the everyday lives of historical Native American tribal people in the absence of records of specific individuals. In our projects the creation of composite narratives and characters overlapped, and many traits of our resulting characters were developed through our composite narrative work. Willis (2018) identifies four key practices that she sees as crucial to the assembling of composite narratives from data. These are that: (1) each narrative is based on, and condensed from, transcripts of around three to five interviews; (2) quotations are direct from transcripts; (3) details from the research and interactional context are taken from one account; and (4) researchers are careful to not impose motivations or feelings, ensuring narratives are reflective of the interviewees themselves. With these four factors in mind, Willis (2018, p 475) suggests that this allows for a clear link between original transcripts and final narratives. However, they stress a distinction between 'narratives' and 'fiction', emphasising that narratives are derived directly from raw data and not simply 'made-up'.

Willis' caution here is indicative of the tensions that can arise in working at the borders of analysis and fiction, where researchers may feel an unease at intervening in individuals' accounts, assimilating them with others', or somehow disconnecting them from the 'whole' participant. However, we found that inhabiting these spaces engendered creative opportunities for crafting knowledge. Watson (2022) highlights how the sociological imagination may be created in, and with, fiction. In particular, she details how the language, the mode of storytelling and voices with which a story is told, and the imagery and sensory qualities that bring the sociological story to life are methodological processes through which to work. While the artefacts we generated from our projects were not straightforwardly works of fiction, the process of creating them sparked our analytical imaginations. This analytical work also produced artefacts able to spark public imaginations, and the narratives at their heart ensured we did not present what Duneier (2006, p 554) terms 'a bunch of disembodied thoughts without developing characters to show people as full human beings'; '[you need to] have a character that lives in the text'. Indeed, what 'characters' and 'narratives' have in common is their capacity to be translational tools for communicating research and public sociology (Watson, 2016), and here the links between analysis, dissemination, and sociological fiction reside (Inckle, 2010).

In this chapter, we unpick how we came to see the process of assembling composite narratives to create characters from our research as a deeply analytical endeavour which, in addition to facilitating the communication of our 'findings' to non-academic audiences, deepened our knowledge of our data in ways we had not expected. We focus on three ways in which this dissemination process became analytical work and trace how creating our characters fed further iterations of the analytical process: (1) through the processes of storying narratives and assembling themes; (2) in accounting for difference and representation; and (3) by tapping into (in)tangible knowledge in our attempts to capture the 'spirit' of our research projects and participants (Mason and Davies, 2009).

Storying narratives and assembling themes

The analysis of qualitative data is often thought of in two central ways: thematic analysis (to identify themes occurring across a data set) or narrative analysis (to engage in the details of one participant, place, or period of time to understand processes and account for context). Many researchers also combine these approaches. In both our projects, creating composite narratives and accompanying characters involved moving iteratively between thematic and narrative analysis in new and exciting ways.

Brexit Britain timeline

To create the timeline of stories in Brexit Britain, Adam and Katherine drew upon thematic analysis to identify points in 'Brexit Time' that were felt by their participants as significant – for instance the prorogation of parliament in September 2019 or the Brexit election in December 2019. These, among others, would come to be the major time stamps of the timeline, aiding the thematic analysis of the temporal 'ebbs and flows' experienced by families in Brexit Britain. Adam and Katherine then used their narrative analysis of family cases, bringing together interviews, communications, diaries, and videos from members of each family group involved in the study. This aided them in spotting similarities and differences in how different families experienced key points in 'Brexit Time'.

For instance, Christmas 2019, which immediately followed the UK's general election of 2019, emerged as a key timestamp in the Brexit experiences of the participants, but its significance was felt differently. For some, Christmas was a time to remain silent on the subject, often following previous disagreements and hurt. This experience is contrasted to others, who experienced Christmas as a time of rapprochement. At one point, two participants (siblings) were not on speaking terms. However, their positions had thawed, and so coming together at Christmas and talking about Brexit was experienced as a form of reconciliation rather than relational threat. By creating narratives out of these experiences and linking them to more universally experienced moments, such as the theme of Christmas, Adam and Katherine were able to bring into focus both the general and particular, and the structural and the biographical. They moved not just between thematic and narrative analysis, but between different units of analysis – Christmas Day became a case study alongside the families themselves. Considering these two case types as different but related 'units of analysis' within their data brought into focus how 'thematic' events in Brexit time were weaved into the life stories and experiences of the participants.

'Everyday Life with IBS' animation

In beginning to consider how to communicate the central themes of the research and assemble a narrative thread for the animation, Lauren began initial conversations with the team around the topic of IBS, its key factors and findings, and the relationship of these to the pitch and tone of the animation. Lauren selected a series of participants' accounts which captured a diversity of experience deriving from their individually produced narrative diaries and follow-up interviews. Following this, Lauren assembled a document of key themes and findings that were centred around four thematic and analytical areas (negotiating the social experience of food and diet, accounting for extra time and changing routines, acquiring privacy in personal life, and public toilet access on everyday journeys). This built upon a combination of narrative (Reissman, 1993) and thematic (Braun and Clarke, 2006) analysis in the stages of post-data collection and before the formalised write-up. In selecting quotes and particular participants, the intention was not to prioritise or 'cherry-pick' cases, but rather to engage with a range and to follow what felt like somewhat instinctive decisions based on research relationships and individual narratives. The project scriptwriter then went through a process of familiarising herself with the research and participants' accounts to assemble a script, including the creation of characters, for the animation. This was a complex process. Challenges included working together to produce a script that was streamlined while maintaining participants' nuanced speech and representing participants in ways that felt authentic.

Reflecting upon the accounts, Lauren and the scriptwriter assembled three central characters to tell narratives of participants while working to capture thematic stories in the assembling of the script. There was an ongoing conversation about how these central characters might lend themselves to telling stories about the whole project, whilst remaining sensitive to the intricacies of individual accounts. Further to this, and perhaps connected with Katherine and Adam's example, was the overarching ontological and epistemological position where the everyday was centred within the script. The ordering and articulation of the narratives, and the visual representation, worked at capturing the routines of everyday life; for example, a morning routine, and the journeys and practices within these routines.

Therefore, creating composite characters in both of the artefacts discussed here involved moving across narrative and thematic forms of analysis. In each example, our thematic analyses helped us to identify and represent practices, experiences, and moments in time, and our narrative analyses helped us to communicate these through 'individual' everyday, relatable tales capable of capturing public imaginations. Creating stories in these creative formats helped us to think differently about our themes, adding texture and context across recurring themes, and about how they were embedded in individual lives and relationships. We now move on to focus on the process of making and assembling our characters and its relationship to capturing difference and representation.

Accounting for difference and representation

Our second reflection relates to the analytical questions we asked when developing our central characters, related to authenticity, difference, and representation. For the Critical Methodologies Collective (2022), there are moments of discomfort in presenting research findings and it is important for us to sit with these as qualitative researchers in the process of knowledge production. Sou and Hall stated that when producing materials for public engagement, it is crucial to avoid 'ethnographic errors' (2021, p 96) that undermine the accuracy and integrity of research findings and participant accounts. This is especially the case when translating into *visual* forms where difficulties arise in representing the nuance of experiences and dialogues (Sou and Hall, 2021). Similarly, we recognise the role of the visual as significant in analysing, translating, and capturing our research and its participants. We next detail some of the discomforts around striving to engage with difference and representation and their relationship to the authenticity of participants' narratives.

Brexit Britain timeline

Adam and Katherine's early analysis indicated that British ethnic minorities and European Union (EU) nationals were experiencing the ramifications of life in 'Brexit Britain' with greater intensity than their White British compatriots. These participants' stories needed to form part of the timeline exhibition. Given the visual nature of the online exhibition, Adam and Katherine inevitably faced issues surrounding the representation of racialised and ethnic minority characters. However, the process of creating these characters, building their narrative, and considering how they could be 'displayed' had much analytical value in itself.

The analytical starting point was to collate data from ethnic minority and EU nationality participants who mentioned the key Brexit events, and then identify similarities in their stories, including challenges applying for settled status in the UK and shifts in their conceptualisation of 'home'. These formed the main points of the composite narrative for one of Adam and Katherine's constructed families. This was a married couple – a British East-Asian man and a female EU national from Germany.

Early discussions with the timeline web designers centred on the issue of how to visually represent these characters. For the overall 'look' of the timeline, representations would need to be drawn in 'cartoon' form. However, having cartoonish characters depicting German and British East-Asian people, Adam and Katherine feared, could veer into crude caricature. Also, the composite narrative was assembled from a range of participant groups and stories – to overspecify the visual look of the characters would not do justice to the variety of perspectives contained in the narrative.

It was here that the use of emoji was raised – a visual language based on simplistic cartoon faces. The design team produced custom-made emoji – similar to those

Figure 16.1: Representation through emoji

Note: The original has a grey/blue background with two emoji figures. In the original, both faces are emoji yellow; the emoji on the left has pink hair; the emoji on the right has brown hair and a brown beard and moustache.

Source: Thanks to the University of Sheffield for permission to reuse this image.

found in social media, but that included visual hints at diversity and difference in their detail (see Figure 16.1). This is an imperfect solution to the issue of racial representation in visual sociology, given the association between the 'race-neutral' emoji and Whiteness (Robertson et al, 2021). Yet the decision to not over-rely on the visual in the representation of our characters meant that Adam and Katherine had to be absolutely sure that the composite narrative attached to the characters was sound. They revisited the data, checked their analysis, and clarified that the central point of the narrative – that racism had been a part of the lived experience of Brexit for a portion of our study participants – came through.

Adam and Katherine could, of course, be accused of avoiding further nuances of Brexit's impact on different ethnicities and nationalities, subsuming too much into a narrow set of narratives and 'neutral' emoji. However, they maintain that grappling with questions over incorporating a diversity of stories in a fundamentally visual format led them to approach the data from a novel perspective, and deeply analyse the stories provided by their participants.

'Everyday Life with IBS' animation

In the creation of the three characters for the animation, 'Anna', 'Jim', and 'Sheila', there were questions and anxieties about how to capture differences and represent all of the research participants. Lauren and the scriptwriter regularly engaged in dialogue about 'who' these characters should be. Each character in the animation had a voice-over accompanying it, and so the diversity of accents was a key point (see also Mannay, 2019). These discussions involved critical sociological questions about representing age, gender, race, social class, and sexuality, and their relationships to experiences of IBS. Lauren worked hard to strike a balance between maintaining the authenticity of participant accounts, while honouring the public-facing nature of the animation and making efforts to speak to diverse audiences.

One of the ways in which matters of difference and representation came to be felt and questioned most intensely was through the visual element of the animation itself. Questions about what the central three characters should look like were troubling, given the role of bodies and their visual representation. Since the project focused on the everyday, including spaces, places, routines, and objects, it was decided that visual representations of these elements rather than bodies could be a useful compromise. This demonstrates again how thinking through the politics of visual representation can aid the analysis of data in refocusing the analytical eye on questions of difference. An example of this is illustrated in the animation still in Figure 16.2.

In this section, we have sought to draw attention to the necessary and critical questions of representation. We have considered how creating characters with visual accompaniments evokes discomfort in decision-making and analytical

Figure 16.2: 'Everyday Life with IBS' animation still image

Note: Image still from the 'Everyday Life with IBS' animation. The original shows a green background with five illustrations. The central image is a handbag. Around this are a circle of images that include a box of tissues, a pair of underwear, a box of Imodium, and a collection of coins.

Source: Thanks to Fettle Animation for permission to reuse this image.

processes. There are ethical questions to consider in the assembling of composite characters and narratives – 'who' gets represented, 'how' do they get represented, and 'for whom'. From our examples, the role of the visual in the creative public engagement process drew careful analytical attention to these crucial questions about representation, and improved our analysis. This section also touches on seeking to authentically represent our participants and their stories, hinting at the analytical issue of generalisability – which we will reflect on together in our next section.

Uncanny knowledge and capturing the spirit

Textbooks on research methods often emphasise the limitations of generalising from qualitative data, pointing to the usually small number of cases in a sample and the need for restraint in the claims that are made on their basis. However, it is of course possible for qualitative researchers to generalise, by pointing to patterns, causality, or making particular comparisons. Mason (2008) argues that generalising from qualitative data can be conceived of differently, as an act of evocation and creativity, and that generalising about 'real life' *requires* attention to the particular, as this is after all where everyday life resides (Neal and Murji, 2015). When faced with the challenges of creating knowledge for public audiences through representative stories and characters in our artefacts, we found that the process of creating these composite characters and stories nudged us towards grounded and robust generalisations about the nature of everyday life as lived with IBS or in Brexit Britain.

In both projects, when making decisions about which features to include in the lives of our characters and which events to incorporate into the narratives we created about them, we found ourselves striving to capture the 'spirit' or 'tone' of the data in our representations. We had to think carefully about how we could *know* and articulate this spirit during the process of creating materials. Knowing this 'spirit', we argue, encapsulates a lived, embodied knowledge and experiential 'close' connection to data coupled with systematic thematic and narrative analytical work. This is, of course, entangled with our own researcher positionalities and research relationships in which the 'spirit' became known or shared. For Adam and Katherine, their spirit involved capturing the ways that everyday lives and relationships are lived alongside the shadow of political events such as Brexit. Adam and Katherine were also living with Brexit themselves and their own experiences of Brexit shaped their relationships with their participants and the data they generated (see Davies and Carter, 2021). For Lauren, the tone comprised how participants negotiated structural barriers and personal struggles, while also sharing stories of getting by and capturing shared solidarities as a researcher also living with a bowel condition. Inckle (2010, p 39) describes the value of fiction and its power to represent and embody the diversity of social life by conveying the 'complexity, contradictions, pains, pleasures and politics' of research. Drawing on our data to harness the spirit and tone of our findings was indeed a process of generalisation in the sense that we were representing a large amount of data through a much smaller set of stories and ideas. However, it would be wrong to conceive of this as somehow 'boiling down' or 'reducing' the detail of our data. For us, representing the spirit and tone of our data meant highlighting particularities – in individuals' biographies, day-to-day lives, and relationships – because the spirit of our participants' accounts and understanding them as a whole, and our epistemological position, lay in the details of the everyday. As such, both artefacts included examples of mundane practices (watching television or waiting for the family bathroom to become free), which worked to highlight the significance of everyday domestic space for the experiences of Brexit and IBS through the playful telling of relatable stories.

There is also an element of intangibility to this knowledge, and it is here that we experienced the process of generalisation as a slow and painstaking practice (Smart et al 2014). As well as utilising our systematic analysis, we also drew upon more intangible knowledge that included sensory aspects of the data as well as elements that seemed less easy to capture and somehow beyond the sensory (see Mason and Davies, 2009). We relied strongly upon our ethnographic knowledge of having 'been there' in the field. In some ways this is uncomfortable ground from which to generalise and was necessitated by the goal of representing our data in a form that would captivate and interest public audiences but, coupled with our more systematic analytical work, it took our analysis to more creative places.

It also worth noting that even in the event where direct quotes or 'narratives' may have involved a degree of 'tinkering' to facilitate a 'smoother' prose to aid the visualised process of the animation (Sou and Hall, 2021), capturing the general spirit meant that audiences could engage or identify with the central characters and the everyday stories that they articulated. This perhaps works at engaging with the sometimes confusing interplay between composite 'characters' or 'narratives', demonstrating the playful spectrum of creativity as part of analytical processes. And yet, regardless of the 'closeness' to the 'raw data', if the spirit of the research findings are reflected, then sociological endeavours are achieved in telling important stories.

Conclusion

In this chapter, we have reflected upon our experiences of creating characters and assembling composite narratives for creative outputs in two qualitative research projects. In so doing, we have emphasised how the increasing inclination to engage wider publics in academic research offers new opportunities for researchers to revisit and enrich tried and tested analytical practices, necessitates critical thinking about difference and representation, and ultimately invites more creative ways of knowing and generalising with data (Hall and Sou, 2021). This highlights the connections between research processes that are often conceptualised separately, disrupting the assumption that dissemination is an end point in a research project, demonstrating how analysis, writing, and engaging public audiences are all iterative processes within the craft of knowledge (Smart et al, 2014).

We have shared examples of navigating creative data analysis in three key ways. Firstly, through the assembling and combining of thematic and narrative analysis in the creation of characters. Secondly, through critically engaging with 'who' and 'how', difference and representation come through, and into, our analysis, and thus into our finalised (and significantly, visual) artefacts. And, thirdly, how the 'spirit' or 'tone', and the ethnographic knowledge of such, is an important creative sensibility to pay attention to in communicating and disseminating our research. With each of these in mind, we offer a series of questions, prompts, and provocations for those who may be interested in developing composite narratives and characters for public engagement artefacts and would like to critically engage in the creative and analytical processes that accompany their formation.

Key questions, prompts, and provocations

- Have you considered non-researcher involvement, for example, research participants and creative partners, and their role in collaborating with you to describe, interpret, and creatively engage with data?
- How might you think reflexively about creating composite narratives and/or characters for the purposes of your research, its analysis, and its outputs?
- How might you take care with individual narratives while building central themes through storying and character approaches?
- Ask yourself critical questions – who are you representing? Is anyone missing? How might composites negotiate this and are there particular ways to represent across differences?
- What is the role of the visual in your creative and analytical process?
- What is the 'spirit' of your research? How might you capture and communicate this? What matters to the people in your research and how might you represent the diversity of their experiences?

Notes

1 Dr Lauren White would like to thank Amy Carter Gordon for her expertise in curation, convening creative partners, and facilitating knowledge exchange. She would also like to thank Charlie Barnes (Creative Director, Dead Earnest Theatre Productions, https://www.deadearnest.co.uk/) for co-producing content and script and facilitating voice-overs. Thanks are also due to the team at Fettle Animation (https://www.fettleanimation.com) and to the ESRC (ES/V009397/1) and University of Sheffield Public Engagement for Funding. Finally, she would like to thank the 25 people who kindly participated in the research.

Dr Katherine Davies and Dr Adam Carter would like to thank Lynette Hodges of the University of Sheffield Public Engagement Team for her organisation and liaison prowess. They would also like to thank the digital creatives at Sheffield-based Joi Polloi (https://joipolloi.com) for their role in assembling the My Year in Brexit Britain website, as well as the ESRC (ES/S006362/1) for funding, and the Festival of Social Science for providing the online exhibition space. Additionally, Adam would like to acknowledge the support of The Leverhulme Trust Early Career Fellowship. Finally, they would like to thank those who took part in the research.

2 Information on the 'Brexit and Family Life' project available to view here: https://www.sheffield.ac.uk/socstudies/research/current-projects/brexit-and-family-life.

3 *Gogglebox* is a popular UK TV programme, where families are filmed watching television together and having conversations with each other.

4 'Everyday Life with IBS' animation; available to view here: https://player.sheffield.ac.uk/events/everyday-life-ibs.

5 Both research projects referred to in this chapter received ethical approval through the University Research Ethics Committee at the University of Sheffield.

References

Arjomand, N.A. (2022) 'Empirical fiction: composite character narratives in analytical sociology', *The American Sociologist*, [online] 23 September. Available from: https://doi.org/10.1007/s12108-022-09546-z

Back, L. and Duneier, M. (2006) 'Voices from the sidewalk: ethnography and writing race', *Ethnic and Racial Studies*, 29(3): 543–65.

Braun, V. and Clarke, V. (2006) 'Using thematic analysis in psychology', *Qualitative Research in Psychology*, 3(2): 77–101.

Critical Methodologies Collective (eds) (2022) *The Politics and Ethics of Representation in Qualitative Research: Addressing Moments of Discomfort*, London: Routledge.

Davies, K. and Carter, A. (2021) 'Research relationalities and shifting sensitivities: doing ethnographic research about Brexit and everyday family relationships', *Families, Relationships and Societies*, 10(1): 169–77.

Edmunds, D.R. (2010) 'People's lives: some thoughts on composite characters and Native American biography', *Journal of the West*, 49(4): 38–44.

Inckle, K. (2010) 'Telling tales? Using ethnographic fictions to speak embodied "truth"', *Qualitative Research*, 10(1): 27–47.

Mannay, D. (2019) 'Revisualizing data: engagement, impact and multimodal dissemination', in L. Pauwels and D. Mannay (eds), *The SAGE Handbook of Visual Research Methods* (2nd edn), London: SAGE Publications, pp 659–69.

Markham, A. (2012) 'Fabrication as ethical practice', *Information, Communication & Society*, 15(3): 334–53.

Mason J. (2008) 'The art of generalisation in real life research', Paper presented at the Vital Signs: Researching Real Life conference, University of Manchester, 9–11 September 2008.

Mason, J. and Davies, K. (2009) 'Coming to our senses? A critical approach to sensory methodology', *Qualitative Research*, 9(5): 587–603.

McElhinney, Z. and Kennedy, C. (2022) 'Enhancing the collective, protecting the personal: the valuable role of composite narratives in medical education research', *Perspectives in Medical Education*, 11(4): 220–7.

Neal, S. and Murji, K. (2015) 'Sociologies of everyday life: editors' introduction to the special issue', *Sociology*, 49(5): 811–19.

Piper, H. and Sikes, P. (2010) 'All teachers are vulnerable but especially gay teachers: using composite fictions to protect research participants in pupil–teacher sex-related research', *Qualitative Inquiry*, 16(7): 566–74.

Riessman, C. (1993) *Narrative Analysis*, London: SAGE.

Robertson, A., Magdy, W., and Goldwater, S. (2021) 'Black or white but never neutral: how readers perceive identity from yellow or skin-toned emoji', *Proc. ACM Hum.-Comput. Interact.*, 5: CSCW2.

Smart, C., Hockey, J., and James, A. (2014) *The Craft of Knowledge: Experiences of Living with Data*, Basingstoke: Palgrave Macmillan.

Sou, G. and Hall, S.M. (2021) 'Communicating crisis research with comics: representation, process and pedagogy', in H. Kara (ed), *Qualitative and Digital Research in Times of Crisis*, Bristol: Bristol University Press, pp 93–109.

Watson, A. (2016) 'Directions for public sociology: novel writing as a creative approach', *Cultural Sociology*, 10(4): 431–47.

Watson, A. (2022) 'Writing sociological fiction', *Qualitative Research*, 22(3): 337–52.

Wertz, M.S., Nosek, M., McNiesh, S., and Marlowe, E. (2011) 'The composite first person narrative: texture, structure, and meaning in writing phenomenological descriptions', *International Journal of Qualitative Studies on Health & Well-Being*, 6(2): 1–10.

Willis, R. (2018) 'The use of composite characters to present interview findings', *Qualitative Research*, 19(4): 471–80.

Wolseth, J. (2019) 'Writing after betrayal: *desahogarse*, street outreach and ethnography', *Ethnography*, 20(3): 342–58.

17

Beyond the brick wall: transdisciplinary and creative research through Scholarly Personal Narrative and Lilyology

Dawn Wink

A spider created an intricate web in the outside corner of my writing room window. I see her perched in the centre when I look up from my desk. That the delicate legs support the large, round abdomen the size of a small stone amazes me. Throughout the year, she goes about her business, and I go about mine here in the high desert of Santa Fe, New Mexico, in the south-western United States. The seasons of the year serve as a backdrop to the web's silhouette. The web frames the cumulus clouds of summer as they drift across bright blue sky and the bright green leaves and clusters of magenta flowers of a New Mexico locust tree just beyond the fence. The leaves of that same tree turned gold, rust, and red with the crispness of autumn beyond the web. When snow begins to fall in soft flurries to announce winter's arrival, I wondered at the spider's ability to survive. I still do not understand; just know that I am grateful when the air warms again in the late-arriving spring of this altitude. The spider's web maintained its shape, structure, and strength through the seasons. As the frame of web connects the seasons outside the window, so connective power has much to teach us as we delve into transdisciplinary and creative research methods.

Transdisciplinary and creative methods

Transdisciplinary research and creative methods honour complexities. These complex dynamics encourage us to move beyond discipline-specific and isolated analyses and into rich and deep research, as 'only a thoroughly transdisciplinary perspective can navigate such issues, which are at once technological, cultural, ethical, political, economic, and ecological' (Wells, 2013, p 126). This chapter explores the creative method of Scholarly Personal Narrative (SPN) (Nash and Bradley, 2011) and theoretical framework of Lilyology (Blair, 2015). SPN falls under the overarching umbrella of narrative inquiry (Creswell and Poth, 2017). The narrative method 'begins with the experiences as expressed in lived and told stories of individuals' (p 70). Clandinin and Connelly (2004) detail 'experience as central to the epistemology and ontology of narrative inquiry' (p 38). Using narrative methods, researchers weave their lived experiences and understandings into the research. Narrative and story convey the findings of qualitative research. Qualitative research reflects an interpretive approach to make sense of the meaning (Denzin and Lincoln, 2011). Dewey (1980) describes narrative inquiry as a way to make the 'tangled scenes of life more intelligible in aesthetic experience' (p 290).

The epistemology of SPN falls within narrative inquiry as, when done well, it 'combines scholarship, personal stories, and universalizable themes and insights in a seamless manner' (Nash and Bradley, 2011, p 24). Essential aspects of SPN highlight the writer's intentionality to identify broader themes and insights and connect these universalisable ideas within non-academic and academic prose. SPN explores research, not through a positivistic paradigm that assumes researcher objectivity, but rather by assuming researchers bring their whole selves and experiences to the research. This concentration on connective knowledge differentiates SPN from interdisciplinary research, as it transcends the divides that remain and embraces research that weaves and fuses together to create a new whole. Interdisciplinary research can be represented as a braid. The disciplines remain distinct, even when woven together, as the distinct sections of hair remain separate while woven within the greater whole. Within transdisciplinary research, the disciplines come together, swirl, interweave, and create a new whole, much as recycled Indian saris are cut into ribbons that are twisted and fused together so seamlessly that the separate strips are indistinguishable within the new whole.

This embrace of lived experiences grounds the *axiology* of SPN, reflected in key terms used by SPN researchers: *vigour* rather than *rigour*, *subjective experience* rather than *experimental design*, *personal testimony* instead of *evidence*, *perspectives* over *data*, *introspective questions* instead of *interview questions*, *universalizability* rather than *replicability*, *according to my experience* that acknowledges *limitations*, *plausibility/honesty/coherence* replaces *hypothesis*, and *illustrative, embedded references* replace *literature reviews* (Nash and Bradley, 2011, pp 82–5). These paradigmatic foundations of SPN reflect with transparency where the researcher is situated within the inquiry. The relevance of SPN as the methodological framework lies in its expansiveness not only to allow, but also to encourage, a meaningful, energetic, and genuine relationship between the researcher/writer and the work.

In *The Offense of Theory* Maggie MacLure (2010, p 277) asserts that the 'value of theory lies in its power to get in the way: to offend and interrupt'. MacLure invites us to consider the power of theory to resist the reproduction of the status quo and be open to new possibilities. MacLure (2010, p 278) argues that 'theory stops us from forgetting then that the world is not laid out in plain view before our eyes … it stops us from thinking that things speak for themselves'. SPN and Lilyology interrupt the status quo and open research up to creative potential and possibilities.

Lilyology

Reflecting Indigenous perspectives from Australia, the theoretical framework of Lilyology, created by Nerida Blair (2015), offers the potential to connect across spaces in-between through the strong, transparent threads of *spider webs*. Storytelling as a way to share human experience, convey wisdom and knowledge, and connect with others and the world spans the long arc of human experience. Stories have been shared around fires and throughout daily rhythms long before written words, long before books, and long before distinctions of disciplines. This theoretical framework embodies the connective wholeness of story throughout time. The connective wholeness does not marginalise according to hierarchy but rather embraces all stories of all people. This lens opens the analytical aspects of this creative method to include the nuances and subtleties of data collection and interpretation where the essence of that data lives, yet has often historically been dismissed. Herein lies the importance of creative research methods and an embrace of a wider and more holistic lens of theoretical frameworks and methods, to allow researchers to shine light on invisibilised data through scholarship.

This creative research integrates Lilyology as a framework and storytelling practice to decolonise critical ethical spaces for academic research (Blair, 2019). The theoretical framework embraces ecosystems, natural landscapes, and Indigenous languages as vital, integral, and necessary elements for global sustainability. Lilyology, as a reflection of Indigenous scholarship, draws directly from the natural ecosystem and lifts Indigenous language worldviews into the greater body of work focused on creative research methods. Lilyology embodies, what hooks (1995) describes as 'learning to see and appreciate the presence of beauty is an act of resistance in a culture of domination' (p 124).

Storytelling is deeply ingrained in many cultures as an integral way to impart wisdom. Blair (2019) highlights the significance of story within methodology. Lilyology 'as a decolonizing framework … reflects a re-imaged concept of the waterlily. Lilyology metaphorically engages waterlilys, sweet potatoes, spiders, and brick walls, creating a space to reflect on and privilege Indigenous Knowings in the academic context' (Blair, 2019, p 203). Blair invites the reader to imagine a drawing of a square, a frame around images of waterlilies, a brick wall, a bush potato (sweet potato), and lightning. Tendrils of a spider web span and connect all. The petals of the *waterlily* are open, a yellow stamen rising from the centre and holding the petals together. Waterlily flowers come in as many distinct colours, shapes, and textures as there are stories. Every element of the waterlily connects to 'Country' (Blair, 2015, p 6) in its most comprehensive sense, encompassing the whole of the ecological environment, comprising 'water land, soil, air, sun, light, dark, atmosphere. … If we all stop our imagining and listen, we can almost hear the different languages being spoken by all of these elements to each other' (Blair, 2019, pp 209–10).

The *brick wall* represents Western academia. Each brick reflects a specific philosophy or methodology as an academic frame of reference. Brick order and construction reflect the linear hierarchy inherent in conventional Western academia. Western knowledge serves as the mortar holding the bricks together. The wall represents boundaries and divisions. *Bush (sweet) potatoes* represent 'Spirit

Figure 17.1: Lilyology

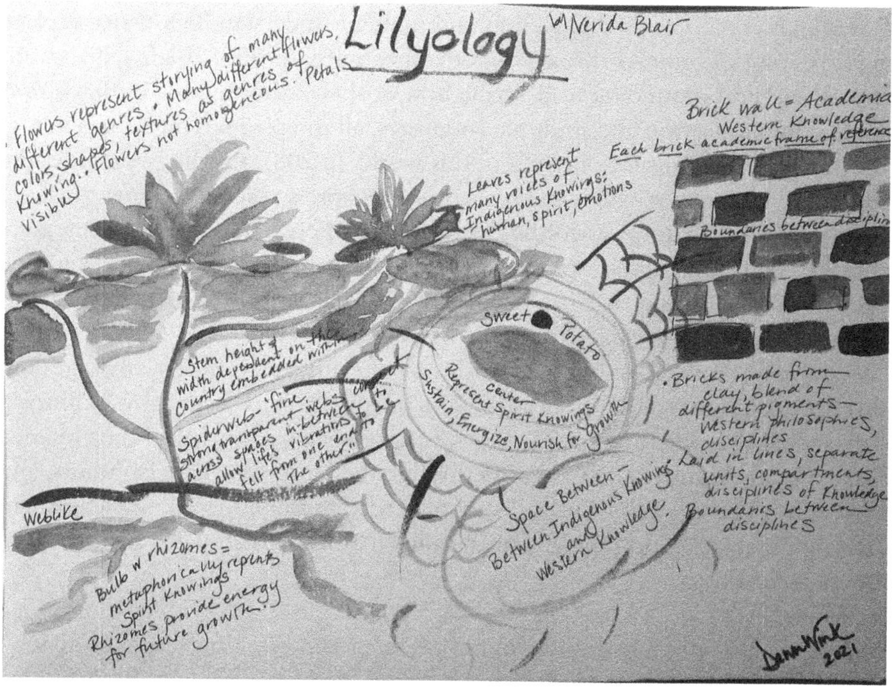

Knowings' (Blair, 2015, p 38) that radiate from a nourishing, sustaining, and growth-producing central core. The strong, transparent threads of *spider webs* represent the potential to create connections across the spaces in-between. The final element of Lilyology is the *space in-between* 'Indigenous Knowings' and 'Western Knowledge' (Blair, 2019). When I first read the details of Lilyology, I wanted to visualise this method. I pulled the watercolour paints and paper from the shelf and painted my way through my understandings of each of the elements. Figure 17.1 reflects my understandings.

Lilyology is grounded in story, stories that are iterative rather than linear. The 'storys [sic] are told in a circular or spiral theme, with each thematic repetition or spiral adding a little' (Blair, 2019, p 213). The iterative and circular Indigenous expression of story contrasts to the traditional linear trajectory of the Western story. Blair refers to Hokari's (2003) expression of storying:

> Maintaining the knowledge did not mean finding a 'right' story but widening the possibilities of storys [sic]. Information of different variations is preferred, pooled and maintained as a bundle of possibilities without judgement. Different storys which contract each other, do not conflict, but simply co-exist. (Blair, 2019, pp 8–9)

The Indigenous worldview of co-becoming with 'Country' is essential to Lilyology. Blair (2019, p 207) describes '[Country as] the Aboriginal English word which encompasses this vibrant and sentient understanding of space/place which becomes bounded through interconnectivity' (p 207). Blair expands on the philosophical grounding to describe how co-becoming 'with Country storys diverse communication and multiple languages, all intertwined into a tapestry of the cycle of life and death, of being in this world' (p 208). Lilyology provides the structure to integrate research and life stories to express voices often marginalised within academic writing. Accordingly, Lilyology offers a framework to integrate the stories of all, an expression of how 'justice may rule on the side of deep time and a storied landscape that has been held in place by the roots of Indigenous wisdom' (Tempest Williams, 2019, p 17).

Within my research exploring stories at the intersection of language and landscape through wildness, beauty, and imagination, I began by exploring experiences in my life through 25 years' worth of personal journals. I then braided these experiences into five SPN essays through the lenses of wildness, beauty, and imagination. The five SPN narratives are presented through the acknowledgement of the thoughts of Maori scholar Linda Tuhiwai Smith (2012, p 24) as she reflects on post-colonialism:

> There is also, amongst Indigenous academics, the sneaking suspicion that the fashion of post-colonialism has become a strategy for reinscribing or reauthorizing the privileges of non-Indigenous academics because the field of post-colonial discourse has been defined in ways which can leave out Indigenous peoples, our ways of knowing and our current concerns.

The brick wall deserves further exploration. Lilyology reflects Indigenous experience expressed with wisdom by Tuhiwai Smith (2012, p 1):

> From the vantage point of the colonized, a position from which I write, and choose to privilege, the term 'research' is inextricably linked to European imperialism and colonialism. The word itself, 'research', is probably one of the dirtiest words in the Indigenous world's vocabulary.

Historically, research as experienced by Indigenous peoples is a brick wall. Lilyology provides the rigorous scholarly structure to weave together research and life stories that have often been silenced by the brick wall within academic writing. Conventional research values a linear, European approach to writing styles and content. The brick wall historically excluded the spiral, intuitive, nature-based stories and wisdom of Indigenous and marginalised communities as lacking quantifiable evidence necessary for academic credibility. The spider web of Lilyology, founded on holistic Indigenous wisdom and rigorous scholarship, connects the brick wall and the spaces in-between of 'Indigenous Knowings' and 'Western Knowledge' (Blair, 2015, p 38). Rather than a binary, Lilyology offers integration and connection. This expansion of what constitutes data deepens the potential for analytical analysis, allowing creative qualitative researchers to interpret findings and connect their meaning within a greater context. Creative research methods do not analyse data in a vacuum, but rather within the greater whole, and then draw conclusions on the meaning of the interpreted data. This can cause cognitive dissonance for those long familiar with research methods historically utilised and accepted within academia. The brick wall invites further musings, as expressed by my Professor Emerita mother, who, when presented with these ideas, called me and exclaimed, 'My brick wall keeps getting in the way of the nourishing sweet potato!'

The narratives presented intend to embody the essence of 'illustrative, embedded references – integrating supportive background scholarship as it comes up ad hoc throughout the writing ... illustrative, timely, coherent. They deepen, enrich, and enlarge the thematic motifs of the writing throughout the writing ...' (Nash and Bradley, 2011, p 85).

Procedure

Nash and Bradley (2011) suggest a structure to research through four main steps: '(a) Pre-Search: How do I get started? (b) Me-Search: What is my personal narrative regarding the ideas emphasized in my writing? (c) Re-Search: What scholars and researchers have informed my writing? (d) We-Search: What are the implications for my profession, or field of study, that can be generalized from my SPN?' (pp 6–7) These structured prompts guide the approach and engagement.

Pre-search: how do I get started?

The pre-search for this inquiry began with pulling the containers filled with years of journals out of the garage and bringing them into my writing room. As I unpacked each container and lifted bundles of journals out to place on the bookshelves, the years and experiences expressed in these pages washed over me. I decorated each journal with quotes, images, and beautiful paper. The aesthetics of each journal visually reflect the resonances of that time in my life.

I started this SPN inquiry by reading through these journals to discover what themes emerged around my relationship with language, landscape, wildness, beauty, and imagination. I codified discovered themes through colour coding: (1) language: blue; (2) landscape: green; (3) wildness: red; (4) beauty: purple; and (5) imagination: yellow. I was also aware of other themes that emerged through the texts. I was grateful for the spectrum of colours within my set of markers for these emergent threads. I discovered the themes: (1) creativity: orange; (2) writing: poppy; (3) story: sky blue; and (4) the tension between academic and creative life: chartreuse.

After identifying the themes, I reread each themed area to select a narrative kernel that holds the most potent energy to reflect the essence of this idea for the SPN. The narratives presented here reflect a holistic arc to address the inquiry question in full. Nash and Bradley (2011, p 6) ask scholars to reflect on the following question during the pre-search stage: 'Is there some pivotal belief, ideal, or conviction I hold with such energy and passion that I am willing to take the time and make the effort to research and write about it in a sustained way?' I taped this question to my desk as I engaged with reading my journals.

Nash and Bradley (2011) also ask the scholar to consider at this stage, 'Who is my audience, and have I made this clear from the start?' (p 6). My intended audience groups are the readers, researchers, and writers of ecolinguistics/linguistic human rights (LHR) in the realms of academia and landscape literature. My specific intention behind my selection of SPN as a methodology is for the narratives to meet the scholarly rigour criteria for academic writing and the creative voice criteria for landscape literature.

Me-search: what is my personal narrative regarding the ideas emphasised in my writing?

Nash and Bradley (2011) ask the scholar to reflect on the following question during the me-search: 'Where exactly is the "me" in my writing, and why is the "me" important in advancing my ideas to others?' (p 6). I recognised years ago that I wanted to express knowledge and research of ecolinguistics and LHRs through creative prose within landscape literature. Nash and Bradley (2011) also ask the scholar to self-pose the question 'How do I write from the core of my life experience in order to inform my trained writer's intellect?' (p 6). As I reflected on this question, what emerged for me was how intimately intertwined my life experience and my trained writer's intellect have become over the years of journal writing, combined with academic and creative writing and publication. My journals reflect the initial spark that would become a flame when a friend mentioned the book *The Artist's Way* by Julia Cameron (1992) to me as we pushed our babies on the swings in the local park. The baby in the swing was my youngest child, Wynn, who was then 6 months old. My other children, Luke (2 years old) and Wyatt (then 4 years old), played on the jungle gym.

The reading and writing explorations presented in *The Artist's Way* forever transformed my life. As I read and wrote, I realised that I wanted to write professionally. After *The Artist's Way*, I dove headfirst into reading and learning all that I could about the art and craft of writing. As my kids' bookshelves evolved from cardboard books to children's stories to novels, the bookshelves in my writing room filled and overflowed with books on the art and craft of writing. I dedicated the early morning hours before the kids woke and naptimes when they slept to reading and learning about writing – and writing.

Wynn, Wyatt, and Luke are now 23, 24, and 26. Novels of weight and heft long ago replaced the children's books on their shelves. My bookshelves bend with the weight of writing books from those early years. Writing fiction and non-fiction filled the years since that initial spark of writing energy. My epistemologies reflect the whole of my life and are deeply influenced by scholarly readings and life experience. They have emerged over time to reflect an integration of readings with my lived experiences. My epistemologies include all knowledge that I have learned within the brick wall of academia. They are also deeply grounded in my developed sense of trusting intuitive knowings. My epistemologies include both the head and the heart.

Re-search: which scholars and researchers have informed my writing?

Nash and Bradley (2011) ask the researcher to reflect on 'What are the primary and secondary scholarly sources that ground my research?' (p 7). This stage of creative research within SPN is akin to a literature review. Nash and Bradley (2011) also invite the scholar to define 'In what creative ways do my documents redefine, as well as support, conventional understandings of research?' (p 7). The focus of this inquiry lies in its intention to create new understandings among and within literary landscapes. SPN meets the criteria of both academia and creative prose.

Nash and Bradley (2011) ask the scholar to answer 'How do I identify and convey to my readers the implications for their own fields of study?' (p 7). The translation of academic research into creative prose opens new implications for the academic and creative writing professions and fields of study. Nash and Bradley (2011) encourage the researcher to identify 'How do I generalize (or universalize) my findings to others in such a way that *I* can co-exist with the *we* who are reading my document?' (p 7). The *we* who are reading the work and narratives of this inquiry are very important to me. SPN provides the methodological framework to provide such universalisability. The essential components for universalisability in ways that *I* can coexist with the *we* who are reading this work are grounded in academic, scholarly rigour translated through creative expression. Only this combination will allow me to coexist with the *we* who read this work. Without the scholarly and academic rigour, I could not coexist with the *we* of my academic colleagues in ecolinguistics and LHR. The work must have the foundation of rigorous academic scholarship to be credible.

Procedure details

1. First, I listed the required elements at the end of each narrative: *language*, *landscape*, *wildness*, *beauty*, and *imagination*.
2. Next, I printed out each narrative and bundled individually. I spread these out on the table, along with the printed draft of the whole of my dissertation.
3. I intentionally brought in the research and writers specific to the topic and the journal focus to speak most resonantly with the readership.
4. Next, I identified the element (*language*, *landscape*, *wildness*, *beauty*, *imagination*) within the narrative and listed that sentence, paragraph, or section alongside the specific element.

5. Then, I identified the elements of Lilyology within each narrative: waterlily, brick wall, bush (sweet) potato, spider web, and spaces in-between, and listed each component of the specific narrative.
6. At the end of this stage, I had ensured that each narrative contained all required elements specific to this inquiry and Lilyology.
7. I then began to write the narratives and wove the required components into the whole of the piece.

Data analysis

Data analysis in SPN follows the emergent themes that unfold throughout the research, as the researcher explores 'ways in which our researcher relationship to ongoing participant stories – the stories being lived and told – shapes the nature of field texts as well as the kinds of field texts that may be composed' (Creswell and Poth, 2017, p 92).

> Within narrative inquiry researchers' personal, private, and professional lives flow across the boundaries into the research site. … The consequence of this fluidity for making field texts is that there is a virtually endless list of life experiences that might be and frequently are turned into field texts of value to the inquiry. (Clandinin and Connelly, 2000, p 115)

Research journeys reflect the intertwining relationship that flows across boundaries of private and professional lives recognised and honoured by SPN and expressed through the framework of Lilyology. I began each day with the reading and writing research dedicated to my inquiry and described in one of the narratives within this inquiry. I moved then into the early morning run, a time for musing, ruminating, and reflecting on what I had read or written that morning and how it fitted into the greater whole of the inquiry. This was a time that I remained intentionally open and receptive to ideas and connections that somehow coalesced and took shape during my runs. I ran with a tiny notebook and pen in my pocket, so I could capture these, which would later become another narrative within this inquiry. After my run, I moved into my professional life with the college and all that goes with collegiate life. As I engaged in the meetings, emails, events, and issues of collegiate life, my inquiry was never far from my mind, just set on the back burner of the stove on low to simmer until I could return. Under SPN, limitations are experienced *according to my experience*. The expected manner for a writer to acknowledge limitations with SPN is to be completely honest throughout the study regarding the strengths and limitations of personal narrative scholarship. Limitations that unfolded are addressed with transparency.

Within narrative inquiry and SPN, validity is understood 'through four primary criteria: credibility (Are the results an accurate interpretation of the participants' meaning?); authenticity (Are different voices heard?); criticality (Is there a critical appraisal of all aspects of the research?); and integrity (Are the investigators self-critical?)' (Creswell and Poth, 2017, p 258). Validation strategies include: (1) corroborating evidence through triangulation; (2) discovering a negative case analysis or disconfirming evidence; (3) clarifying researcher bias or engaging in reflexivity; (4) generating rich, thick descriptions; (5) enabling external audit; and (6) prolonged engagement and persistent observation in the field (Creswell and Poth, 2017, p 260). These validity criteria were also applied to the act of translation, a primary focus of this inquiry. The act of translation, and naming it explicitly, offers an interpretative lens that encompasses experiential details within the analytical interpretation of the data.

Narratives as an act of translation

SPN offers the means to bridge the brick wall of academia identified in Lilyology through translation into narratives ultimately published in landscape literature. Translation is understood here as a way to help 'make common sense of our conflicted and fragile world, and to join together in a mutual search for constructive human responses to the serious crises and challenges we face as inhabitants of the same small planet' (Gillespie, 2021, p 2).

Lilyology integrates into the methodological context of SPN the four principles of respect, responsibility, reverence, and reciprocity, which act as ethical guides for the researcher to work with Indigenous people, their Indigenous knowledges, and stories, through 'engaging in holistic meaning-making [that] involves using the heart (emotions), mind (intellect), body (physical actions), and spirit (spirituality), as well as recognizing the relationships of these realms to oneself, family, community, land/environment, and wider society' (Blair, 2019, p 1).

The structure of the brick wall of Western academia provides the essential structure of research, knowledge, and information on ecolinguistics and LHR. Lilyology elevates Indigenous worldviews and knowings:

> Lilyology, grounded in Country, reflects the organic and dynamic space that is Indigenous Knowings, the cyclic and holistic patterning of life. It presents a storytelling practice for strategic interventions into operational spaces within academic, which in the model of Lilyology is storyed as a brick wall. (Blair, 2019, p 203)

Lilyology decentres SPN from the Western lens, to address the idea that 'until the lions have their storytellers, the story of the hunt will always glorify the hunter' (wa Thiong'o, 2017, p 5).

The inspiration of this inquiry is to support the knowledge and research of LHR to an even broader readership, as

> The cartilage of history is precious and perishable: millions of family chronicles of millions of people lost in transit or … lost in translation or

> untranslatable, abridged out of fear or neglect or unsuitability, so that we are displaced not merely in place but also in time, our ancestral narratives mislaid or missing or hidden from our sight, inaccessible. (Badkhen, 2020)

These translative narratives offer what Gillespie (2021) suggests as an alternative way to think about the binary of translation: as it changes 'the original work into new language, it also changes the language of the translator in both meanings of this genitive phrase: it hangs both the language that the translator "has", and the language that "has" the translator' (p 57).

Narrative arc of the SPNs

Each narrative created reflects the framework structure of Lilyology (Blair, 2019) – each story narrative reflects the multiple voices and stories represented by the petals of a *waterlily*, each story integrates into the *brick wall* of academia, each story holds a central, sustaining, energising, nourishing core necessary for growth represented by the *bush potato*; the overarching, connective narrative arc of the stories reflects the transparent *spider web* connecting across *spaces in-between* and life's vibrations. As Vuong (2020) said in an interview reflecting on how people tell stories, 'Stories are carried in the body, and it's edited each time the person tells it. And so what you have, by the time someone tells a story, is a masterclass of form, technique, concision, imagery.'

These stories maintain the spirit of Lilyology and narrative inquiry as they are not intended to find a single 'right' answer but to bring experiences to life. The stories are linked together chronologically to create a narrative arc. Each story builds on the story before. The culmination of the five SPN essays reflects a holistic perspective in sequential order.

Stories as seed bombs

The intentions of these narratives are to serve as story seed bombs. A seed bomb is a little ball composed of compost, clay, and seeds. The mixture of compost and the clay holds the seeds together and serves as a carrier for the seeds to be spread to wider areas, just as these narratives are being published – tossed – into the wider area of landscape literature. The compost and the clay hold the seeds together until they reach a particular area. Seeds (ideas) cannot be thrown loose into these distant spots as they would not arrive due to distance or being blown away in the wind. Companion planting utilises seeds that encourage each other's growth (Jeffrey, 2011). These story seed bombs are composed with the intention of companion planting between the landscape of academic writing and that of creative prose, each enriching and strengthening the other. Jeffrey (2011) proposes the beauty of seed bombing as a little revolution in the palm of a hand that can change the face of the earth. Herein lies the heart of the correlations between physical seed bombs and the story seed bombs of this inquiry – with each publication, the readers hold a little revolution in the palm of their hands.

Conclusion

The spider web outside of my writing room window glistens in the burnished bronze light of late afternoon. It is winter now, and snow lines the tops of the branches of the tree beyond the web. I rise from my desk to study the web closely, just on the other side of the window. Each spider web is unique, much like fingerprints. I wonder at the infinite possibilities expressed in each design. Creative research invites each of us to create unique webs. SPN and Lilyology offer potential structure and form to spin connective webs of research beyond the brick wall.

References

Badkhen, A. (2020) Once I Took a Week-long Walk in the Sahara. *Orion Magazine*. Available from: http://www. https://emergencemagazine.org/essay/once-i-took-a-week-long-walk-in-the-sahara

Blair, N. (2015) *Privileging Australian Indigenous Knowledge: Sweet Potatoes, Spiders, Waterlilys, and Brick Walls*, Champaign: Common Ground.

Blair, N. (2019) 'Lilyology as a transformative framework for decolonizing ethical spaces within the academy', in J. Archibald, A.A. Xiiem, J.B.J. Lee-Morgan, and J. De Santolo (eds), *Decolonizing Research: Indigenous Storywork as Methodology*, Sydney: Bloomsbury, pp 203–23.

Cameron, J. (1992) *The Artist's Way: A Spiritual Path to Higher Creativity*. New York: Tarcher/Putnum.

Clandinin, D.J. and Connelly, F.M. (2000) *Narrative Inquiry: Experience the Story of Qualitative Research*, San Francisco: John Wiley & Sons.

Clandinin, D.J. and Connelly, F.M. (2004) *Narrative Inquiry: Experience and Story in Qualitative Research*. San Francisco: Jossey-Bass.

Creswell, J. and Poth, C. (2017) *Qualitative Inquiry and Research Design: Choosing among Five Approaches* (4th edn), Thousand Oaks: SAGE.

Denzin, N.K. and Lincoln, Y.S. (2011) 'Introduction: the discipline and practice of qualitative research', in N.K. Denzin and Y.A. Lincoln (eds), *The SAGE Handbook of Qualitative Research* (4th edn). Thousand Oaks: SAGE, pp 1–20.

Dewey, J. (1980 [1934]) *Art as Experience*, New York: Perigee Books.

Gillespie, S. (2021) 'The possibility of translation', *Boundary 2*, 48(1), 49–63.

Hokari, M. (2003) 'History happening in/between body and place: Journey to the Aboriginal way of historical practice' in *Proceedings of the Habitus: A Sense of Place. Proceedings of the 2000 Conference*. Perth: Curtin University of Technology, p 2.

hooks, b. (1995) *Art on My Mind: Visual Politics*, New York: The New Press.

Jeffrey, J. (2011) 'What is a seedbomb?' *The Ecologist*, 4 May. Available from: https://theecologist.org/2011/may/04/what-seedbomb

MacLure, M. (2010) 'The offence of theory', *Journal of Educational Policy*, 25(2), 277–86.

Nash, R., and Bradley, D. (2011). *Me-Search and Re-Search: A Guide for Writing Scholarly Personal Narrative Manuscripts*. Charlotte: Information Age.

Tempest Williams, T. (2019) *Erosion: Essays of Undoing*. New York: Sarah Crichton.

Tuhiwai Smith, L. (2012) *Decolonizing Methodologies: Research and Indigenous Peoples*. Dunedin: Otago University Press.

Vuong, O. (2020) 'A life worthy of our breath', *On Being with Krista Tippett*, April 30. Available from: https://onbeing.org/programs/ocean-vuong-a-life-worthy-of-our-breath/

wa Thiong'o, N. (2017) *Devil on the Cross*, London: Penguin Classics.

Wells, J. (2013) *Complexity and Sustainability*, Oxford: Routledge.

Section VI

Creative arts-based analysis

18

Slow stitch: reflexive creative analysis and meaning making

Naomi Clarke[1]

Introduction

While there are an increasing number of texts exploring creative research methods (Gauntlett, 2007, 2018; Smith and Dean, 2009; Mannay, 2016; Kara, 2020), few provide detailed explorations of the in-practice use of creative analysis. My use of the term creative analysis, in this chapter, refers to analysing data creatively rather than analysing creative data.

Accordingly, this chapter focuses on one element of my doctoral research, namely stitching as a tool within creative, reflexive, thematic analysis. Stitching is threaded throughout every stage of my work, from the literature review through to analysis and dissemination, because it is an inherent part of my identity as a crafter as well as a researcher. There are plentiful 'how-to books' on creative research (as will be discussed) but few that go into detail about the in-practice lived experiences of creative analytical approaches. This chapter aims to offer:

- a combination of theoretical and reflexive positionings;
- some 'how-to' guidance in using sewing as a tool for creative analysis;
- discussions and thoughts on what worked well (or did not) and why;
- and, finally, some potentials for developing this stitching approach.

As well as sharing my reflections on this approach, I am keen to encourage reflection on the part of the readers. Consequently, I encourage readers to utilise reflexive approaches during, and after, reading this chapter (whether through stitching or another method) in relation to their own responses and thoughts.

Background to the research project

When the COVID-19 pandemic took hold, popular media reported global increases in everyday making and creativity (Brignall, 2020; Kurutz, 2021). The 'Crafting during Coronavirus' research project utilised diary methods over a nine-month period in 2020,[2] to explore everyday making during crisis, as recorded, understood, and explained by makers themselves (Clarke and Watson, 2020). The study involved 313 makers, aged 21–84 years, from 19 countries. Creative diary data were collected, comprising over 23,000 files, including textual diaries (physical and digital), images and videos of making, and tangible crafted items. These materials were explored using reflexive thematic analysis via both textual and textile means (Braun and Clarke, 2022b).

This research, and the following discussion of the stitch-based creative analysis approach, is grounded within broader feminist and social constructionist approaches. These approaches seek to challenge and disrupt patriarchal views of 'knowledge' while acknowledging that 'knowledge' and 'knowers' are not static end points, but social constructs impacted by various external factors (Wilkinson, 1988; Harrison and Ogden, 2021). Utilising creative approaches allows researchers to step away from dominant paradigms (Mannay, 2016; Harrison and Ogden, 2021); and using textile-based methodological approaches further challenges dominant paradigms (Harrison and Ogden, 2021). Fibre-based activities, such as knitting and sewing, carry long associations of women and women's voices being subjugated within research, wider society, and mainstream cultures (Parker, 2012; Harrison and Ogden, 2021). The very act of using stitch as a form of creative analysis can, within research, offer new and valuable insights as well as non-linguistic knowledge forms while also challenging traditional patriarchal knowledge production positionings (Hawkins, 2017; Harrison and Ogden, 2021).

The role of creativity, creative methods and theorising this space is the focus of the following section.

Theorising craft and researching creatively

Our lived realities are social constructs that are influenced by, and should be examined alongside, the historical, cultural, economic, and political contexts in which they occur (Butler-Kisber, 2010). The ongoing act of meaning making is co-created between participants and researchers using activities and processes (Stelter, 2010), therefore we make, rather than conduct, research (Kara, 2020). Lives are represented through stories that are themselves performances and ongoing acts of co-creation, meaning they do not hold inherently stable meanings or intentions (Denzin, 2014). Given the complexities, and multilayered nature of our social world and human experiences, there are arguments suggesting that these complexities cannot be fully captured or portrayed through numbers (Butler-Kisber, 2010).

Over recent decades there has been increasing usage of creative and arts-based methods within qualitative research, including participatory textile workshops (Shercliff and Twigger Holroyd, 2020) and knitting focus groups (Harrison and Ogden, 2021). Playful methods allow space for knowledge production that is not reliant on verbal language (Henwood et al, 2019), while using a range of creative methods provides opportunities for different kinds of knowledges, which may more accurately reflect the multitude of meanings within our social world and contribute to building a richer tapestry of understanding (Gauntlett, 2007; Kara, 2020). Creative and visual research methods thus offer and 'promote a "slow science" built upon the opportunity for reflection, reflexivity and the aim of better understanding the subjective worlds of our research participants' (Mannay, 2016, p 104).

With no singular definition of reflexivity, the various aspects of reflexivity (including personal, functional, introspective, disciplinary, social, and discursive) can often overlap (Wilkinson, 1988; Finlay, 2002). Informed by a range of these approaches, I am not aligning with a single definition of, or approach to, reflexivity. Throughout this chapter, I speak about reflexivity as a tool for: examining the impact of the researcher's presence and perspective; supporting rich understandings through noting personal responses; being clear about the researcher's implicit biases; evaluating the research process; and supporting research integrity with a methodological record of research processes and decisions (Finlay, 2002).

If visual, creative, and playful thinking opens up space for slow thinking and reflexivity (Kahneman, 2011; Wellesley-Smith, 2015), how can we adopt and build upon such approaches within the analysis process itself?

The slow-stitching creative analytic process

This slow-stitch creative analysis approach is informed by reflexive thematic analysis (Braun and Clarke, 2022b) and trajectory thematic analysis (Spencer et al, 2021). Thematic analysis is a flexible method with no inbuilt theoretical positionings or presupposed specific philosophical assumptions (Braun and Clarke, 2022a). This slow-stitch analysis approach is underpinned by a constructionist research approach, acknowledging participants and researchers as engaged in 'a dialogue of knowledge-constructions' (Wilkinson, 1988, p 495). It is also informed by a material-oriented ontology that understands our world is multilayered, multifaceted, multidimensional, and multisensory (Woodward, 2020).

A material-oriented ontology positions materials and things as pivotal, and entangled, elements of our social world entwined with our relations to others, things, understandings, and knowledge (Woodward, 2020). People, things, and relations are multifaceted and coexistent in their active roles and the meanings ascribed to them.

Fabrics cannot just have stories imposed on them; instead, through working with fabric, using different threads, stitches, and patterns, the fabric comes to carry the maker's intentionality. The fabric can also 'act back'; if stitches come undone, or seams fall apart, then the maker's intentionality has been thwarted (see Woodward, 2003).

There are, currently, eight sections in this slow-stitch analysis approach, which I will now outline and discuss. Each section includes both the physical act and the thinking behind it. These are outlined and discussed together, as the making and the thinking inform, and are entangled with, one another, as is so often the case within material method approaches.

For colour versions of the figures/images, please see Clarke (2023).

Step 1: The data familiarisation process began as I read the diaries, upon their receipt, during COVID-19. This initial reading was not structured but, reflecting the messiness of daily life as well as of data collection, was an immersion in situ during the crisis. Once I had initially read each diary entry, I saved it (with the date as the title) and placed it into a folder (one folder per participant).

Step 2: In addition to deep immersion, I wanted to get an aerial view of the diaries and their trajectories (Spencer et al, 2021). I created a spreadsheet with a table that enabled me to see participants on the x axis (the columns, left to right) and the dates diaries were submitted on the y axis (the rows, top to bottom) (see Table 18.1).

Step 3: I worked my way through the folders and reread each diary. With each diary, I highlighted that entry date on the spreadsheet and entered '1'.

By inputting '1' into the cell, I could see both the 'sum' of cells per participant, to see how many diaries they had made, and the 'sum' of cells per date, to see if some dates gathered more diaries than others (Table 18.2).

Highlighting the cells did not aid the calculations but it did brighten up the spreadsheet. While that sounds trivial, it was an important part of my process of moving from digital data to tangible, material, and creative analysis. Data immersion is often very textual and the visual of the spreadsheet highlights enabled me to start reflecting on the data pictorially. Each participant, and their story, is more than a digital '1' on a spreadsheet: it is a multilayered story.

Table 18.1: Initial spreadsheet

	Sarah	Lee	Rachel	Clare	Beth	Tom
23/06/20						
24/06/20						
25/06/20						
26/06/20						
27/06/20						
28/06/20						
29/06/20						

Note: These columns and rows later fitted with the slow-stitch material representation of this table.

Table 18.2: Populated spreadsheet

	Sarah	Lee	Rachel	Clare	Beth	Tom
23/06/20	1	1	1	1		1
24/06/20	1	1	1		1	1
25/06/20		1	1	1	1	1
26/06/20	1		1	1	1	1
27/06/20		1	1	1		1
28/06/20	1	1	1		1	1
29/06/20	1	1	1	1	1	

Step 4: With the spreadsheet completed, I slowly reread each person's diary entries and reflected upon what they had shared. At the end of each day, I noted the number of entries read for each person that day and recorded that in my non-digital research journal before stepping away from my computer. Braun and Clarke (2022a) talk about the dual processes of deep, prolonged data immersion *and* time away from the data with space for reflection. In the evenings, I immersed myself in sewing as part of my reflective time away from the data while still holding the data, the people, and their stories in mind. The people who took part in my research shared their stories with me during a global time of crisis and I personally needed to create something that acted as a visible acknowledgement that their voice had been heard, that it was important, and that they had been held in mind.

Stitching is an activity that encourages the researcher to pause. This act of 'pausing' encourages the researcher to step back and have space for reflexivity and to generate new ideas, connections, and dissonances (Fletcher, 2013).

Step 5: I chose a different fabric for each person, from which I handmade a 0.5″ hexagon for each diary entry submitted. Here, the spreadsheet supported the calculations as I needed to ensure I had enough of each fabric to make the required hexagons per person (Figure 18.1).

Each hexagon was handmade using a paper template, folding fabric over the template and taking small stitches through both layers in the corners to hold in place. Slow-craft movement encourages slow processes of handmaking and time for reflection – whether that be reflecting on the object, the properties,

Figure 18.1: A 0.5″ hexagon representing a single diary entry

the stories, the processes, or a combination of them all. Time, thought, and energy goes into my making of each hexagon, and it ties in with the slow process of thinking, reflexivity, craft, and how the three are intertwined (Ravetz et al, 2013).

Fletcher (2013) talks about metaphors as shifting from standardised 'languaging' expression modes to something more creative and open. Building upon this, each hexagon that I made individually represented a single diary entry, an individual's story, someone's experience, someone's time, someone's making. On their own, they are fragile yet striking. They are a tangible representation of someone's story.

I continued making hexagons using an individual's fabric until I had the total number of hexagons required for that participant and their diary entries (Figure 18.2).

I placed these hexagons into an A4 wallet, labelled with the participant's name or pseudonym (depending on their preference[3]), and kept them in a ring binder until all hexagons were completed.

Step 6: All hexagons were made with a paper template that needed removing before the next stage of stitching. This was another slow process; I ironed the hexagons individually (so the fabric folds were crisp), unpicked the basting

Figure 18.2: Hexagons representing different diary dates for a single participant

stitches holding the template in place, removed the paper template, and re-ironed the hexagon. This was a repetitive task that enabled my hands to keep busy while allowing my mind to reflect on the data.

I used stitching as a tool for fighting familiarity; it is an activity that makes me slow down, be in the present moment, linger, and fight automatisation (Gurevitch, 1988; Mannay, 2016). I also used stitching as a tool for familiarising the strange for the same reasons: by slowing down, lingering on the reading, and representing it within sewing (a medium I feel comfortable using to express, share, and visualise), I engaged with texts on a deeper level.

Step 7: I appliquéd each hexagon individually onto a background base fabric in a similar manner to the spreadsheet: each participant had a column, and each date had a row (headings were later embroidered on for visual reference) (Figure 18.3).

Each hexagon was appliquéd with a thread matching the background fabric so the stitching did not detract from the hexagon (or the next stage). Handstitched into place, each hexagon represented a diary entry for that individual on that date, with spaces left when individuals did not make a diary entry on that date. This very act of slowing down and handstitching a predefined hexagon onto fabric allowed me to gain an element of control in reflecting upon my own experience,

Figure 18.3: Hexagons appliquéd onto background fabric for stability

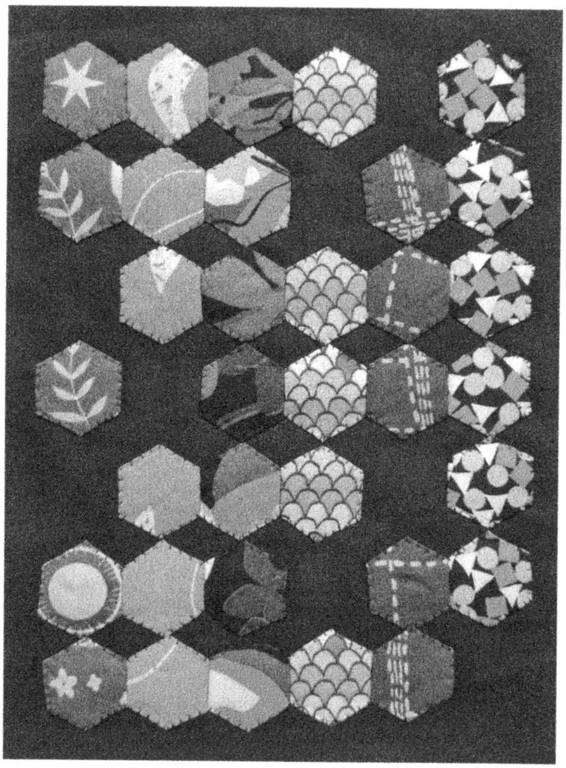

as textiles have the capacity to act as record keepers of events (Hemmings, 2012). By stitching, I was making sense of what I read and familiarising the unfamiliar. Similarly, by stitching, I was also making the unfamiliar familiar (Gurevitch, 1988; Mannay, 2016).

Having a base fabric on which to appliqué the hexagons meant that those small, fragile shapes had stability for the next stage of the slow-stitch creative analysis: stitching directly *onto* the hexagon with stitches representing initial coding.

Analytically, this step enabled me to begin seeing initial visual patterns within the data in relation to each participant's contributions, each date's contributions, and the cross-section between the two.

Step 8: With the hexagons appliquéd onto the base fabric, I returned to the computer for initial, digital coding. This meant rereading diaries while using NVivo (released March 2020) to mark text sections with colour codes, with each representing a different initial code.[4] For example, texts marked in

- yellow represented a sense of control;
- purple represented feelings of pride and accomplishment;
- green represented escapism.

My use of NVivo was personal preference; it allowed me to organise the data in a single project, integrating all materials and enabling connections. The ability to move, merge, and group codes supported the stages of reflexive thematic analysis. This is another way of initial coding in the way we might use different colour highlighters, representing initial codes, and tangibly mark texts or use a highlighter on Microsoft Word to mark-up initial codes in the data.

The colours used for this initial digital coding correlated with the embroidery thread colours that I had, rather than being informed by art-therapeutic approaches of colour meanings (Liebmann, 1990) (see Figure 18.4).

The NVivo software enabled me to organise the colour codes so I could see which diaries were highlighted in which colours. Informed by Braun and Clarke's (2022a, 2022b) reflexive thematic analysis, I stepped away from the raw data, using my time and space in the evening to reflect on that day's reading by moving from the digital initial coding to re(presenting) these initial codes through tangible, slow stitches on the hexagons.

Through these steps, I was involved in an ongoing cyclical coding process of moving from the digital data to the tangible stitched material, before returning to the digital data and so forth. This is a lengthy process and one that could be questioned. Digital software can be incredibly high-tech and clever, so what does the combination of the two approaches add to creative analysis? Coding involves looking at the data set as a whole (zooming out) *and* looking deeply at each data item methodically (zooming in) (Braun and Clarke, 2022a). The systematic process of coding involves working through the data set more than once as analytical meanings develop through ongoing interaction with the data (Braun and Clarke, 2022b).

Figure 18.4: Hexagons with handstitched straight lines correlating with digital coding

Note: The handstitching was done in different colours, correlating with that day's initial digital coding within the corresponding diaries.

Braun and Clarke (2022a) argue that good research practice is multilayered and I wanted to reflect the multilayered-ness within my stitching analysis. Having the base fabric supported this; the hexagons alone are too fragile to easily withstand, and hold, slow-stitch initial coding. Accordingly, the background fabric provides strength and stability for multilayered initial coding. There are also analytical parallels in that this base piece of stitching comprising the background fabric represented the research as a whole, with individual hexagons representing diaries, dates, and diarists (Figure 18.5).

Moving beyond the coding

For now, I have not written further steps beyond this point as I remain at this early, initial coding stage in the research process. Slow stitching is, as the name suggests, slow; it is a process that I do not want to rush so as to group the initial codes into themes. Coding is not simply identification; it is interpretation and researcher subjectivity greatly impacts this process. Themes do not pre-exist

Figure 18.5: Hexagons with increasing stitches representing early coding

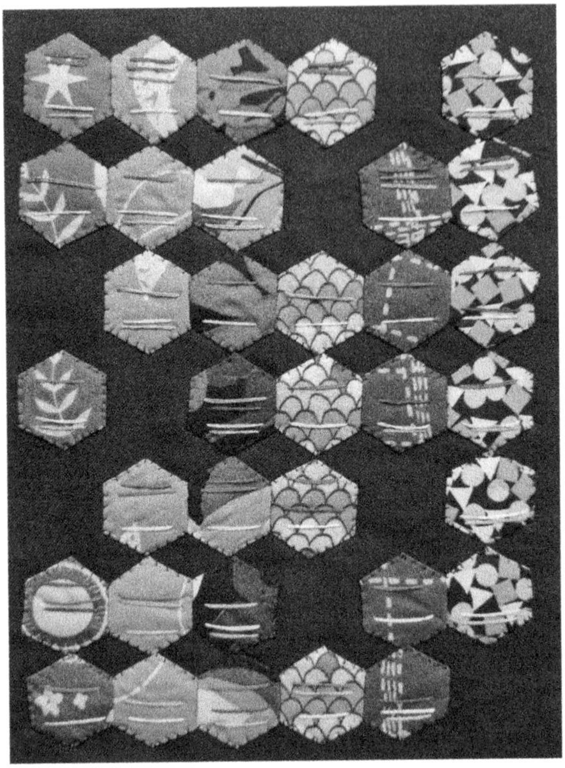

within the data to be unearthed but are *produced* by the researcher's engagement with data and their own positionings and experiences (Braun and Clarke, 2022b). Therefore, generating themes from codes early on could restrict the analysis findings to a superficial level (Braun and Clarke, 2022a).

Given the vast amount of data I am working with, I continue to use this approach of combining NVivo with slow stitching for early coding so that I am engaged in an iterative, ongoing process of zooming in and zooming out of data and reflexivity. Naturally, there will be stitch-based progressions of this approach when I reach a stage of generating themes and dissemination.

This textile approach to early coding can continue being used throughout this research, from evolving the coding, generating initial themes, reviewing themes, and beyond. There is no fixed, singular end point for coding. Further data engagement with the material (both textile and digital diaries) could make possible new insights as knowledge is generated and constructed; thus there will always be possibilities for new understandings (Braun and Clarke, 2022a).

Visuals often carry a multitude of meanings that are subject to continuing interpretation (interpretations of which will be subject to the viewers' own experiences, culture, histories, and positionings); visuals are not static and finite objects (Mannay, 2016).

Where do I go from here?

Reflexive thematic analysis requires researcher deliberation and thoughtfulness (Braun and Clarke, 2022b) and the same can be argued for stitching. These processes (reflexivity, thematic analysis, slow stitching, coding) should be reflexive and unhurried. There is no quick undo or backspace button when we stitch in the way there is when working on a computer. Each mark with a needle and thread is a deliberate and thoughtful one so weight should be given to that very act. The meanings ascribed into stitching are those of re(presentation), of metaphors, and of reflexivity. The lives of the people for whom the hexagons represent are entwined for a specific context as this item is made and the act of stitching is a 'constant flux of activity' comprised of micro-events (Shercliff, 2019, p 351).

The hexagons, and their stitches, act as physical representations of voice. They are not static and further stitches can be added as well as cross-stitches (in different directions, overlapping, weaving through, and wrapping around) to represent interactions of codes and subtheme development.

These materials act as a visual, tangible prompt for further and additional ideas to flow and for the development of lines of enquiry. They can also aid discussions beyond typical academic boundaries as they provide creative space for playful resistance against typical oppressive views of knowledge. Critical reflexivity is an ongoing process, not an end point, and something continually strived for (Fletcher, 2013). Similarly, slow stitching for reflexivity within the analytic process is not an end point but something that is part of the cyclical process of theory, analysis, reflexivity, and stitching, all feeding into one another throughout the ongoing, meaning-making process.

Strengths and challenges of the creative analytical method

Haseman and Mafe (2009) argue that when practice-led research is conducted, the research should take place *through practice*. As I read, think, analyse, critically engage with, and write, I am inspired to stitch. Through this stitching, I am rethinking, re-analysing, reflecting and re-understanding. This iterative, cyclical process can develop different forms of knowledge and understandings, with creative practice contributing to academic research as well as academic research contributing to creative practice (Smith and Dean, 2009). For Hawkins (2017), creativity supports new knowledge production, both individually and collectively, in ways that challenge traditional disciplinary norms. This resonates with my own use of the slow-stitch analysis process as a creative act that enables me to engage in a practice that both inspires creativity while also pushing the boundaries of traditional disciplinary analytic approaches.

However, using a creative stitch approach to analysis and reflexivity is not without challenges. Space for in-depth reflexivity is often limited by word counts within academic writing (Finlay, 2002). There are also increasing factors (time restraints, financial restraints, institutional regulations, efficiency and impact measurements, publishing deadlines) that restrict the ability to conduct creative and visual research (Mannay, 2016; Kara, 2020) and exploratory, longer-term research (Mills and Ratcliffe, 2012). Making is, in itself, an investment of time and resources, and it requires consideration and contemplation (Turney, 2012). Given that stitching is a slow process, these factors need to be considered carefully.

Nonetheless, it could be argued that stitching's slow nature and multilayered approach supports the depth and richness of data that reflexive thematic analysis seeks (Braun and Clarke, 2022a, 2022b). Visuals in themselves can offer great depth and richness (Mannay, 2016) and there are visual capabilities within the creative stitch approach to analysis that support research depth and richness.

(Re)presenting the participants' stories, and the researcher's reflexivity, through stitch creates space for a flow of ideas while also creating something tangible that can stand alone, welcoming wider viewer/reader interpretation with the viewer/reader invited into research spaces (Ward, 2011). There is power in using tangible textile methods within analysis; materials have the capacity and power to provoke thoughts, emotions, and understandings (Tobin, 2009; Woodward, 2020). Accordingly, everyday creativity offers positive and meaningful ways of connecting with others, ourselves, and the world generally (Gauntlett, 2018).

The word creativity, however, can be off-putting. For some, creative activities feel threatening, anxiety provoking, and disempowering, especially when people feel vulnerable or lacking in creative ability (Mannay, 2016). Added to this is the daunting fact that researchers feel they have to confront their own *perceived* methodological and creative inadequacies, which can be experienced as personal fragilities and can be difficult to handle (Finlay, 2002). Yet, if we move away from 'abilities' (which is a problematic term itself) and consider creativity as another tool in a creative research toolbox, then this approach has far-reaching applicability and flexibility. It can be adapted to be used according to the individual researcher, with a variety of materials including buttons (each button representing a story), sticks and rocks (painting and colour-coding them), pens, cross-stitch, collage, found materials, and more – as other chapters in this volume show. Ever-evolving digital spaces have shifted creative practice sites away from specialised spaces, opening this up to communities, homes, and amateurs, meaning more of us can engage in creative practice (Hawkins, 2017). Creative practice and academic research are interwoven; by acknowledging and utilising this, there are possibilities for 'distinctive methodological approaches, as well as exhilarating findings and artworks' (Smith and Dean, 2009, p 5).

Conclusions

Throughout this chapter, I have discussed my slow-stitch approach to creative analysis (more specifically, coding). Using NVivo allows me to zoom in (colour-coding data) and zoom out (organising, remodelling, and querying based on colour codes) of the data through multiple rounds of coding. Using stitching also enables me to zoom in (through slow-stitching colour codes) and zoom out of the data (looking across the hexagons in columns, rows, and colour patterns), albeit in a different and material manner. Combining both approaches offers opportunities for prolonged, creative, in-depth engagement with the data and creative analysis. Wilkinson (1988) asserts that, within reflexive research, we should utilise strategies encouraging space for new understandings and expose strategies continuing to restrict research practice; I believe this was made possible by combining these analytic approaches. Thematic analysis involves making the research elements that need to be united for successful qualitative analysis visible, and it also provides space to reflect upon the elements' connections with one another (Braun and Clarke, 2022a). This chapter has made visible the research elements and creative approaches that have informed, and continue to inform, my work. I also hope that this chapter has supported the increasing use of visuals and textiles as a way of speaking out (Parker, 2012; Nickell, 2015; Porch et al, 2022).

Notes

1 This chapter would not have been possible without Dr Sarah Bulloch's support in structuring this chapter and teaching me NVivo's potential. My supervisor, Professor Debbie Watson, has expertly guided me in writing about my approach, encouraging me to be creative and push traditional academic boundaries. Finally, my work rests upon my identity as a maker and stitcher, so I owe a massive debt of gratitude to my mum, who taught me all that I know about sewing.

2 Part of Naomi's PhD. Ethical approval for the study was received from the University of Bristol.

3 All participants were given the option of whether to be anonymised or to be identified by their first name. The labels that I used on the A4 wallet were then in line with the participant's preference. The A4 folder was kept in my home sewing room due to the nationwide 'stay at home' message.

4 https://www.qsrinternational.com/nvivo-qualitative-data-analysis-software/home

References

Braun, V. and Clarke, V. (2022a) 'Conceptual and design thinking for thematic analysis', *Qualitative Psychology*, 9(1): 3–26.

Braun, V. and Clarke, V. (2022b) *Thematic Analysis: A Practical Guide*, London: SAGE Publications.

Brignall, M. (2020) 'Hobbycraft reports 200% boom in online sales since start of pandemic', *The Guardian*, [online] 3 August. Available from: https://www.theguardian.com/business/2020/aug/03/hobbycraft-reports-boom-in-online-sales-since-start-of-pandemic-crafts-coronavirus

Butler-Kisber, L. (2010) *Qualitative Inquiry: Thematic, Narrative and Arts-Informed Perspectives*, London: SAGE Publications.

Clarke, N. (2023) '"Slow stitch: reflexive creative analysis and meaning making" images', *Wordpress Blog*, 1 December. Available from: https://naomialicecm.wordpress.com/2023/12/01/slow-stitch-reflexive-creative-analysis-and-meaning-making/

Clarke, N. and Watson, D. (2020) 'Crafting during coronavirus: creative diary approaches for participant-centred research', in H. Kara and S.-M. Khoo (eds), *Researching in the Age of COVID-19: Volume 3: Creativity and Ethics*, pp 19–28, Bristol: Policy Press. Available from: https://doi.org/10.46692/9781447360438.003

Denzin, N.K. (2014) *Interpretive Autoethnography* (2nd edn), London: SAGE Publications.

Finlay, L. (2002) 'Negotiating the swamp: the opportunity and challenge of reflexivity in research practice', *Qualitative Research*, 2(2): 209–30.

Fletcher, G. (2013) 'Of baby ducklings and clay pots: method and metaphor in HIV prevention', *Qualitative Health Research*, 23(11): 1551–62.

Gauntlett, D. (2007) *Creative Explorations: New Approaches to Identities and Audiences*, London: Routledge.

Gauntlett, D. (2018) *Making is Connecting: The Social Power of Creativity, from Craft and Knitting to Digital Everything* (2nd edn), Cambridge: Polity Press.

Gurevitch, Z.D. (1988) 'The other side of dialogue: on making the other strange and the experience of otherness', *American Journal of Sociology*, 93(5): 1179–99.

Harrison, K. and Ogden, C.A. (2021) '"Knit 'n' natter": a feminist methodological assessment of using creative "women's work" in focus groups', *Qualitative Research*, 21(5): 633–49.

Haseman, B. and Mafe, D. (2009) 'Acquiring know-how: research training for practice-led researchers', in H. Smith and R.T. Dean (eds), *Practice-Led Research, Research-Led Practice in the Creative Arts*, Edinburgh: Edinburgh University Press, pp 211–28.

Hawkins, H. (2017) *Creativity*, Oxford: Routledge.

Hemmings, J. (ed) (2012) *The Textile* Reader, London: Bloomsbury Academic.

Henwood, K., Dicks, B. and Housley, W. (2019) 'QR in reflexive mode: the participatory turn and interpretive social science', *Qualitative Research*, 19(3): 241–6.

Kahneman, D. (2011) *Thinking, Fast and Slow*, London: Allen Lane.

Kara, H. (2020) *Creative Research Methods: A Practice Guide* (2nd edn), Bristol: Policy Press.

Kurutz, S. (2021) 'What we learned from a year of crafting', *The New York Times*, [online] 13 May. Available from: https://www.nytimes.com/2021/05/13/style/craft-boom.html

Liebmann, M. (ed) (1990) *Art Therapy in Practice*, London: Jessica Kingsley Publishers.

Mannay, D. (2016) *Visual, Narratives and Creative Research Methods: Application, Reflection and Ethics*, Oxford: Routledge.

Mills, D. and Ratcliffe, R. (2012) 'After method? Ethnography in the knowledge economy', *Qualitative Research*, 12(2): 147–64.

Nickell, K. (2015) '"Troubles textiles": textile responses to the conflict in Northern Ireland', *TEXTILE*, 13(3): 234–51.

Parker, R. (2012 [1984]) *The Subversive Stitch: Embroidery and the Making of the Feminine*, London: I.B. Tauris.

Porch, L., Canavan, K., Treadaway, C., and Cazeaux, C. (2022) 'Caring through cloth: textiles and the trauma of stillbirth', *TEXTILE*, 21(2): 491–501. Available from: https://www.tandfonline.com/doi/full/10.1080/14759756.2021.2021727

Ravetz, A., Kettle, A., and Felcey, H. (2013) 'Introduction: collaboration through craft', in A. Ravetz, A. Kettle, and H. Felcey (eds), *Collaboration through Craft*, London: Bloomsbury Academic, pp 1–15.

Shercliff, E. (2019) 'Making memories in storage: the aftermath of site-specific, collectively made textiles', *TEXTILE*, 17(4): 347–56.

Shercliff, E. and Twigger Holroyd, A. (2020) 'Stitching together: participatory textile making as an emerging methodological approach to research', *Journal of Arts and Communities*, 10(1–2): 5–18.

Smith, H. and Dean, R.T. (2009) 'Introduction: practice-led research, research-led practice – towards the iterative cyclic web', in H. Smith and R.T. Dean (eds), *Practice-Led Research, Research-Led Practice in the Creative Arts*, Edinburgh: Edinburgh University Press, pp 1–38.

Spencer, L., Radcliffe, L., Spence, R., and King, N. (2021) 'Thematic trajectory analysis: a temporal method for analysing dynamic qualitative data,' *Journal of Occupational and Organizational Psychology*, 94(3): 531–67.

Stelter, R. (2010) 'Experience-based, body-anchored qualitative research interviewing', *Qualitative Health Research*, 20(6): 859–67.

Tobin, B.F. (2009) 'Consumption as a gendered social practice', in M.D. Goggin and B.F. Tobin (eds), *Material Women, 1750–1950: Consuming Desires and Collecting Practices*, Surrey: Ashgate Publishing, pp 1–13.

Turney, J. (2012) 'Making love with needles: knitted objects as signs of love?' *TEXTILE*, 10(3): 302–11.

Ward, A. (2011) '"Bringing the message forward": using poetic re-presentation to solve research dilemmas', *Qualitative Inquiry*, 17(4): 355–63.

Wellesley-Smith, C. (2015) *Slow Stitch: Mindful and Contemplative Textile Art*, London: Batsford.

Wilkinson, S. (1988) 'The role of reflexivity in feminist psychology', *Women's Studies International Forum*, 11(5): 493–502.

Woodward, S. (2003) 'Looking good, feeling right: aesthetics of the self', in S. Kuchler and D. Miller (eds), *Clothing as Material Culture*, Oxford: Berg, pp 21–40.

Woodward, S. (2020) *Material Methods: Researching and Thinking with Things*, London: SAGE Publications.

19

The analogue journey method

Karen Gray and Emma Lazenby

Introduction

In this chapter we – Emma Lazenby, a filmmaker, and Karen Gray, a researcher – introduce the analogue journey method. The term 'analogue' has dual meanings, and both are relevant. An 'analogue' is a thing that is similar to, or that is used to represent, something else through the process of comparison or analogy. The analogue journey method is a novel creative analysis tool that can be used with almost any kind of qualitative or mixed-methods research data, making sense of such data by visualising them in the shape of a journey, with the end goal of communicating this sense to others. The word 'analogue' is also now commonly used to denote things whose means of representation is through the quantities and qualities of the physical world. This contrasts to the 'digital' world, in which physical quantities are processed through and represented by electrical signals. For example, analogue time is told through the movement of the minute and second hands of a clock, rather than through changing numbers on a computer screen. The analogue journey method requires analogue tools, such as paper, pens, scissors. However, its results are intended to be translated into either words or images using any media to support wider and creative dissemination of research findings. The word 'journey' in the name indicates that it involves information being organised and presented in a form that, while linear, is mobile and mutable – open to change.

The analogue journey method demands thoughtful connection and reconnection with the data. Its activities encourage cross-disciplinary collaboration, opening up different perspectives or helping to form different constellations of information. As illustrated in Figure 19.1, this process can also help to bring data and how we think about it into a relationship with time.

The origins of the method

In 2009 Emma was making *Mother of Many* – an award-winning animated film about midwifery and childbirth. Many ideas were whirling round in Emma's head, including extracts from books, birth stories, YouTube videos, and the pages of her own sketchbooks. All of these ideas and materials became simply too much to hold or to process. To make sense of them, Emma began sticking the research notes, information, images, and extracts of interviewees' words across her living room wall. When these ideas were collated across the wall, Emma found

Figure 19.1: A relationship with time: data and our thoughts

Source: Created by Emma Lazenby.

she could 'see' and organise each element. In doing this, Emma was applying techniques learned and adapted from film storyboarding but drawing also on a phenomenon that will be familiar to anyone who has ever watched film or TV detectives creating an 'evidence wall'. As Emma made more films, the technique developed and Emma began to use it whenever she needed to see everything together in one place and when she wanted to take both herself and anyone else with whom she was working on a journey through the material. In using this organisational and curatorial technique, Emma could examine intention, duration, instruction, words, narrative, imagery, and inspiration – and, when working with others, they could make this journey alongside each other.

In 2018, Emma and Karen worked together. Initially this was for a commission whose aim was to disseminate the findings of a research project. We planned to create a set of visual assets that would combine illustration with the words of interview participants, forming a 'graphic narrative' to communicate Karen's research findings. The subject of the research was new to Emma and there was a lot of material for her to consider. The task of remaining true to the data and being sensitive to the intentions and rights of those who had been interviewed while presenting and re-presenting those data visually was difficult. The resulting graphic narrative was a useful (re)presentation of research findings. However, the collaborative process of using the analogue journey technique to create it, quite unexpectedly (since the primary analysis was complete) introduced a whole new analytical life to the research project itself. Through creating the graphic narrative, we found ourselves engaging and re-engaging reflexively with the data together and, as a result, asking some important ethical and theoretical questions of this data. Since then, and in writing this chapter, we have reflected on what elements of the process contribute to analysis, and why. It is therefore both this foundational experience and subsequent thinking that informs this chapter. We will outline the method here, but primarily we think of this chapter as an invitation to others to use, to tinker with, and to build on these foundations.

Analogue journey method draws on ideas and theory from visual research methodologies. The kinds of knowledges it can produce are constructed and subjective. In brief, the method involves turning a mix of abstracted data items – which may or may not themselves be visual – into something through which knowledge can be made visible, shared, and usable; creating an analogy for that knowledge. The process of the analogue journey method borrows ideas from creative, lateral, and design thinking, which are approaches that deliberately connect analysis and creative production. Accordingly, these approaches would suggest that turning data into a visual analogy can be a helpful way to generate insight, particularly when working with others. In *Steal like an Artist*, the writer and artist Austin Kleon advised anyone wishing to harness the power of creative thinking to 'Use your hands' (Kleon, 2012, pp 50–61) and, as we suggest later in this chapter, the physicality of the analogue journey process is key. It may be useful to think about what we are proposing as aligned to thinking through drawing (Magnani, 2013).

It is also important to recognise that the method was born out of the applied practice of a filmmaker whose work aims to communicate health issues through animated short films, illustrations, and audio stories based on people's real lives and experiences. Therefore, storyboarding was a key influence. As illustrated in Figure 19.2, in filmmaking, storyboarding is a process used to visually represent a film sequence. In a storyboard, strictly linear, rough sketches break down the action into individual panels, laying out, shot by shot, how the director or cinematographer imagines the video and its story will unfold, in a way that can convey this message to the production team. Applied by silent filmmakers such as Georges Méliès, its origins as a method are commonly claimed by Disney Studios. Storyboarding was used in the design of Disney's major animated films from the 1930s onwards. It redefined the way in which films such as *Snow White and the Seven Dwarfs* were produced and subsequently influenced other genres (Pallant and Price, 2015). Beyond Disney, and outside of its use in filmmaking, storyboarding is also commonly used within collaborative design, often as part of a process of innovating new solutions and services. For example, a storyboard might be brought to a design or user experience workshop. There it can function as a boundary object through which design knowledge is able to cross into the social world, informing discussion around any problem it is hoped that a design solution might address. In research involving participatory video-making, creating storyboards together has been described as a tool for communicating experience, enabling collaboration and supporting problem-solving (Labacher et al. 2012, pp 149–63).

A further underlying visual trope informing the analogue journey method is the characterisation of a pinboard in film or TV, usually, but not always, in crime drama, as an 'evidence wall' (see Figure 19.3). Ganzert (2020) coined the term 'serial pinboarding' to explore instances in contemporary television series

Figure 19.2: Film storyboarding

Source: Created by Emma Lazenby.

Figure 19.3: An evidence or 'murder' wall

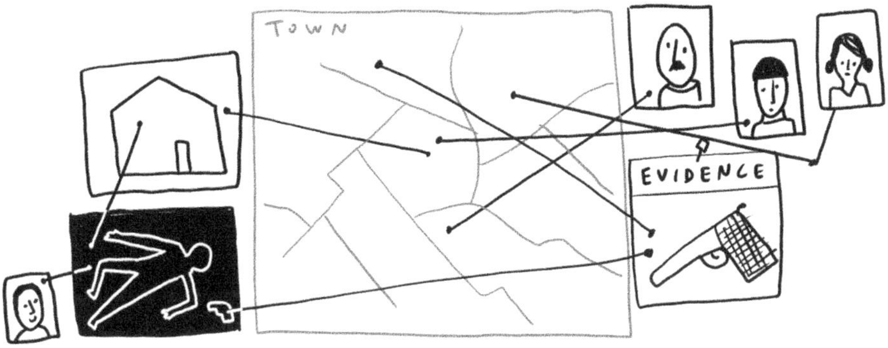

Source: Created by Emma Lazenby.

where pinboards have been applied not just as a prop or storytelling device but as an aesthetic or semiotic system that helps to guide viewers temporally through a series and its narrative. For Ganzert (2020), the kinds of reasoning undertaken by characters we see creating the pinboards, and the reasoning that is simultaneously being encouraged in the viewer, can be inductive, deductive, or abductive, depending on the pinboarder and on the demands of the story. Serial pinboards open up spaces on which characters play with material objects and with abstract ideas, configuring and reconfiguring them in the hope and expectation that the physical activity involved, and its visible results, help to signpost a solution. They are places on which possible narratives and connections are played out and made visible. They can also be full of dead ends as they tell the story of the analysis process itself.

An example: Away from the research context, Emma applies the analogue journey method each time she makes a film. For example, *A Little Deep Sleep* is an animated short she made to explain to children and family members the process of receiving a general anaesthetic in hospital.[1] Emma did her own desk research into the subject, including multiple conversations with her anaesthetist collaborators on the project. Emma spent time in the hospital, recording sounds, taking photographs, shadowing staff, making sketches in operating theatres, and recording interviews with children before and after they were given a general anaesthetic. Emma returned to her studio with this material to plan the film. On a noticeboard her basic structure started with a timeline (the length of the film) and its aims (to ease anxiety and inform children going in for a general anaesthetic). Emma expanded this structure to include an introduction, basic linear journey, and the message she wished to leave viewers with at the end of the film. Using this structure, Emma filled the board with the words, sketches, images, and chosen segments of interviews that would help her to understand and then represent the clinical and the personal experiences her research had uncovered. The noticeboard grew organically and overfilled; she then pruned and edited the content so that it eventually conveyed the aims and told the narrative journey of the film along with the experiences that were central to it. All were clearly plotted against the necessary constraint imposed by the length of the film. Emma then shaped the results into a more traditional sketched storyboard, her ideas moving from analogue into editing software and the final digital animation.

Having described the key origins of and influences on the analogue journey method, the following section offers a more detailed overview of how to apply the method to the analysis of research data.

Applying the method

This section provides some basic principles that can be useful when applying the analogue journey method. These steps are not set in stone and the process is open to adaptation, but we have found them particularly helpful.

Identify your fellow travellers

While it could be taken by a single researcher, this analysis journey is best made with others. Ideally, your fellow travellers will bring different skills and perspectives to the process. We would recommend including at least one person who does not know the project at all, since the task of explaining your data and your aims afresh can be illuminating. You could also include one or more research participants. Consider what the outputs of the analysis will be at the point when you choose your analysis team. If you are looking to develop a creative means of sharing the results, you might want an artist, designer, filmmaker, dramatist, writer, or a technologist on board. If you are thinking of developing materials to influence policy or practice, you might need an appropriate stakeholder. You could involve the group in a series of at least three collaborative engagements with each other to explore understandings of the project, its aims, and the aims of your analysis, and to work directly with data together.

Gather your tools

What you use as tools will depend on the space and resources available to you, the needs of your fellow travellers (consider accessibility), and the nature of your project. You might want to give the process a particular 'aesthetic', but what that is really does not matter. The following points and Figure 19.4 offer some pointers for collating the necessary tools.

1. *To hold the journey together*
   You can use a noticeboard, a table, a wall, a whiteboard, or a roll of blank paper – the bigger the better. You must be prepared to live with your journey for a while as it will not be disassembled until the process is finished.

Figure 19.4: Gathering your tools

Source: Created by Emma Lazenby.

2. *To communicate*
   You will need pens, pencils, brushes, chalk, and coloured paper. You might be tempted to use post-it notes; these do work but the glue is often poor quality and the notes can fall off the board, particularly if you are going to be moving them about a lot. It may seem more time-consuming, but consider writing or printing on coloured paper instead, cutting to shape, and sticking using some other method. Use colour or shape to indicate different kinds of things or to suggest different kinds of meaning. These tools might also be used to highlight, place boundaries around, or indicate a thematic cluster or group of things.
3. *To tailor*
   Scissors, knife, or a guillotine can be used to tailor pieces of information that might have been part of a bigger whole but which you want to be able to move about separately or align with other things.
4. *To pin*
   Pins, magnets, Blu Tack®, or sticky tape will hold items on your board. They might also indicate a node, a point at which several other things connect. They should hold things firmly in place but not create a permanent fix – you will want to move things around.
5. *To connect*
   Thread, tape, or string can be used to connect items, groups, or clusters of meaning together.
6. *To print and document*
   The use of a computer or printer is optional, but if you are going to be analysing data such as interview transcripts, documents, or notes, you may want to print out text or images. It is likely you will also want to use a camera, taking pictures at regular intervals so that you can document how the board has developed.

Beginnings

In this stage you will share understandings of the research project, agree the aims of the analysis, and develop an outline structure for the journey. The stage could be completed through a single collaborative Beginnings workshop, held either when data collection is complete or as part of the process of guiding data collection. If held early, this might help to identify gaps in your plans or bring to the surface an ideal way of sharing your findings when they are complete. Each person should have an opportunity to outline what they wish to contribute. This could be experience, a skill, or playing a role, such as that of a questioner or information-gatherer.

It is useful to identify a place-holder name for the journey – one that clearly conveys its nature to everyone involved. You will need to agree on the end point of the journey: the essence of what the audience for your final research output needs to know. What your journey looks like will depend on the nature of your research, the data you are bringing to the method, and what this end goal is. It will also be informed by what you are going to do with the analysis when you have completed it. Do you want to write a paper? Create a poster or presentation? Make a film? The demands of this format will give you another essential element of structure to inform your analysis process. For example, in drawing on the analogue journey method for filmmaking discussed earlier in the chapter, Emma had the basic timeline and aims of a film in mind alongside the need to chart different kinds of experience.

You should now transfer the bones of this structure to your board, so that it is ready to hold your data and research material.

An example: When this chapter's authors first met to work together, Karen explained her research, its aims, and initial findings in response to probing questions from Emma. The research examined the methodological challenges involved in evaluating the impact of engaging in arts and creative activity for people living with dementia. We decided on a name for the project ('What IS it about arts and dementia?') and identified a structure that suited the data and our dissemination goal of creating a graphic narrative to share the research findings. The data consisted of interviews with researchers and evaluators, artists, and funders, discussing their experiences of capturing and seeking evidence that might demonstrate the effects of engaging people living with dementia in arts and creative activity. We used the simple structure of a beginning, a middle, and an end, but within that we also agreed that we needed to describe a context and introduce some characters (composites of the different kinds of interviewees). Then, for each of the characters, we wanted to examine what the data revealed about their relationships to and perspectives on the problem, as well as potential solutions to it:

1. Beginning: the overarching problem addressed by the research.
2. Middle:
   - the setting;
   - the 'characters' (researchers and evaluators, artists, and funders of activity);
   - their relationships to and perspectives on the problem;
   - the solutions suggested by the data.
3. End: recommendations and summary.

Middles

Having agreed an initial shape or structure for your journey, you can begin to add your research material to it. Depending on what this is, you may be pinning up participant quotes or sections of interview transcripts, photographs, parts of documents, or extracts from your observational notes. You can divide up longer texts or pull out and highlight individual words. You will need to have a visual system for categorising and identifying different types of information, but once this is in place you could also throw many other things at the board around the data: reflections, illustrations, asides to yourself, gaps or issues you have spotted, jokes or silly comments that occur to you or others. You or others can add to the board at collaborative meetings, or in-between them.

At collaborative meetings with your analysis team, you can then move items around and use the thread or markers to indicate links between different elements. Together you will be identifying where there are nodes of connection, where there are thematic similarities, and where there may be gaps. Very little is forbidden – except perhaps that, if you are working with others, it is best not to simply remove something that is on the board without discussion. If there is disagreement – note it on the board. You could document the board at the beginning and at the end of each collaborative meeting so that you have a record of any major changes in your thinking over time. Each analysis team member may want to quickly note and share their own reflections of the meetings.

This phase can be as lengthy as you need it to be, with the board growing organically throughout.

An example: When we first worked together, and after agreeing the structural elements outlined, Karen revisited the original interview transcripts from her research using the structure as a guide. She extracted verbatim quotes, cutting and pasting them into tables on a document, sorted according to the character type of the speaker. As illustrated in Figures 19.5 and 19.6, the quotes were printed onto

Figure 19.5: An analogue journey line of interview extracts (1)

Source: Created by Emma Lazenby.

Figure 19.6: An analogue journey line of interview extracts (2)

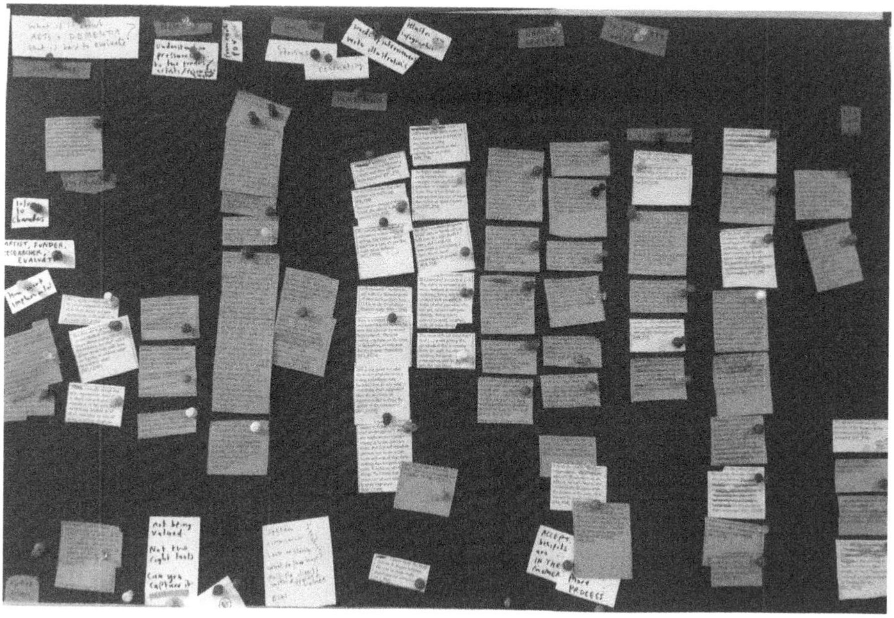

Source: Created by Emma Lazenby.

paper that indicated different character type by colour; these were then cut up to create a store of individual quotes. Then, together, Karen and Emma considered each of the quotes and positioned them along the analogue journey line.

Ends

Eventually you will need to start editing. This is the point at which the boundaries that come with knowing how your analysis is going to be shared can be helpful in giving a shape and structure for this process. Continually remind yourself of the aims of the project and the shape and structure that you agreed at the start. Agree on the one or two quotes or other extracts that exemplify each point on the journey and identify where ideas and themes might be useful combined or elided. Less is more but do keep any material that you remove. Work towards a board that is streamlined, and which makes sense to everyone involved.

It is from this point that the board can begin its move into different media.

An example: In our first project, we sifted through participant quotes and then moved the board into a simple 'script' on which we highlighted the key quotes from the board (Figure 19.7). Emma then used these to inform development of a set of illustrative rough sketches and eventually the graphic narrative itself (Figure 19.8).

Figure 19.7: Sifting through participant quotes from research project 'What IS it about arts and dementia?'

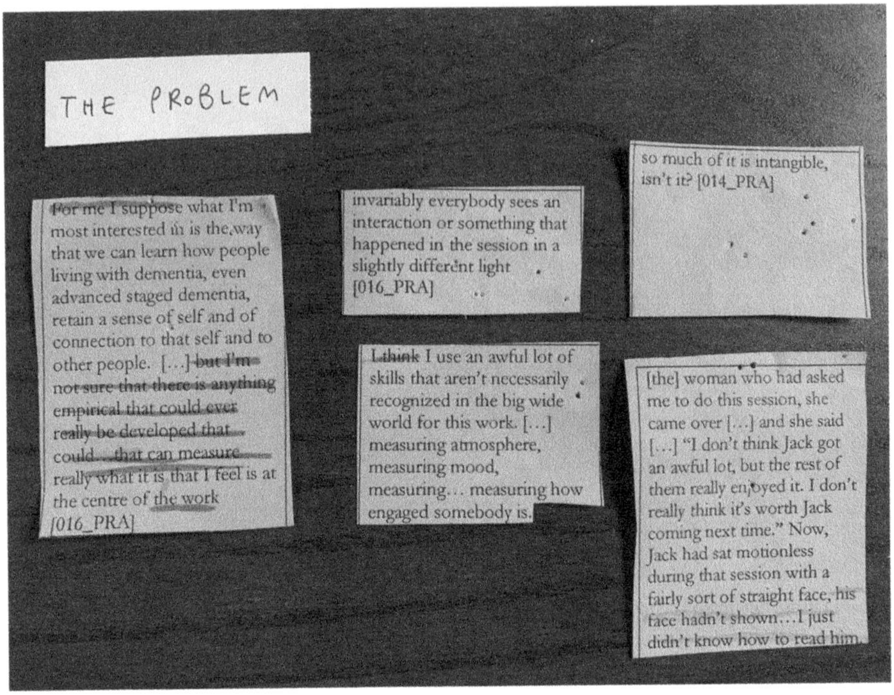

Source: Lazenby and Gray (2019).

Benefits and challenges

As a way of making sense of information, the analogue journey method thrives on a balance between structure and playfulness. It creates and celebrates a space for 'mess'. It demands that researchers make repeated returns to the data and surrounding material; it works on the basis of an understanding of the world that is producing this data as being inherently slippery and difficult to grasp (Law and Urry, 2004). At the same time, the method places useful parameters around this messy space, encouraging the researcher to consider directly how their analysis and its findings will eventually be communicated. The method celebrates both the journey and its destination.

The analogue journey researcher and analysis team are not neutral observers; rather, they are actively shaping the material. Along with the repeated returns to the data, the emphasis on working collaboratively to co-produce knowledge both with stakeholders in the work and with those who can provide outside perspectives is intended to help ground the eventual interpretation. Like other visual methods involving participants and other research stakeholders, it may also encourage a useful process of defamiliarisation – making the familiar strange again (Mannay, 2015) and helping to overcome researcher preconceptions and

Figure 19.8: Extract from final graphic narrative *What Is It about Arts and Dementia?*

Source: Lazenby and Gray (2019).

assumptions about the data. These aspects of the process align themselves with calls for a 'reflexive turn' in research, through which attention can be drawn to the researcher's role in shaping both the research process and its outputs (Mauthner and Doucet, 2003). The method also supports cross- or interdisciplinary working; all kinds of data can be made equally visible on the board.

While researchers do, of course, still engage directly with the physical world when they collect data, some of us are finding ourselves doing this less and less often: it is increasingly common for data to be gathered, created, stored, analysed, and disseminated in digital formats. Introducing analogue methods at the analysis stage intentionally disrupts this tendency. For many people, it will be refreshing to stand up and move away from the computer keyboard, to think alongside others and with technologies and objects that are tactile and have aesthetic properties. In an interview, Kleon talks about encouraging 'the dance between analogue and digital' (Glei, 2017), and his description of a creative process that involves physical movement between different modes of production have been key influences on our thinking around this method.

As a result of the method's generous celebration of messiness, its collaborative elements, and its physicality, those involved get a chance to see in the process everything that feels of importance to them – even if some things need to be streamlined for the ideas to be eventually communicated. They can discuss and share what they see, move things around over time in response to reflection or exposure to new information, and have a hand in finally editing and shaping the material. While the work is in progress, what is on the board stands in for a working understanding or analogue of how the research data might potentially answer an issue, idea, or problem. And whatever form it takes afterwards will be built on and out of shared knowledge.

However, the method can be challenging to use. Although the materials required are not in themselves costly, physical space, sustained commitment, and time are needed. The method works best when all those involved in the analysis are physically in the same room. We have considered whether there is scope for introducing digital tools (such as Trello, Jamboard, or Padlet) into the process to allow for remote collaboration. While these might work, the risk is that some of the method's essential physicality is lost. A further challenge is that for collaborative activity to be effective, one member of the team may need to be skilled in ensuring that different voices are balanced and equally represented; remote collaboration can further complicate this. It is also important to consider that any creative professional involved will usually need to be paid appropriately for their contribution, as will research participants and other stakeholders.

Additionally, as with all interpretative methods of analysis, it requires care to ensure that researchers are aware of how internal and external influences are shaping their interpretation of the data. And any choices made as part of the analysis process will need to be ethical in their representation of individual research participants' voices and experiences.

An example: In the work Emma and Karen did together, in the initial structuring, some composite characters were identified to 'stand in' for research participants. This is one of the ways that we made sense of and performed the task of editing the vast mass of material available to us. Illustrative quotes were attached to each of the characters. In reflecting their experience on the board, and through the final graphic narrative, we had to work hard to ensure that some of the important differences and contradictions that existed between individual participants remained – even if we could only use a limited number of quotations. The work involved in doing this proved a valuable reminder that while participant experience is often comparable, it will not be universal; in reality there was no one simple or coherent story. In analysis we necessarily strive to make complexity less confusing, but rather than concealing the fact that complexity exists, the analogue journey method helped us to acknowledge this complexity.

Next steps

In an essay of 1986, the philosopher of science and technology studies, Bruno Latour reflected on the power and history of processes that visualise complex information – what he called 'thinking with eyes and hands':

> And, without moving more than a few inches, I unfold in front of your eyes figures, diagrams, plates, texts, silhouettes, and then and there present

> things that are far away and with which some sort of two-way connection has now been established. I do not think the importance of this simple mechanism can be overestimated. (Latour, 1986, p 14)

Latour reminds us that the power of any mechanism that makes confusing things less confusing, that is capable of transforming data into something more easily discussed with others, or that can be placed into other contexts to aid understanding should not be underestimated. It is an appreciation of this that has encouraged us to introduce a method whose aim is to harness this kind of power for analysis and whose processes also foreground the vital importance of considering at an early stage how the results of analysis will be communicated.

As we wrote earlier in this chapter, we see this work as being at a foundational stage. We would now like to see others using and adapting the analogue journey method summarised in Figure 19.9 – as we will do ourselves.

Figure 19.9: The analogue journey method: this chapter

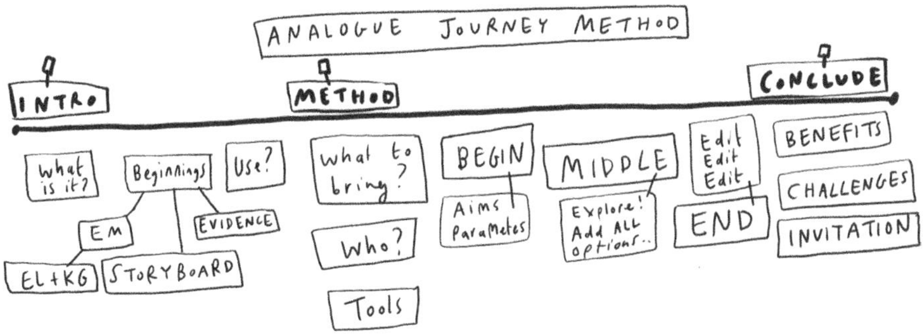

Source: Created by Emma Lazenby.

Note

1 https://www.formedfilms.co.uk/alittledeepsleep.

References

Ganzert, A. (2020) *Serial Pinboarding in Contemporary Television*, Cham: Palgrave Macmillan.

Glei, J.K. (2017) 'Austin Kleon: pencil vs computer', *Hurry Slowly*, 7 November. Available from: https://hurryslowly.co/005-austin-kleon/

Kleon, A. (2012) *Steal Like an Artist: 10 Things Nobody Told you About Being Creative*, New York: Workman Publishing.

Labacher, L., Mitchell, C., de Lange, N., Moletsane, R., and Geldenhuys, M-M. (2012) 'What Can a Visual Researcher Do with a Storyboard?', in E.J. Milne, C. Mitchell, and N. de Lange (eds), *Handbook of Participatory Video*, Plymouth: AltaMira Press, pp 149–63.

Latour, B. (1986) 'Visualization and cognition: thinking with eyes and hands', *Knowledge and Society: Studies in the Sociology of Culture Past and Present*, 6(1): 1–40.

Law, J. and Urry, J. (2004) 'Enacting the social', *Economy and Society*, 33(3): 390–410.

Lazenby, E. and Gray, K. (2019) 'What is it about arts and dementia that makes it so hard to evaluate?' *Formed Films*. Available from: www.formedfilms.co.uk/artsanddementia

Magnani, L. (2013) 'Thinking through drawing: diagram constructions as epistemic mediators in geometrical discovery', *The Knowledge Engineering Review*, 28(3): 303–26

Mannay, D. (2015) *Visual, Narrative and Creative Research Methods: Application, Reflection and Ethics*, Abingdon: Routledge.

Mauthner, N.S. and Doucet, A. (2003) 'Reflexive accounts and accounts of reflexivity in qualitative data analysis', *Sociology*, 37(3): 413–31.

Pallant, C. and Price, S. (2015) *Storyboarding: A Critical History*, Basingstoke: Palgrave Macmillan.

20

Using creative mapping methods to analyse multimodal data

Erin Roberts, Merryn Thomas, Karen Henwood, and Nick Pidgeon

Introduction

Mapping and map making are creative methodologies that are widely used in both quantitative and qualitative research practice. Whether cartographical or metaphorical in form, maps are powerful multimodal artefacts that are rich resources for communication (Clark, 2011; Powell, 2016). Their ability to visually communicate complex information in an accessible manner has made maps and map making an invaluable part of many an academic's methodological toolbox.[1] While they are often used for research dissemination, whether to demonstrate the process of research in flow charts or to provide visuals to communicate findings, far less attention has been paid to the use of mapping as an analytic technique, with the notable exception of McKinnon and McCallum Breen (2020). This is rather perplexing, not least because creative and multimodal methods have gained much traction among qualitative researchers since the mid-1990s, but also because by their very nature – that is, being inherently multimodal – maps of various kinds are seemingly well positioned to aid researchers in their efforts to organise and make sense of multimodal materials.

As such, this chapter sets out to explore how mapping in its various forms can be used to make sense of data generated from the use of creative methods in interpretive research. Drawing on our own research practice, we will discuss how we used multiple mapping techniques to collect, analyse, and (re)present rich, multimodal data collected as part of an interdisciplinary study of well-being at two estuarine case sites in Wales. We will then take you along our analytic journey – broken down into three steps – detailing the decisions we made and why we made them, in the hope that this might help others as they develop their own analytic practice. But first we will introduce our project.

The CoastWEB project: 'Valuing Nature for Wellbeing'

Our story begins with the research team's involvement in the interdisciplinary CoastWEB project from 2017 to 2020. The project aimed to better understand the well-being benefits of saltmarsh habitats, and our team were tasked with exploring the 'intangible' cultural values associated with these vulnerable environments. We developed a bespoke, qualitative multimodal approach to elicit a broad range of values and a rich understanding of our diverse sample's (n = 26) entanglements, interactions, and relationships with the coast – and with saltmarshes in particular – at two case-study sites: the Taf Estuary in Carmarthenshire, south Wales; and the Mawddach Estuary in Gwynedd, north Wales. The study was granted ethical approval by the School of Social Sciences Research Ethics Committee at Cardiff University in 2018. As is common practice in the reporting of qualitative data, participant names were replaced with pseudonyms (apart from one participant, who requested that we use their real name).

Our methodology consisted of five tasks across three distinct phases (see Figure 20.1), purposefully designed to help our participants to express the 'inexpressible' (Satterfield et al, 2013, p 112). For the first task, which was to be carried out before the interview, participants were asked to take photographs of anything that they felt was significant to them when using the coast/estuary in any way,[2] to help them discuss their relationships with the coast over time during the interview. For the second task, participants were invited to guide us on a walk around the case site via a route that is meaningful to them, to explore values relating to the place in situ. We adopted narrative and open-ended techniques to explore participants' senses of well-being, which are best elicited by asking people about their experiences, feelings, and interactions with the world (see Roberts et al, 2021; Thomas et al, 2021, 2022).

On completing the mobile interview, participants were invited to a sit-down interview, where they took part in three further tasks. These were a photo

Figure 20.1: Mapping interview methodology

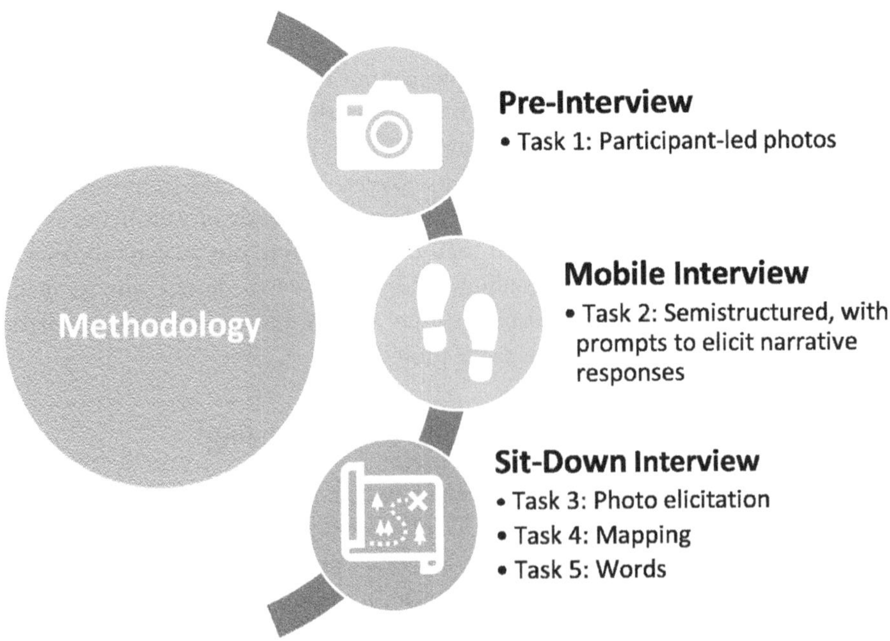

elicitation task, a map-annotating task, and a word association task. Each task was designed to provide ample opportunity for expression by engaging with the subject matter from multiple angles, which is particularly relevant when interested in the multifaceted qualities of human–nature relationships. While the use of multimodal methods was primarily for the purpose of stimulating conversations (Trombeta and Cox, 2022), we also recognised that the resulting materials were rich sources of information that ought to be analysed alongside transcripts. As such, we view the material and the discursive as inextricably intertwined, providing context for one another, as well as different lenses on the same topic (Kara, 2020).

When returning from fieldwork, we were faced with making sense of vast amounts of rich multimodal material, and promptly set out to find productive ways to organise and interpret the data. While the research team had experience of using creative methodologies to elicit talk about various topics and had used multiple strategies for analysing textual data before (see Henwood et al, 2016), we had not applied these to analysing multimodal data. As such, we were eager to see whether some of these skills were transferable, and whether they could address the aims of the research.

Step 1: transforming and familiarising (some of) the data

Whether research is qualitative or quantitative, issues of data transformation and analysis are embedded within the research process. Along with the theoretical and methodological frameworks that underpin the research, more practical constraints, such as limited budgets and timescales, will influence the analytic process (Tobias, 2014, p 298). After deciding on the theoretical and methodological underpinnings of a study, researchers must begin thinking about ways of processing and making sense of the resulting data that will address the aims of the research. When dealing with multiple qualitative methods and modalities, deciding on what transcription convention to use, whether to employ professional transcribers, or even whether transcription is necessary or appropriate in the first place, will all influence the researcher's interpretations of the data.

Transcription is often likened to the 'translation' of one mode into another – that is, from speech into writing – which suggests that modes are like different languages; and no two languages are the same. As such, both transcription and translation involve making decisions that have a direct impact on how words and their meanings are interpreted and later reported (Bird, 2005). As a selective and situated act, transcription therefore runs the risk of 'flattening' rich multimodal data into a purely written form. That is to say, some meaning will always be 'lost in translation'. This is not to devalue academic conventions that are predominantly text-based; rather, we are suggesting that researchers should consider and reflect on this issue early in the research process, so that they may develop strategies to retain the unique insights enabled by each mode.

The question of who does the transcribing – whether it is the research team or whether the work is outsourced to professional transcribers – also has a bearing on the resulting data (Bird, 2005). Many scholars highlight the value of researchers doing their own transcribing, as it enables them to build an intimate knowledge of the data (Bazeley, 2021). Despite the obvious merits, constraints on researchers' time and resources, particularly in large research projects such as ours, often result in the outsourcing of transcription. In such cases, it is essential that researchers provide transcribers with guidelines for transcription (ie convention and formatting), so that the resulting text is as true to the conversation as possible, while being 'pragmatic in dealing with the data' in ways that align with the purposes of the research (Bazeley, 2021, p 101).

In the case of our study, we were interested not only in the meanings underpinning the stories created and shared during the research encounter, but also the role that different people–place interactions played in generating meaning. These interactions included direct, in situ interactions, as well as indirect interactions such as relived memories, and interactions with photos and maps. As such, we developed a denaturalised transcription convention with some verbatim features that conveyed affect and emotion – such as meaningful pauses, gasps, laughter, and so on – and gave the transcribers a detailed guide to aid them in translating recordings to text. Even when such precautions are taken, however, it is important to remember that transcription is an interpretive task (Bird, 2005). Given that the transcribers were not present during data collection, we were aware that their distance from the material was likely to result in unintentional misinterpretations and errors. As such, it was essential that each transcript was reviewed and edited by the researchers who carried out the interviews. We approached this by reading the transcript while carefully listening to the corresponding recording, paying special attention to the emotional overtones of speech, while simultaneously checking accuracy (ie missed words/misinterpretations/inaudible sections) and punctuation; both of which can change the intended meaning of a sentence (Bazeley, 2021). During this process, we also drew on our field notes and researcher-generated photographs – which were taken during the interview to facilitate our recall of the encounter – enabling us to communicate what occurred more accurately by providing important contextual information that would have been impossible for anyone not present during the interview to know. This was particularly useful in the case of speech that responded to sensory stimuli during the interview encounter; including sights, smells, the weather, sounds, images, haptics, or ground-feel (Brown, 2017), and in the case of one participant, even taste:

<< Walking through a wooded area (pine, uneven ground due to roots) along an upland walk overlooking the Mawddach estuary. Trystan had previously been talking about his work engaging young people with nature when we came across some Wood Sorrel and paused for a moment>>

[Trystan]: Yes. But you know I told you about, uh, bringing the children to have experiences? Well, this plant here, this plant <<*bending down, Trystan picks up a plant for me to see*>> … It's Wood Sorrel. A white flower, three leaves, heart shaped.

[ER]: Oh yeah?

[Trystan]: Eat it. <<*offering me the plant*>>

[ER]: <<*hesitantly reaching for the plant*>> There hasn't been a, uh, dog here recently, has there? [laughing]

[Trystan]: Um not that I know of! [laughing] … Bite it into small pieces, right … <<*Trystan pauses expectantly, waiting for my response*>>

[ER]: <<*I take a bite of the sorrel*>> … Like lemon.

[Trystan]: That's it, yes! <<*in excitement*>> Yes. Apple skin is what I tell them. Or if I have –

[ER]: [overlapping] – Yes, apple skin! <<*in excited agreement*>> You're right, like a really green apple.

[Trystan]: That's it. Yes, like a Granny Smith's kind of apple –

[ER]: [overlapping] – Yes … yeah. It's a quite nice taste, isn't it?

[Trystan]: Yes, yes, it is, yeah. And then when I bring kids this way, you know, everything is in packages, isn't it? They love being able to pick something up off the floor and eat it, and getting such a nice, strong taste […] And these experiences are quite – You know I was saying in the car earlier about the soft sciences … the soft skills, isn't it, of working together and solving a problem, figuring something out. A little bit of logic instead of facts, facts, facts. It does so much – It's worth – it's worth – There's a lot of value to something like that.

[Trystan, Mawddach – *translated from Welsh*]

This excerpt demonstrates the importance of contextual information in the interpretation of transcribed data; information that would have otherwise been lost if the person who conducted the interview had not reviewed the transcript. Researcher notes interspersed with the transcripts were integral to understanding how the interactions between researcher, participant, and environment generated meaning. The discussions that took place during these encounters were 'profoundly informed by the landscapes in which they take place' (Evans and Jones, 2011, p 849), with the environment able to prompt and interject as interviewee and researcher moved through it. The ensuing discussion of the taste of wood sorrel created a fleeting moment of affect, a shared atmosphere of wonder and appreciation of the natural world and excitement at the ability to share it with others, both in that moment and beyond.

Step 2: making sense of it all: thematic mapping with a multimodal twist

After familiarising ourselves with the data, we found it useful to reflect on the sort of insights that could be gleaned from each method and mode (see Table 20.1). Each method had produced a different type of qualitative data, and while these can be brought together to enhance researchers' interpretation of the data set, we also needed to think critically about how this could be done and what it would achieve. The process of bringing different forms of qualitative data together can be characterised as data integration, which is described by Cronin et al (2008) as producing a special *relationship* between different methods and methodologies (p 573). With this in mind, we developed a table that laid out the relationship between the methods used and the type of insights that we could glean from them. In doing so, we were better placed to recognise how best to approach analysing each data type, the degree of alignment between different data types, and what sort of claims could be made from the data.

Following this reflexive process, we began thinking of ways to organise and interpret the data. First, we adopted a thematic approach, like that of Trombeta and Cox's (2022) textual-visual thematic analysis, where we looked for patterns in the

Table 20.1: Method insights

Material	Mode	Insight
Transcript	*Text*	A (selective) written record of the interview that contains a co-produced discussion of participants' engagement with the coast in situ, as well as reflections of their past experiences. Contains stories of participants' emplaced experiences and sensory engagement with the case site (during interview).
GPS recordings	*Spatial, image*	A spatial (and visual) record of the mobile interview. Augments transcript by enabling researchers to pinpoint when and where things were said.
Researcher photos	*Image*	Researcher-generated visual snippets of serendipitous moments during (mobile) interviews. Purposefully generated to enhance visual record of mobile and sit-down interviews.
Interview recordings	*Audio*	Co-produced auditory snippets of serendipitous moments during the interviews, purposefully selected by the researcher. Adds to transcript by providing a record of things that are not easily translated (eg soundscapes).
Participant photos	*Image*	Participant-generated without the presence of the researcher. A visual representation of participants' meaningful engagement of the coast. Not always purposefully generated for the research (as in the case of 'found' photos).
Participant map annotations	*Multimodal (spatial, images, text, symbols)*	A (selective) multimodal record of past experiences at the coast. Both a prompt, and providing context, for interview discussions.

transcripts and began sorting them into themes. As with most thematic approaches, this was an iterative process, partially grounded in the data and partially informed by theory, and the ensuing emic–etic flip-flopping helped us to develop and refine our themes throughout this stage of the analysis. Eager to produce an analysis that was more than the sum of its (modal) parts (Clark, 2011), we then began thinking of ways of integrating or aligning the different research materials to develop richer insights (Kara, 2020). When bringing together different types of qualitative data, it has been argued that researchers should be mindful of retaining the character of each method used and treat them equally (Cronin et al, 2008). When it comes to bringing insights from multiple methods into conversation with one another, Cronin et al suggest that researchers try to interface insights from each method through some kind of 'designed juxtaposition' (p 573). As such, we began going back and forth between transcript excerpts and their corresponding photographs, maps, GPS plot, and word-sorts, all the while taking note of their content and asking ourselves;

1. What does this method/mode do differently?
2. Does its content support or contradict the corresponding excerpt?

Each method focused on a different spatial and temporal scale, which allowed our participants to think about their well-being and about the saltmarsh from different perspectives. To summarise, the mobile interview captured particularly well fleeting moments, chance encounters, and small minutiae. The photo task tended to focus on historical time, whether special events or images that show how the landscape and relationships have changed. The maps also elicited events and favourite places but focused on those that we were unable to visit with the first two tasks. Finally, the words task provided a 'quick-fire' route to new ideas and stories, as well as links between different themes. We also found that different types of photographs tended to elicit different types of insight (Thomas et al, 2021).

Given these differences, we were keen to explore how the different types of data 'speak' to one another and how this might enhance our interpretation of each theme. Keeping abreast of the development of themes and how they related to each other across the data set, however, became increasingly difficult to manage as time went on. It quickly became apparent that we needed to reorganise the data in a clearer, more accessible way.

Mapping is a commonly used technique for exploring relationships of various kinds, and for helping researchers to recognise connections between modalities as well as the themes within them (Marx, 2022). Creating and using visual artefacts such as maps can not only enable researchers to integrate and navigate multimodal data, but also provides a mechanism for demonstrating participants' knowledge, understanding, and perspectives of phenomena while also displaying relationships between concepts (Tobias, 2014, p 296). Mapping out, or visually organising, thematic data not only enabled us to explore how themes were discussed across the sample, but also provided a different means of highlighting the unique insights provided by each method (as seen in Figure 20.2).

Figure 20.2: Thematically organised multimodal data

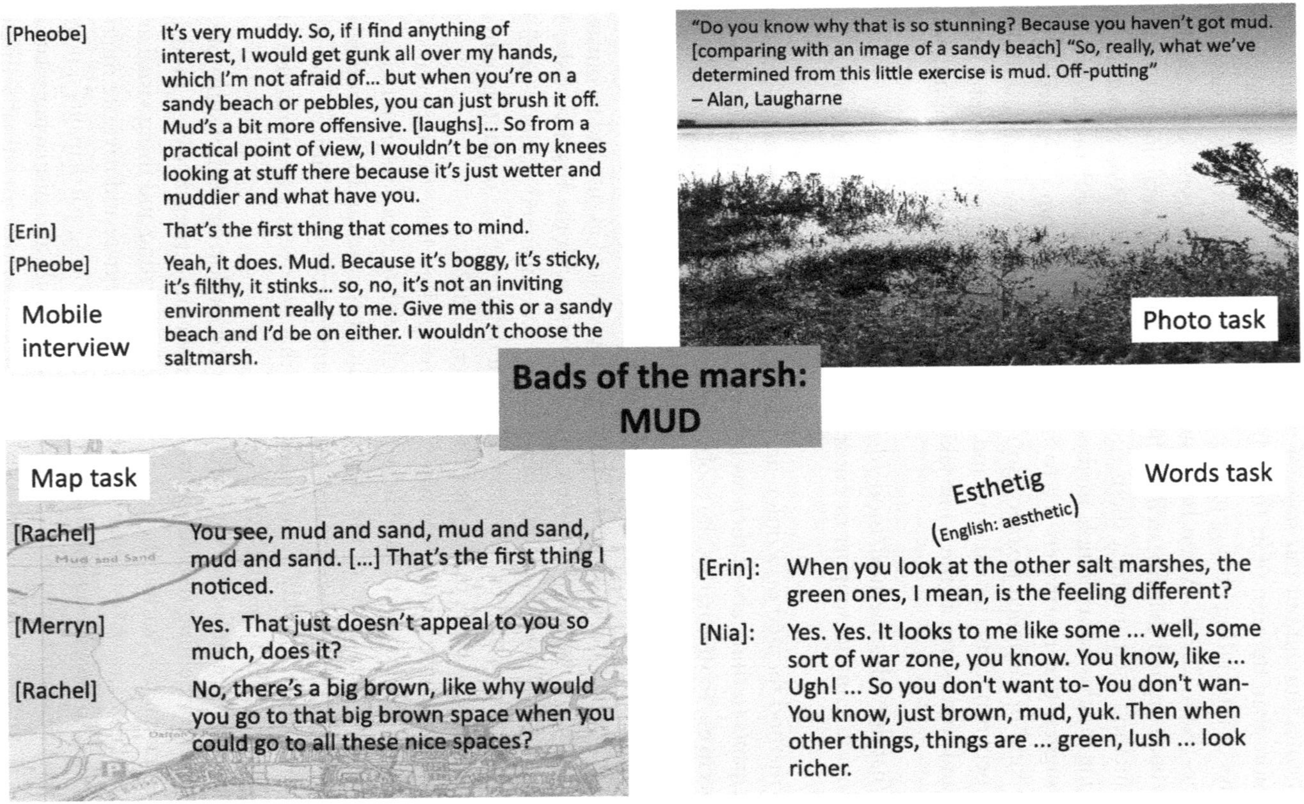

Source: Photo credit for photo task: Merryn Thomas.

Figure 20.2 demonstrates only a very small snippet of a single subtheme (mud) under the broader category of 'Bads of the marsh'. Each participant's data displayed here has two related elements – one text-based and the other image-based – which allowed us to compare between related text and image, as well as between different participants. For example, the juxtaposition between Alan's excerpt from the photo task and the content of the (researcher-generated) photograph he was discussing is particularly interesting. Here, Alan defines the beauty of the image not by what it contains, but by what it does not. Rachel also discusses mud in a negative light; however, her engagement with the method shaped the discussion in a different way. Rachel's excerpt and image come from the mapping task, where she was immediately drawn to the colours of the map (browns, blues, and greens) and the repeated labelling of 'mud and sand' along the estuary. Here, the content of the image directly shaped the ensuing discussion and highlighted the role of symbolic representations – and particularly the symbolic use of colour – in shaping meaning making.

Step 3: (re)presenting participant stories with geo-narratives

So far, mapping had enabled us to enrich our understanding of key themes across the data set, but we had yet to unlock the full potential of mapping as an analytic approach. One of the key strengths that drew us to mapping as an analytic technique lay in the inherent multimodality of maps and their ability to represent the development of meaning. As such, maps are useful communicative tools for visualising spatial relationships, and, as noted by McKinnon and McCallum Breen (2020), they allow audiences to 'see' or understand the data in new ways. As one of our aims was to explore how a sense of place was storied among our participants, we wanted to utilise a mapping technique to visualise and explore this aspect of people–nature/place relationships through time.

Initially, we approached spatialising transcripts on a case-by-case basis by layering data from each part of an interview and spatially plotting it on a map of the case site. Our early attempts at spatialising the transcript (see Figure 20.3) involved using Google Maps, which possesses a level of interactivity, as researchers/audiences can zoom in and out of the data by choosing to expand certain nodes along the map. However, due to the sheer quantity of data, the task of spatialising transcripts quickly became overwhelming. Once again, this prompted us to reflect on what insights could be gleaned from this exercise, at which point it become apparent that this approach was creating a multimodal record of the interview rather than having a specific analytic purpose.

To produce something valuable for meaning making we needed to be more selective in what we included in the map, as it was impossible to include every last detail from the different types of data that we had collected. We were inspired by Kwan and Ding's (2008) geo-narrative approach (as well as Bell et al's (2015) adaptation of the geo-narrative for coastal research), which as the name suggests, is concerned with spatialising *narratives*. As such, we drew on

Figure 20.3: Early spatialised transcript of a canoeing interview using Google Maps

Source: Base map: Google Maps.

conventions commonly used in biographical and narrative research traditions that enabled us to make sense of the data both thematically, across the sample, and on a case-by-case basis (Thomson, 2007; Butler et al, 2014). While the thematic analysis has been largely discussed in relation to the thematic mapping exercise (see Figure 20.2), making sense of participants' changing relationships with the coast through time required us to construct case narratives by using what Saldaña (2003, p 151) terms 'through lines', which 'describe, connect and summarize the researcher's primary observations of participant change'.

This process entailed identifying the 'sequence' of participant stories (Cronin et al, 2008), while taking note of key moments or turning points that prompted thinking about how and why particular paths were followed while others were not (Butler et al, 2014, p 173). Doing so enabled the analysis to move beyond merely identifying past events and experiences, and instead focus on understanding the contemporary significance they hold for the individual telling the story (Cronin et al, 2008, pp 575–6). Here, participant-generated map annotations and photos, along with field notes and researcher-generated photographs, were important in providing further context for these through-lines and were brought alongside any related text during this process.

While we were interested in the story of how participants' relationships with the coast had changed over biographical time, we were also interested in better understanding how in situ encounters had shaped meaning making during the mobile interview. As such, we also paid attention to key moments during the mobile interviews, such as surprise encounters, references to sensory and social engagements along routes, and places where the researcher and participant paused or bimbled. In anticipation of this, and as suggested by Bell et al (2015), we synchronised our GPS and audio recorders at the start of every interview, so that we could pinpoint when and where things were said along with the pace of movement, and thus spatially identify important interactions and moments during the interviews. Images taken during the interview by the researcher to visually represent the topic of discussion were then placed on the GPS map to enhance the record of the encounter and to provide further visual and spatial contextualisation.

The purpose of bringing together insights from these two analytic approaches was not to create a 'holistic' or 'complete' data set: indeed, that would neither be possible nor practical. Instead, the research team sought to weave together a spatio-temporal narrative of participant–place relations that enhanced our understanding of the significance of interactions with – and connections to – the more-than-human world. Selecting what to include in each geo-narrative was done on a case-by-case basis and was largely driven by the narrative through lines identified in each participants' data. While the resulting geo-narratives were qualitatively rich, it is important to note that they represented a condensed retelling of stories co-produced during the interviews (see Marx, 2022). The (re)presentation of the stories was curated solely by the research team, which limits the kind of claims that can be drawn from these (re)presentations. Furthermore, while this method was particularly effective at visualising/spatialising people–place relations, we quickly realised that the success of this approach was dependent on the extent to which participants had engaged with the methodology as a whole. Not every participant felt comfortable annotating the maps that we provided them with, for example, and participants varied significantly in their responses to the photo-taking task; while some provided none, others provided as many as 60 photographs. As such, when producing geo-narratives, we focused our efforts on participants with the most 'complete' data sets, to better communicate our research to a variety of audiences.

As the project came to a close, we had the opportunity to explore geo-narratives' potential to go beyond data analyses to inspire others to engage with their own meanings of place (see Figure 20.4). At two end-of-project events – one in London and one in Cardiff – we presented two examples of participants' geo-narratives to generate interest in our findings. We then used printed large-format maps of each event location (along with post-its and felt pens) as stimuli for event attendees to create their own collaborative geo-narratives of London and Cardiff during the event. To this end, the geo-narratives were successful in engaging audiences and getting them to reflect on their own connections and interactions with meaningful places. Event attendees interacted with the maps in various ways, with some deciding to use post-it notes, others directly scribbling on the maps, and some preferring not to leave a mark but instead choosing to animate their discussion by pointing to various meaningful locations on the map as they talked. We were particularly struck by how readily event attendees related to the geo-narratives on display and how quickly the maps acted as a springboard for discussions about their own place relations – both in the event setting and beyond. Consequently, we would argue that geo-narratives are an effective means of demonstrating the value of creativity in qualitative research practice and communication, as they not only bring together various forms of meaning to create new meaning (McKinnon and McCallum Breen, 2020), but

Figure 20.4: Example of a participant's geo-narrative

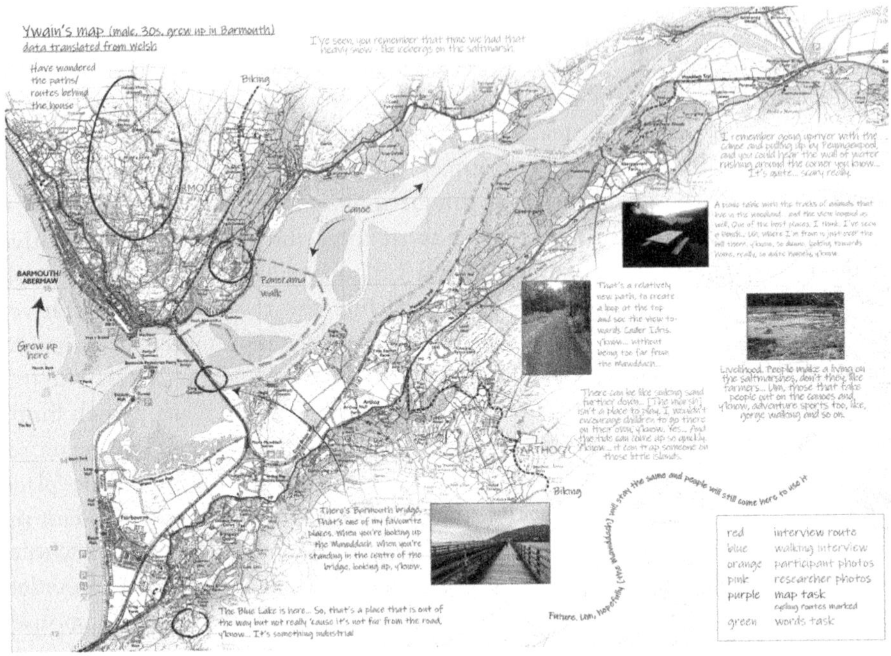

Source: Base map: Ordinance Survey.

also inspire audiences to reflect on their own place relationships and what is at stake when it comes to place change.

Concluding reflections

In this chapter, we have shared one of many potential ways for bringing together insights from multiple qualitative methods to enrich the processes of data analysis and reporting findings. We have demonstrated that even when drawing on commonplace analytic techniques, such as thematic analysis, there are opportunities for purposeful creativity in analytic practice, particularly when integrating insights gleaned from multiple modes of engagement. In our case, both cartographical and metaphorical mapping techniques were used to visually organise, spatialise, and (re)present data; due in large part to their ability to visually communicate complex socio-spatial information and highlight all kinds of processes and relationships – even unexpected ones. While integrating insights from multiple methods and modes can be both challenging and enriching (often in equal measure), maps being familiar cultural artefacts that are inherently multimodal in nature had sparked our interest in mapping as an analytic technique. However, this is not to say that analytic mapping is without limitations.

Underpinning all research are decisions on what is relevant for the purposes of research and thus what is included in the analysis. Along with multimodal data come a whole swathe of ethical considerations regarding the inclusion/exclusion of images and audio in one's analysis. Such considerations include, but are not limited to, participants' informed consent; researcher transparency with regard to how different types of data will be used, including for analysis, journal publications, and presentations; issues of ownership of participant-generated or collected images; and how best to protect participant anonymity. Consequently, researchers must be clear about the intended applications of their research activities as early as possible in the research process. Moreover, while data from qualitative interviews are co-produced by researchers and participants, how researchers intend to analyse data, and whether they will involve participants in the process or not, is also an important consideration. Consequently, researchers should be careful about any claims they make resulting from their analyses, particularly when they (re)present their interpretations of participants' narratives. Finally, challenges remain for the academic dissemination of multimodal data, particularly when it comes to data that is not easily incorporated in conventional journalistic format, for example, audio, video, and a plethora of other sensory engagements. Nonetheless, this is steadily changing with the proliferation of wearable technologies as well as the abundance of digital tools now at our fingertips.

With the conclusion of our reflective account, we hope that readers have found this chapter useful, and encourage them to draw on their creative side during data analysis. After all, who knows what you might find?

Notes

1 A claim that is often made in mapping literature; one that is underpinned by an assumption of a 'general' level of 'map literacy'.

2 Eg visiting seaside attractions, walking, boating, foraging.

References

Bazeley, P. (2021) *Qualitative Data Analysis: Practical Strategies* (2nd edn), London: SAGE, pp 100–14.

Bell, S.L., Phoenix, C., Lovell, R., and Wheeler, B.W. (2015) 'Using GPS and geo-narratives: a methodological approach for understanding and situating everyday green space encounters', *Area*, 47(1): 88–96.

Bird, C.M. (2005) 'How I stopped dreading and learned to love transcription', *Qualitative Inquiry*, 11(2): 226–48.

Brown, K.M. (2017) 'The haptic pleasures of ground-feel: the role of textured terrain in motivating regular exercise', *Health & Place*, 46: 307–14.

Butler, C., Parkhill, K.A., Shirani, F., Henwood, K., and Pidgeon, N. (2014) 'Examining the dynamics of energy demand reduction through a biographical lens', *Nature and Culture*, 9(2): 164–82.

Clark, A. (2011) 'Multimodal map making with young children: exploring ethnographic and participatory methods', *Qualitative Research*, 11(3): 311–30.

Cronin, A., Alexander, V.D., Fielding, J., Moran-Ellis, J., and Thomas, H. (2008) 'The analytic integration of qualitative data sources', in P. Alasuutari, L. Bickman, and J. Brannen (eds), *The SAGE Handbook of Social Research Methods*, London: SAGE, pp 572–84.

Evans, J. and Jones, P. (2011) 'The walking interview: methodology, mobility and place', *Applied Geography*, 31(2): 849–58.

Henwood, K., Pidgeon, N.F., Groves, C., Shirani, F., Butler, C., and Parkhill, K. (2016) *Energy Biographies Research Report: Project Report*, Cardiff: Energy Biographies.

Kara, H. (2020) *Creative Research Methods: A Practical Guide* (2nd edn), Bristol: Policy Press.

Kwan, M.-P. and Ding, G. (2008) 'Geo-narrative: extending geographic information systems for narrative analysis in qualitative and mixed-method research', *The Professional Geographer*, 60(4): 443–65.

Marx, S. (2022) 'Mapping as critical qualitative research methodology', *International Journal of Research & Method in Education*, 46(3): 285–99.

Mckinnon, I. and McCallum Breen, J. (2020) 'Expanding cartographic practices in the social sciences', in L. Pauwels and D. Mannay (eds), *The SAGE Handbook of Visual Research Methods* (2nd edn), London: SAGE Publications, pp 154–72.

Powell, K. (2016) 'Multimodal mapmaking: working towards an entangled methodology of place', *Anthropology & Education Quarterly*, 47(4): 402–20.

Roberts, E., Thomas, M., Pidgeon, N., and Henwood, K. (2021) 'Valuing nature for wellbeing: narratives of socio-ecological change in dynamic intertidal landscapes', *Environmental Values*, 30(4): 501–23.

Saldaña, J. (2003) *Longitudinal Qualitative Research: Analyzing Change through Time*, Walnut Creek: AltaMira Press.

Satterfield, T., Gregory, R., Klain, S., Roberts, M., and Chan, K.M. (2013) 'Culture, intangibles and metrics in environmental management', *Journal of Environmental Management*, 117: 103–14.

Thomas, M., Roberts, E., Pidgeon, N., and Henwood, K. (2021) 'Using photographs in coastal research and communication: reflections from two case studies', in M. Gustavsson et al (eds), *Researching People and the Sea: Methodologies and Traditions*, Cham: Palgrave Macmillan, pp 181–207.

Thomas, M., Roberts, E., Pidgeon, N., and Henwood, K. (2022) '"This funny place": using affordances to explore the value of saltmarshes for human wellbeing', *People & Nature*, 4(3): 181–208.

Thomson, R. (2007) 'The qualitative longitudinal case history: practical, methodological and ethical reflections', *Social Policy and Society*, 6(4): 571–82.

Tobias, E.S. (2014) 'Collecting, generating and analyzing multimodal and multimedia data', in C.M. Conway (ed), *The Oxford Handbook of Qualitative Research in American Music Education*, Oxford: Oxford University Press, pp 288–306.

Trombeta, G. and Cox, S.M. (2022) 'The textual-visual thematic analysis: a framework to analyze the conjunction and interaction of visual and textual data', *The Qualitative Report*, 27(6): 1557–74.

Section VII

Existing methods adapted in creative ways

21

Embracing creativity in familiarisation

Louise Couceiro

Introduction

Familiarisation happens throughout the research process: 'The process of becoming aware of what is interesting and significant about the information and data you are collecting begins during the first moments of conceiving the research project and continues throughout' (Silver and Lewins, 2014, p 230). In other words, becoming familiar with or getting to know your data is something researchers do continuously. However, familiarisation is also a phase that has been formalised in approaches to analysis. For example, in their work on reflexive thematic analysis Braun and Clarke (2021, p 35) write that familiarisation is when 'you become deeply and intimately familiar with the content of your data set, through a process of immersion'. This definition seems broadly applicable to other types of data analysis – regardless of the approach one might take there is almost always a phase of familiarisation at the outset, requiring researchers to 'get to know' their data. Although the case example I present in this chapter focuses on familiarisation as a formal phase of data analysis, my chief contention – that consciously embracing a creative sensibility during familiarisation can open up the data in exciting ways – is pertinent for any moment where data familiarisation is occurring.

Familiarisation is integral to data analysis. It allows researchers to 'get a feel' for the data, ultimately benefiting other phases of analysis. Though more creative approaches to familiarisation are appearing (Culshaw, 2019; Bazeley, 2020; Thomson, 2020), guidance for familiarisation often includes writing memos while reading, listening, or watching the data repeatedly (see Cohen et al, 2018; Harding, 2018; Byrne, 2022). This is not to suggest that reading, listening, watching, and writing are not creative practices. As the chapters in this book illustrate, creativity takes many forms. However, when I talk about consciously embracing a creative sensibility, I am referring to the deliberate intention of looking beyond the norm, to experiment and explore. In this case, this means going beyond the commonly cited methods of repeated reading, listening, or watching, to invent new ways of doing and knowing.

The study

After providing a brief overview of the limited scholarship on creative approaches to data familiarisation, I draw on my doctoral research to illustrate how I embraced a creative sensibility during the formal phase of familiarisation.

This research explored how a group of eight children in the UK, aged between 7 and 10, responded to and engaged with collective biographies about women published between 2016 and 2020, including *Good Night Stories for Rebel Girls: 100 Tales of Extraordinary Women* (Favilli and Cavallo, 2016) and *Little Leaders: Visionary Women Around the World* (Harrison, 2020). The study was granted ethical approval by the University of Glasgow.

The data I gathered was multifarious, including transcripts from interviews and group reading sessions I conducted over Zoom, and copies of creative artefacts participants produced through engagement with a reader response toolkit (see Couceiro 2020). The toolkit, a handmade reference book containing suggestions for arts-based activities, allowed participants to choose their own method(s) for responding to the texts. Children's artefacts included comic strips, PowerPoint presentations, drawings, dances, and baked goods. For the purposes of this study, I did not analyse the artefacts as distinct data objects. Rather, the artefacts served as springboards for conversation during interviews. I thus analysed participants' discussions of their artefacts in the transcripts. My overarching approach to data analysis combined reflexive thematic analysis (Braun and Clarke, 2021) with insights from discourse analysis, drawing on the work of James Paul Gee (2011a, 2011b). I started by following Braun and Clarke's (2021) six phases of reflexive thematic analysis, beginning with familiarisation. However, as I did not find reading and rereading the data effective during this phase – the recursive process feeling tedious and tiresome – I decided to experiment with alternative approaches.

My ontological position is that there is no single, observable reality that can be investigated or measured; accordingly, discovering 'truth' is neither possible nor desirable. Epistemologically, my writing is grounded in the belief that meanings are subjective, shifting, context-specific, and intricately intertwined with language. What this means in the context of this chapter on data familiarisation is that no single, pre-eminent approach to analysis is deemed to exist; rather, meanings are constructed by researchers through engagement with the data. Consequently, researchers should offer transparency about their approach.

As with many contributions in this book, this chapter does not, therefore, offer a step-by-step guide on how to 'be creative'. There are many ways one could approach data familiarisation with creativity. As Anuja Cabraal (2020, para 2) advises in a blog post on coding and analysis for qualitative researchers: 'Identify how you think and process information and start exploring the data that way.' Indeed, the example I present is intended to inspire readers to experiment in finding ways to get to know their data that work best for *them*. Just as there are a myriad of different ways to process information, there are a myriad of different ways to 'do' data familiarisation.

Getting familiar: invisible creativity

There is a paucity of literature focusing on creativity during data familiarisation. Anecdotally, I have heard tales of post-it notes flooding living room floors, of whiteboards filled with diagrams, and of notebooks suffused with doodles and drawings. One colleague told me they like to condense interview data into haiku. Another described how they had once spent hours building models out of pipe cleaners. I have also heard of researchers running, walking, and swimming laps while listening to their data; bodily movement seeming to open up space for the data to breathe and seep into one's consciousness more easily. Thus, despite the dearth of formal reporting on creativity during familiarisation, researchers are clearly embracing creative sensibilities as they get to know their data.

While some scholars have called attention to the benefits of using 'alternative' tools and strategies for sense making during data analysis (Jewitt, 2014; Leavy, 2020; Woodward, 2020; Kara et al, 2021), much of this work escapes formal reporting. As Pat Bazeley (2020, p 124) notes, the early stages of analysis 'are where much of the hard work of analysis is done, yet what you do here rarely rates a mention when the work is published'. In response, Bazeley (2020) dedicated an entire section of her book on qualitative data analysis to familiarisation, providing numerous 'unorthodox' strategies. These include writing news bulletins to summarise parts of the data; creating word pictures to profile a person or an event in the data; and translating data transcripts into poems. Bazeley (2020, p 134) describes these strategies as 'purposeful play', positing them as fun and helpful ways to develop familiarity with data sources.

Pat Thomson (2020) also draws on the notion of play in a blog post titled 'Playing about with data', offering ten suggestions to help researchers think about their data in new ways. Thomson (2020, para 6) writes that 'it is sometimes the case that doing something a little out of the ordinary can alert us to other possibilities, and to how our expectations shape what we see in our data'. One of her suggestions involves taking two interview transcripts with two different participants and writing the dialogue for an imagined conversation they could be having. Another requires the researcher to make a list of the emotions they feel while engaging with the data and to do some free writing based on these emotional responses. Recognising that 'none of these exercises are the same as the formal procedures that most of us are expected to use in analysis' (Thomson, 2020, para 5), she advocates for transgressing expectations – at least momentarily – to enable researchers to become more familiar with their data and consider it from different perspectives.

Evidence that researchers are embracing creative sensibilities during familiarisation can also be found in doctoral theses, where the word allowance is considerably higher than in journal articles or book chapters. Doctoral researchers have space to discuss (and are expected to discuss) the details of their analyses. In her thesis exploring what it means to be a struggling secondary school teacher in England, Suzanne Culshaw (2019, p 73) describes how she created 'scrapbooks' for each participant following their interviews. The pages were split in half, with the top half used for drawing, doodling, and writing notes, and the bottom half used for writing narratives of the participants' stories. This process of writing by hand enabled Culshaw to slow down and think about the material.

In her thesis on the development of digital capabilities in engineering and management, Tünde Varga-Atkins (2018) also describes how she approached the familiarisation process. Varga-Atkins (2018, p 75) summarised lecturers' transcripts through poetry, contending that creating these poems helped to 'identify the salient points of each interview and what [she] felt were important points about the discipline or the approach for developing digital capabilities'. In both examples, the researchers' investment in creative approaches to familiarisation enabled them to reflect and begin identifying what they perceived to be the most important aspects of the data.

Finally, some scholars have pointed to the benefits of using digital tools for data familiarisation. For example, Christina Silver and Ann Lewins explain how CAQDAS (computer-assisted qualitative data analysis software) provides a variety of annotation tools, helping researchers to create visual maps and networks, and to write analytic memos that can link to other objects and writing spaces. This approach is particularly helpful when working with large data sets as it 'can significantly help with the process of spontaneously noting observations and thoughts, linking them to the data that prompted them, managing them in systematic and useful ways, retrieving them in order to use at a later stage of analysis' (Silver and Lewins, 2014, p 230). Accordingly, digital tools can not only facilitate creative familiarisation, but early thoughts and reflections can be easily identified later in the analytic process. This is an important point because being creative is not antithetical to being systematic or structured. For familiarisation to be beneficial for other stages of analysis, work undertaken during familiarisation should be readily available and accessible throughout the research process.

Case example: unravelling responses to Rosa Parks' biographies

As noted, my doctoral study focused on children's responses to and engagement with contemporary biographies about women. Data gathering took place in the second half of 2020 in the wake of the Black Lives Matter movement. Participants were acutely aware of Black Lives Matter. They alluded to conversations they had in school and to learning they had undertaken about influential Black figures. Most seemed keen to engage with the biographies of Rosa Parks in *Good Night Stories for Rebel Girls* (Favilli and Cavallo, 2016) and *Fantastically Great Women Who Changed the World* (Pankhurst, 2016). Parks was a prominent twentieth-century civil rights activist in America, best known for her refusal to give up her bus seat for a White man in Alabama in 1955. One participant, Olivia, said that as soon as she received the books in the post she flicked through to see if she could find Parks' picture; Darcy chose Parks' biographies for the group to read in the reading sessions; Mickael created a comic strip in response to what he had read about Parks; and Harry and Sofia both instigated conversations about Parks' biographies in their individual interviews.[1] Needless to say, the participants' interest in Parks resulted in a lot of data centring on her biographies. I undertook some pilot analysis to explore how I might best utilise Braun and Clarke's (2021) reflexive thematic analysis in conjunction with insights from discourse analysis, which I applied once I had developed the themes. The variety and richness of the data pertaining to Parks made it an ideal subset for piloting different approaches.

In hindsight, during this piloting stage I proceeded to phase two of Braun and Clarke's reflexive thematic analysis (coding) without seriously committing to phase one (familiarisation). I read and reread the transcripts, skimmed over my field notes, and collated the data I had on participants' responses to Parks' biographies in an Excel spreadsheet. I then began experimenting with different coding methods (Saldaña, 2021), investing heavily in in vivo coding ('the terms used by [participants] themselves' (Strauss, 1987, p 33)). I began categorising codes under the following headings: 'Rosa's Actions', 'Consequences of Rosa's Actions', 'Racism Today', 'Racism "Then"', 'Who Should Act?', 'Comparing Texts', 'Fact or Fiction?', and 'Illustrations'. After many hours struggling to connect and make sense of these prematurely formed categories, I realised I did not have a clear, overall sense of the participants' responses. I still found it difficult to articulate what their general reactions were, what was important to them, and what kept coming up in our conversations.

To put it another way, I realised I had not sufficiently 'connected' with the data, something Stephanie Toliver contends is crucial before proceeding with in-depth analysis. Reflecting on her research exploring 'how Black girls might use written and oral storytelling to discuss, critique, and subvert experiences with social in/justice', Toliver (2022, p xxv) provides a fairly detailed account of how she spent time connecting with her data before beginning analysis. For example, she engaged with the girls' stories by revisiting television shows and news articles they mentioned, and investigating the history of various cultural aspects they discussed (Toliver, 2022, p xxvi). Comparatively, I had lurched into experimenting with different coding methods without fully grasping the nature of participants' responses. This inhibited me from moving forwards.

I returned to the raw data and, instead of reading and rereading, I experimented with writing analytic memos and poems. As illustrated in Figure 21.1, I also rewrote Parks' biography using participants' words and berated myself for developing what I suspected may have been yet another frivolous procrastination strategy (though, at least this time, helpfully disguised as formal familiarisation). However, as I go on to discuss, rewriting Parks' biography proved invaluable for later stages of analysis. While I had been consistently struck by the richness and multiplicity of participants' responses – they were insightful and reflective, thoughtful and thought-provoking – I struggled to hold them in my mind simultaneously. Re-presenting participants' responses to the biographies in the form of biography provided me with a vantage point from which to view their most salient responses, and to begin familiarising myself with the contradictions and complexities across them.

As I began extracting participants' words from the transcripts and artefacts to create this new biography, I noticed that narrative retellings were especially prevalent. That is, participants often responded to the biographies by recounting

Figure 21.1: Parks' biography rewritten

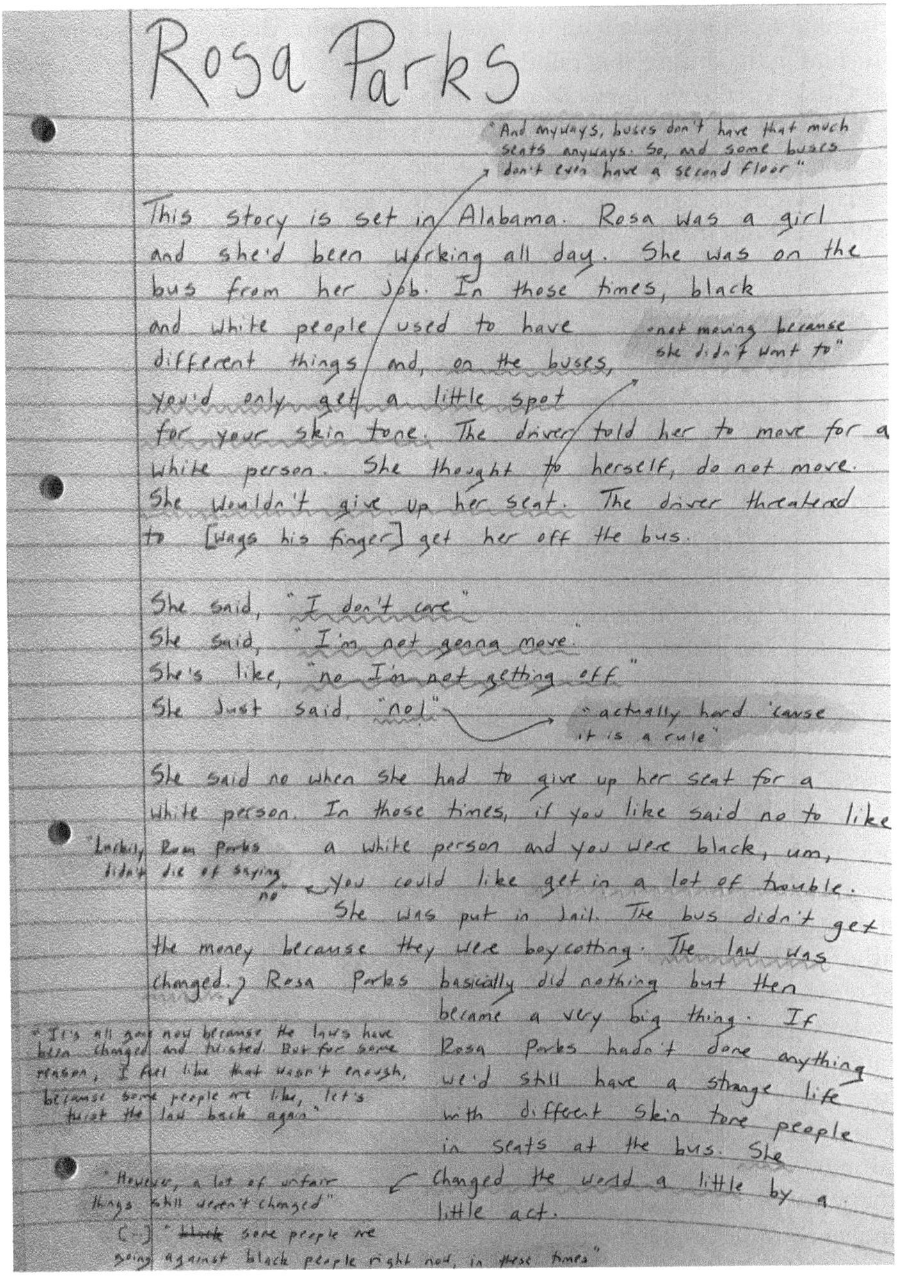

Rosa Parks

"And anyways, buses don't have that much seats anyways. So, and some buses don't even have a second floor"

This story is set in Alabama. Rosa was a girl and she'd been working all day. She was on the bus from her job. In those times, black and white people used to have different things and, on the buses, you'd only get a little spot for your skin tone. The driver told her to move for a white person. She thought to herself, do not move. She wouldn't give up her seat. The driver threatened to [wags his finger] get her off the bus.

"not moving because she didn't want to"

She said, "I don't care"
She said, "I'm not gonna move."
She's like, "no I'm not getting off"
She just said, "no!"

"actually hard 'cause it is a rule"

She said no when she had to give up her seat for a white person. In those times, if you like said no to like a white person and you were black, um, you could like get in a lot of trouble. She was put in Jail. The bus didn't get the money because they were boycotting. The law was changed. Rosa Parks basically did nothing but then became a very big thing. If Rosa Parks hadn't done anything we'd still have a strange life with different skin tone people in seats at the bus. She changed the world a little by a little act.

"Luckily, Rosa Parks didn't die of saying no"

"It's all good now because the laws have been changed and twisted. But for some reason, I feel like that wasn't enough, because some people are like, let's twist the law back again"

"However, a lot of unfair things still weren't changed"

[...] "~~black~~ some people are going against black people right now, in these times" details from Parks' life. For example, when I asked Darcy why she had chosen this biography for the group to read, part of her response included the following: 'So, um, she was on the bus from her job, and somebody told her to move.' Similarly, when I asked Harry if there were any biographies he had especially enjoyed, he mentioned Parks' biography, recalling that 'the driver threatened to, [wags his finger] get her off the bus!'

For visual clarity, I decided to write the main text using participants' narrative retellings, identifying which parts of the narrative were repeated most often by participants and arranging these first. I then annotated the main text using the responses that had more of an analytic quality. Therefore, responses that felt more analytic were intertwined within the narrative retellings. Parks' 'no' was particularly salient, with participants' using a range of expressions to describe this moment.

If the analytic response connected to a particular point in the narrative, I highlighted this connection through shading. For example, as shown in the bottom right of Figure 21.2, I highlighted participants' description of Parks' 'no' and Sofia's reflection, 'it is actually hard [to say no] 'cause it is a rule'.

Annotating the main text with participants' analytic comments helped me to become familiar with the complexities and contradictions within and across their responses. I could literally *see* them, side by side. Some participants' framing of Parks' refusal to give up her seat could be interpreted as them perceiving this as a very simple and straightforward act. For example, Olivia said, 'she just went on a bus and then she just said no!' In this instance, the adverb, 'just', might imply a certain effortlessness, while Sofia said that Parks did not move because 'she didn't want to'. However, some reflected that saying no is not always easy, with Sofia noting that this is especially true when it is in opposition to an established rule. Participants also praised Parks' bravery, expressing that what she did was difficult and must have required a lot of courage. In short, participants framed Parks' refusal as something she did easily, while concomitantly problematising this apparent simplicity.

Similarly, participants described the extraordinary effect that Parks' 'no' had, with Sofia and Olivia both noting that it 'changed the world'. Yet, they also questioned the extent to which the world was changed by Parks' 'simple' actions.

Figure 21.2: Parks' biography rewritten (excerpt)

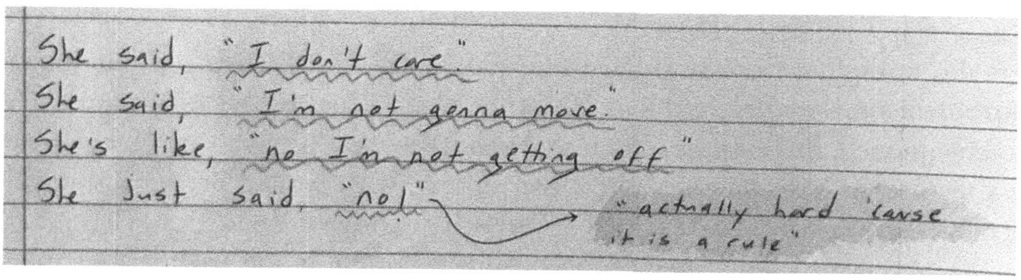

In a beautiful moment of reflection featured in the bottom left of Figure 21.1, Harry pondered the effects of changing segregation laws:

> But for some reason, I feel like that wasn't enough, because some people are like, let's twist the law back again. Let's regain the power of having a strange place to live. And it is already strange and that's wonderful, but some strange can be very awful to have. For example, that's a very mean and selfish one for your skin tone and the other skin tones who will be very sad.

Harry questioned the long-term effects that Parks' actions had, clearly stating that the subsequent change of law was not enough – 'some people are like, let's twist the law back again'. Neither Harry nor the other participants seemed altogether convinced by the simple, straightforward narrative presented. Although, at times, their responses did echo and even champion the overall narrative framing of the texts – that Parks and others achieved extraordinary things with ease and simplicity, they also problematised this narrative. They seemed to question how simple it is to do extraordinary things, and how extraordinary extraordinary things are, anyway.

The process of unravelling participants' responses by creating this new biography was incredibly helpful for my subsequent phases of analysis. Seeing these paradoxes on paper was integral to my analytic process, alerting me to the fact that the patterns of response I was hoping to capture through theme development were not necessarily going to be straightforward or tension-free. Becoming familiar with these paradoxes in particular – participants' echoing of the texts' simplistic narrative framing and their simultaneous problematising of it – ultimately inspired the evolution of the theme 'affirming and challenging inspirational-ness'. This theme encapsulates a repeated occurrence across the wider data set. Namely, that participants affirmed the texts and women's inspirational-ness, while concomitantly challenging it. They seemed to struggle to relate to the women's extraordinariness, perceiving the women as extraordinary yet unrelatable, and their achievements as impressive yet unattainable (at least for now). This finding resonates with the work of Myra Zarnowski (1988) who, in her study into children's engagement with biographies of contemporary women, found that the children perceived the women to be both potentially inspirational *and* overwhelming. There is very little research exploring children's engagement with biographies, and so these findings have important implications for how we frame role models and facilitate children's engagement with these role models.

It also transpired that rewriting Parks' biography with participants' words was instrumental for my development of another theme, 'permissive questioning'. As aforementioned, this exercise illuminated that there were parts of Parks' biographies that seemed to have more salience than others, with certain events and details recounted repeatedly. The example given – repetition of Parks' 'no' – led me to wonder about the significance of participants' focusing on this particular element and the manner in which they engaged with Parks' refusal. Around the same time, I was reading Clémentine Beauvais' (2020) writing on children's biography. Beauvais draws on the work of Roland Barthes to describe how 'the biographeme' is frequently employed by writers of children's biography. She explains that biographemes are small details that decant some of the essence of an individual's life, where certain moments are 'intensified, compressed into a small space, packed with symbolic meaning and narrative force' (Beauvais, 2020, p 68). In Parks' biographies, her refusal to give up her seat for a White person is certainly intensified, compressed, and packed with symbolic meaning. For example, in *Good Night Stories for Rebel Girls*, details of the event are squashed into one short paragraph, and the words 'Rosa said no' are afforded prominence on a separate line (Favilli and Cavallo, 2016, p 166). As this single act of resistance is elevated through the biographeme, details of Parks' other, extensive engagements with activism are eschewed. Given its accentuation in the texts and its prominence in dominant narratives of Parks and Black activism more broadly, it is thus unsurprising that participants tended to focus on this part.

However, rather than noting the salience of Parks' 'no' across participants' responses and moving on, I stayed with it. As participants had not simply recounted Parks' refusal – they had also challenged and problematised it – I wondered what else might be happening. Further reflection and engagement with the data led me to realise that there was a relationship between participants' problematisation of this biographeme (and, as I later discovered, with other devices often utilised in biographical texts) and their expectations of non-fiction. Participants had stringent expectations of what non-fiction should be, which included the qualities of being realistic, detailed, and factually accurate. Therefore, when they encountered elements that they perceived to be in defiance of these expectations – such as the biographeme's highly abridged version of Parks' refusal – they problematised or questioned it. Crucially, they started doing this when I gently intimated that non-fiction could be questioned, which they may have interpreted as my granting permission, hence the theme name, 'permissive questioning'.

Discussing this theme in my thesis, I suggest that participants' expectations that non-fiction should be detailed, realistic, and factually accurate inhibited them from being fully persuaded by the texts' simple yet extraordinary narrative, intended to be inspirational. This conclusion is a long way from my initial observation that participants' responses to Parks' biographies tended to include narrative retellings, and that Parks' 'no' featured heavily among these. This shows that taking the time to familiarise myself with the data, in a way that worked for me, was fundamental to the evolution of my analytical thinking, development of my themes, and subsequent reporting of my findings.

Familiarisation: it's just the beginning

Researchers must take care not to get waylaid in familiarisation, nor allow themselves to hold onto pieces of data that are peripheral to their area of focus or irrelevant to the research questions. 'Holding on' to work created during this phase can result in a premature narrowing of focus. For this reason, Gareth Terry and Nikki Hayfield suggest researchers keep their familiarisation notes removed from the data, perhaps in a separate notebook (Terry and Hayfield, 2020, p 435). Indeed, knowing when to move on from this initial phase can be difficult. In my experience, struggling with the second phase of analysis (coding) was indicative that I needed to revisit the first (familiarisation). It might also be helpful to remind yourself that familiarisation will inevitably continue throughout the process, though in a less concentrated form. That said, letting go is not always easy. Although during later phases of analysis I repeatedly asked myself whether I was paying undue attention to aspects I had initially found intriguing, letting go of certain parts often felt unnerving. What if I was discarding something crucial? What if dispensing with this element would discredit my entire analysis?

These are the types of questions I grappled with as I thought about one of the analytic responses I had noted in my participant-inspired, rewritten biography: 'Luckily, Rosa Parks didn't die of saying no.' In fact, I became utterly fixated on this response, ruminating on the horrifying reality that people of colour continue to be killed for standing up for their rights, and the equally horrifying corollary that those who are not killed are considered the 'lucky' ones. I felt this data item demanded deep discussion, but as I progressed through coding and theme development, I could not find a place for it. I eventually came to realise that my engagement with this utterance and its implications was not bound by the parameters of the research project. Temporarily letting go of the data item in the context of this study (coding it as 'miscellaneous') did not mean letting go of it altogether. I have continued to reflect on it, and I have had discussions with fellow educators about how we can problematise and facilitate discussions on sentiments such as the one expressed in this data item in our classrooms.

Therefore, while I advocate embracing a creative sensibility during familiarisation – that is, experimenting to find methods that work for you – remaining mindful that this is the first phase of many is crucial. It is also important to note that the amount of familiarisation required by researchers depends on a range of factors including the research questions, the particularities of the study, and the size of the data set. It may be neither necessary nor feasible for the researcher to become intimately familiar with all the data that has been gathered. Choices need to be made. I certainly did not rewrite *all* the biographies with which participants engaged. Instead, I focused my attention on those that were most prominent within the data set. I learnt that getting to know the relevant parts of my data through imagining, wondering, and playing was neither futile nor frivolous: it was indispensable to my analytic process.

Note

1 Pseudonyms that were agreed upon with participants are used throughout.

References

Bazeley, P. (2020) *Qualitative Data Analysis: Practical Strategies* (2nd edn), London: SAGE Publications.

Beauvais, C. (2020) 'Bright pasts, brighter futures: biographies for children in the early twenty-first century', in N. op de Beeck (ed), *Literary Cultures and Twenty-First-Century Childhoods: Literary Cultures and Childhoods*, Cham: Palgrave Macmillan, pp 57–79.

Braun, V. and Clarke, V. (2021) *Thematic Analysis: A Practical Guide*, London: SAGE Publications.

Byrne, D. (2022) 'A worked example of Braun and Clarke's approach to reflexive thematic analysis', *Quality and Quantity*, 56(3): 1391–12.

Cabraal, A. (2020) 'Coding and analysis tips for qualitative researchers', *Anuja Cabraal*, [blog] 1 May. Available from: https://anujacabraal.com/2020/05/01/coding-and-analysis-tips-for-qualitative-researchers/

Cohen, L., Manion, L., and Morrison, K. (2018) *Research Methods in Education* (8th edn), New York: Routledge.

Couceiro, L. (2020) 'Disorientation and new directions: developing the reader response Toolkit', in H. Kara and S.-M. Khoo (eds) *Researching in the Age of COVID-19: Volume 1: Response and Reassessment*, Bristol: Policy Press, pp 30–9.

Culshaw, S. (2019) 'An exploration of what it means to be struggling as a secondary teacher in England', PhD thesis, University of Hertfordshire.

Favilli, E. and Cavallo, F. (2016) *Good Night Stories for Rebel Girls: 100 Tales of Extraordinary Women*, Venice, CA: Timbuktu Labs.

Gee, J.P. (2011a) *An Introduction to Discourse Analysis: Theory and Method*, New York: Routledge.

Gee, J.P. (2011b) *How to do Discourse Analysis: A Toolkit*, New York: Routledge.

Harding, J. (2018) *Qualitative Data Analysis from Start to Finish* (2nd edn), London: SAGE Publications.

Harrison, V. (2020) *Little Leaders: Visionary Women around the World*, London: Penguin Books.

Jewitt, C. (ed) (2014) *The Routledge Handbook of Multimodal Analysis* (2nd edn), London: Routledge.

Kara, H., Lemon, N., Mannay, D., and McPherson, M. (2021) *Creative Research Methods in Education: Principles and Practices*, Bristol: Policy Press.

Leavy, P. (2020) *Method Meets Art: Art-Based Research Practice* (3rd edn), New York: Guilford Publications.

Pankhurst, K. (2016) *Fantastically Great Women Who Changed the World*, London: Bloomsbury.

Saldaña, J. (2021) *The Coding Manual for Qualitative Researchers* (4th edn), London: SAGE Publications.

Silver, C. and Lewins, A. (2014) 'Managing processes and interpretations by writing', in C. Silver and A. Lewins (eds), *Using Software in Qualitative Research: A Step-by-Step Guide*, London: SAGE Publications, pp 229–55.

Strauss, A.L. (1987) *Qualitative Analysis for Social Scientists*, Cambridge: Cambridge University Press.

Terry, G. and Hayfield, N. (2020) 'Reflexive thematic analysis', in M. Ward and S. Delamont (eds), *Handbook of Qualitative Research in Education* (2nd edn), Cheltenham: Edward Elgar, pp 430–41.

Thomson, P. (2020) 'Playing about with data', *Patter*, [blog] 24 February. Available from: https://patthomson.net/2020/02/24/playing-about-with-data/

Toliver, S.R. (2022) *Recovering Black Storytelling in Qualitative Research: Endarkened Storywork*, Oxford: Routledge.

Varga-Atkins, T. (2018) 'Designing curricula to develop digitally capable professionals in engineering and management: the case in two UK universities', PhD thesis, Lancaster University.

Woodward, S. (2020) *Material Methods: Researching and Thinking with Things*, London: SAGE Publications.

Zarnowski, M. (1988) 'Learning about contemporary women: sharing biographies with children', *The Social Studies*, 79(2): 61–3.

22

How to be creative when creativity is policed

Shehr Bano Zaidi

Introduction

In this chapter, I chart my journey of creativity spanning the entire length of my doctoral programme. It details how my insistence on being creative resulted in the generation of a discursive model (D/doxa) and a substantive theory (ENGURD) by introducing some changes to the grounded theory method (GTM), my research method. I made a total of four changes to the original model: three creative and one practical.

'Creativity' was (and still is) not a word often used in my academic department – located in a university in the capital of Pakistan. We generally use the word 'new'. A typical question during a dissertation proposal defence is 'What is new?' or 'What new elements are you adding to the existing body of knowledge?' These types of questions are quite easy to answer as every new rendering is 'new' in some way – mostly it is new data within an old and tested theoretical framework. The word 'creative' is generally reserved for creative artists. 'Adaptation' is, however, a word used more frequently, but usually when two theoretical frameworks are drawn together in one study – in the sense of a practical rather than creative adaptation. In the case of adaptation as we use it, candidates are expected to spell out which aspects from their selected frameworks they will use, which they will discard, and the reasons for doing so. Predictably, when I announced my intention to be creative, alarm bells rang and an effort was made to discourage me with comments like 'PhD dissertations are not supposed to be creative because creative endeavours can fail' and 'one can experiment (another word for being creative) after one has achieved the doctorate'. When I insisted and argued that the GTM allowed creativity, I was told in no uncertain terms that it would be 'guided' and 'limited' creativity, and that I should not insist on 'forcing' creativity if there was not any need to do so.

As I planned to write this chapter, I went through some respected theorists of creativity (Lubart, 1998; Kara, 2020; Tow, 2022) to test my ideas – a Global South scholar exploring the work of Global North scholars. This East–West divide as well as convergence is an idea that I take up in several ways in this chapter. My idea of creativity is similar to the somewhat broad postulation by Kara (2020, p 56), who notes that 'thinking creatively helps us to look at the world in different ways'. I also agree with Strauss and Corbin (in Kara, 2020, p 57), who conceive of creativity as an activity that tries to overcome 'normative words and statements, such as "always" or "never"'. Creative activities provide us with alternatives that may or may not be challenged in future. For Tow (2022, p 10), creativity involves both 'divergent thinking and convergent thinking'. When doing creativity within the confines of an academic degree programme, divergence allows us choices whereas convergence helps us with the 'narrowing down' to the most feasible option.

Now that I have completed my doctoral studies, I write with the benefit of hindsight. One thing that was in my favour was that according to the Pakistani university rules, a PhD dissertation can only get a pass if two examiners from technologically (and academically) advanced countries give it a pass. There are also three local examiners. My interactions (via email) with the notable Global North scholars of my field gave me confidence that was hard to miss for the examiners during the public defence. Moreover, my international examiners (one from the US and one from Canada) made positive comments about my dissertation, especially with regard to it breaking new ground in a doctoral work, which proved to be extremely useful.

Writing on creativity also needs to account for the fact that I have engaged with Global North scholarship all my life while living in an Eastern society that has its own ethos, which is quite different from that of the West. Lubart (1998, p 347) encapsulates the difference in these words: 'the Western definition of creativity as a product-oriented, originality-based phenomenon can be compared with an Eastern view of creativity as a phenomenon of expressing an inner truth in a new way or of self-growth'. As Pakistan follows the British/American educational system, it is normal for Pakistanis to interact with the Western scholars. This is advantageous because Pakistani scholars try to stay abreast in their fields as per Western standards, but disadvantageous because they have to undertake academic study by straddling Western and Eastern sensibilities and approaches, which, at times, becomes difficult. Local academic authorities may or may not be comfortable with the views that such scholars accumulate through their interactions with the Western academics. The reasons for this discomfort may range from having a different ethos to not wanting to accommodate new ways of thinking. Needless to say, many scholars are faced with tightrope walking in such an environment.

My broad field is sociolinguistics. A Hudsonian way of defining sociolinguistics is to separate it from sociology of language with the disclaimer that there is some overlap between the two (Hudson, 1996). He defines sociolinguistics as a 'study of language in relation to society' and sociology of language as a 'study of society in relation to language' (Hudson, 1996, p 4). Hudson's somewhat narrow definitions are considered as the only version by many in departments like mine in the Global South. The definition of sociolinguistics that such scholars use is generally variationist and quantitative in nature. Even qualitative/interactional sociolinguistics is mostly underpinned by positivism and leaves little scope for social constructionism and the like. It also leaves no room for creativity as it is based on a logico-deductive model of research that starts with a well-defined hypothesis (Lawson, 2015) and is grounded in a predetermined research methodology. In Western academia, not only has the gap between sociology of language and sociolinguistics been mostly bridged in the sense that both foci are accepted as part of sociolinguistics (Blommaert, 2018), but the presence of purely qualitative sociolinguistic studies is not frowned upon.

When I presented my synopsis, and at my final dissertation defence, I was questioned about not following the set pattern of variationist/quantitative studies. I was also challenged about my aim to theorise, as sociolinguistic studies do not normally have theorising as their explicit aim. Although theorists like Coupland (1998, p 116) mention the need to theorise in sociolinguistics, they do not describe how to go about it. Coupland calls theorising 'a restructuring' of the representative value of the discipline without giving any instructions as to the choice of a theorising method. It was inconceivable for a PhD scholar to be so 'bold' as to go against the grain.

Background and aims of the project

I studied doxa formation in selected opinion articles culled from Pakistani English and Urdu newspapers. An early definition of doxa is presented here. Doxai are opinions that achieve a certain discursive status by virtue of being repeated in texts. The aim of my work in which I developed the D/doxa model of grounded theory, was to understand afresh the relationship between Urdu and English as used in Pakistan – where I was born and live. There are close to 74 active languages spoken in Pakistan, so linguistic controversies are not new to the society.[1] Urdu is the national language of Pakistan and is used as a lingua franca to connect linguistically the length and breadth of the country. The reason that Urdu was selected by the country's founders to be the national language of a multilingual country was that it would not assume a hegemonic status (which it somewhat has) as it was not the mother tongue of the majority of Pakistanis and was already being used as a lingua franca in Imperial India (Khan, 2006).

Notwithstanding Urdu assuming a hegemonic status and sidelining minority languages (Rahman, 2002), Urdu itself is in danger from the English language. The English language in Pakistan has had a troubled past, and has an uneasy present. Pakistan, as part of the Indian subcontinent, was colonised by the British, and English is still viewed as the ex-coloniser's legacy. This is most acutely felt and articulated in the country's 1973 constitution,[2] which gave the government 15 years from 1973 to replace English with Urdu. The current situation in the country is that despite the lapse of over 50 years, English is still used as an official language (Jawad et al, 2021, p 269). The elite/educated are not only fluent in English, but also make sure that neither the language nor its literature die away. Generations of upper- and middle-class citizens have grown up reading English literature, listening to English-language songs, and watching English-language movies. This, coupled with the ordinary masses' inability to afford to learn or use the language, has seemed to create an undesirable stratified society where English is used as a status symbol and a divisive tool.

Paradoxically, the education policy of Pakistan (Ministry of Federal Education & Professional Training, 2018, p 4) has an unclear stance on English. On one hand, it describes English as having a 'glorified status in society' and as being used 'as a de facto official language' (albeit in clear violation of the constitution); on the other hand, it requires standard teaching of English in all educational institutions, underscoring the importance in which English is held by the makers of the document. This 'dual approach towards English' (Zaidi, 2020, p 3) in Pakistan necessitated a fresh look at the relationship between Urdu and English in order to understand this. It also meant that none of the extant theories could sufficiently explain the situation. Most of these theories, formulated in the West, could, at best, explain some parts of the unique linguistic situation in Pakistan, but not the whole. I needed a fresh theoretical rendering, for which I needed a method that would allow me to understand the situation.

Data selection

Since my objective was to understand afresh the relationship between Urdu and English given the confusing approach as mentioned, I decided to select opinion articles from Pakistani newspapers written in both Urdu and English. As extant material, this meant that the articles would be 'more detailed' and 'more representative' (Tight, 2019, p 14) than what I could have collected given my time constraints and resources. Besides, they provided 'stability' and 'coverage' (Bowen, 2006, p 31). As extant documents cannot be modified by the researcher's presence (Merriam in Bowen, 2006), this suited my purpose of making 'repeated reviews' for a grounded theory study.

Opinion articles in Pakistani newspapers discuss different issues at length. I began focusing on frequently mentioned topics (the army, corruption, education, the US–Pakistani relationship, and women) within high-circulation English (*Dawn*, *News International*, *The Nation*, and *Express Tribune*) and Urdu (*Roznama Jang*, *Roznama Express*, *Roznama Nawa-e-Waqt*, and *Roznama Dunya*) newspapers. I finally settled on 300 articles (approximately 380,000 words) from 30 writers (ten articles per writer) written between June 2014 to August 2018. I looked at the formation of opinions (doxa) and found out that every writer in all of their ten articles would use the Pakistan army as the main reference for discussing any one, two, or sometimes three of the following topics: women, the US, corruption, or education (all contextualised in Pakistan).

These opinions, by virtue of the frequency of their repetition, were not just subjective articulations. They seemed to somewhat fall under the realm of doxa, which could either be defined as 'false' and 'mindless repetition' (Barthes, as cited in Bengston, 2017, p 83) or as a source of effective communication and a positive linguistic force (Amossy, 2002, p 374). My job was not just to see how doxa were formed and what they meant for the relationship of Urdu and English as used in the selected opinion articles, but whether I agreed with Barthes or Amossy or neither or both to a certain extent. Thinking creatively about doxa not only gave me a new perspective but also led me to formulating a new discursive model (discussed later in this chapter). A technique that I used in my creative process was to write about all im/possible options and discuss them with (willing) colleagues. I knew that the acceptability of my idea was the lynchpin on which its existence depended.

GTM and its suitability for my project

The seminal text (*The Discovery of Grounded Theory Strategies for Qualitative Research*) by GTM founders Glaser and Strauss (1967) is 'considered to be a bible for grounded theory practitioners no matter which variant they are following' (Zaidi, 2022, p 2). It lays down some steps that are considered to be an integral part of a grounded theory study. Paraphrasing Kathy Charmaz (2006, pp 5–11), they are:

- initial coding
- secondary coding
- raising codes to categories
- theoretical sampling
- theoretical saturation
- writing the draft

Each of these steps is marked by new data collection, if need be, and memo writing – an exercise that allows a researcher to play with different possibilities of the emerging theory. A continuous exercise of constant comparison is performed, which means that codes are continuously compared with each other so that some codes are assigned to one or more categories which not only represent the data but also have some conceptual meaning. The act of conceptualisation is important in grounded theory studies. Codes that have the potential to be generalised are selected at each stage and an explanation is provided. This all leads to the generation of one or more core categories. A core category is supposed to represent the entire conceptual value of the data. This category then undergoes another process, which is called theoretical sampling. A researcher obtains more data and embeds it in the past literature in order for a concept to relate to both old and new field(s). This stage continues (in the form of constant comparison with the existing literature) until theoretical saturation is achieved. At this stage, the researcher is in a position to finally enunciate the theoretical formulation that is called the grounded theory of that specific data set. This theory may be applicable (generalisable) to similar contexts. The generalisation process in a grounded theory is different from logico-deductive studies, as the latter generalise to a large number of cases whereas grounded theory aims at limited generalizability through its conceptualisation process (Charmaz, 2006).

GTM is an inductive approach, which means that it starts with data (Patton, 2002, p 306) and ends at generating a theory. That is why it is referred to as grounded in data. The opposite to inductive is deductive, which aims at testing a hypothesis or theory. The inductive nature of grounded theory means that it relies to a substantial degree on the creative ability of the researcher (Suddaby, 2006). No doubt, it provides a researcher with a clear pathway to conceptualise data for theory generation, but a plausible theory is only possible when the researcher is ready to entertain many different options – even the option to creatively adapt the method to suit their own purpose. This, of course, can be done only if a researcher is allowed to work creatively.

Glaser and Strauss (1967, p 4), on the subject of extant theory, seem to speak directly to my project's needs when they argue that '[g]rounded theory can help forestall the opportunistic use of theories that have a dubious fit and working capacity'. I had the data in front of me but no theoretical explanation for it. Grounded theory's dictum that the 'development of theory starts with data and not an existing theory' (Zaidi, 2022, p 2) was all I required to take the decision to adopt this method. As I delved in grounded theory literature, I realised that the GTM had a number of variants, and most studies made it clear at the beginning which variant they were following. I chose Charmaz's (2006, p 9) variant – called constructivist grounded theory – as she allows researchers to 'adopt and adapt in order to conduct diverse studies', and her philosophical underpinnings of constructivism, especially her exhortations on focusing on the language, aligned with my approach and needs. However, after devising my D/doxa model of grounded theory (as explained later in this chapter), I am of the opinion that this particular model need not be attached to any variant and can belong to grounded theory methodology as a whole.

D/doxa model of grounded theory

Some preliminaries

During the time I was writing my dissertation, I made full use of regular progress seminars and other academic meetings and introduced my work to my colleagues. This ensured that my work did not remain a mystery, and that people formed opinions about it, which was good as it meant that there were supporters alongside the naysayers. I still remember the words of a colleague who said that it was difficult to conceive of sociolinguistics without a quantitative analysis of the data. It was painful and demoralising to listen to such comments, to say the least. Another colleague of mine, who was undertaking a PhD in English literature, had recently come back after spending six months in a university in the USA. He would always encourage me by reminding me that a creative endeavour would work as long as I was able to defend it.

The rhetoric of writing

Charmaz (2006, p 173) advises grounded theorists to rely on those writing 'strategies [that] expedite[s] your work'. She exhorts theorists to 'broaden the range of possibilities' (p 172) when it comes to writing. My area was sociolinguistics and I had decided to study doxai (opinions) as they were formed in the data. The concept of doxa, like most sociolinguistic concepts, is borrowed from sociology, as is GTM (the originators were sociologists). I started defining every term in a manner that would make the link of sociolinguistics with sociology clear, so as to show the examiners that GTM should not be limited to sociology. The strategy worked and no one objected to the use of GTM. Another rhetorical strategy that I employed was to replace a word/term with something that would sound more acceptable to a sociolinguistic study. For example, instead of using sensitising concepts (a regular grounded theory term used for entering the research site), I used sensitising perspectives. This had a twofold utility as it sounded closer to sociolinguistics terminology and the initial insights that I applied to the data were broader than concepts. Such rhetorical strategies helped me build my argument more efficiently.

D/doxa model

The D/doxa model of grounded theory is based on three creative adaptations and one practical adaptation. Using phrases/sentences instead of the usual one or two codes, using in vivo codes throughout the work instead of only during the first stage, devising a general model and then using it to generate my theory, are the creative adaptations. Adding one stage of coding to manage a large amount of data is the practical adaptation.

The unit of analysis in my work was doxa, and as I tried to conceptualise it, I realised that, given the amount and nature of my data, I could make certain generalisations about it; devising a general model was the best option in my mind. So, instead of simply generating a substantive theory from a core category, I used it to first devise a model (D/doxa). I then used this model to generate a substantive theory (ENGURD). A clarification is needed here. It is not uncommon for grounded theory studies to use conceptual models to better explain the theoretical rendering (Soulliere et al, 2001). My model, however, is not conceptual in the sense that a conceptual model only explains the associations of the concepts used in a specific substantive theory. My model adds a safety layer of generalizability to my theoretical rendering. I believe that if a general model is generated in a grounded theory study, it strengthens the argument; it adds a multi/interdisciplinary dimension to the work.

To briefly refer to the content in my data, as the codes are based on the content, I found that there were four main discourses that were mentioned repeatedly: women, corruption, relations with the US, and education. All these four discourses were made with reference to one discourse, the Pakistan Army. The army has ruled (directly or indirectly) over Pakistan for a long time and is deemed to have a controversial role in this respect. The important point to note here is that these articles were written long after formal military rules were over (the last military ruler exited in 2018), and yet there is a strong tendency for

Figure 22.1: D/doxa

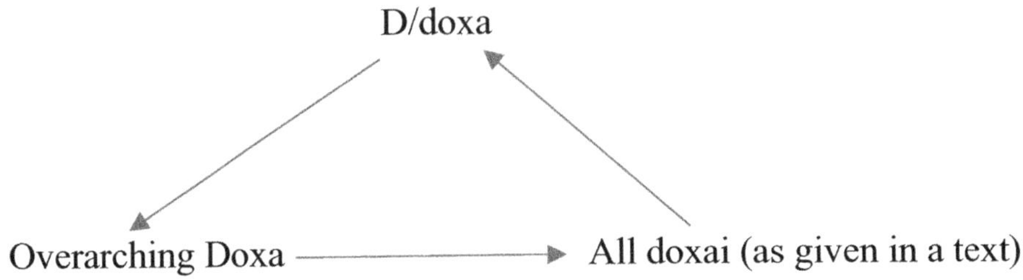

Source: Zaidi (2020, p 200).

writers to make references to the military, which can partially be explained as the country being still under military influence in some ways.

The following list is a detailed account of how I creatively modified GTM to suit my research needs:

1. GTM codes are usually short (two to four words). I used explanatory phrases/sentences as they helped me explain clearly what was happening in the data. They well served the purpose of abstracting, however, which is the basic requirement of GTM. A list of phrases/sentences that I used to code the discourse on women is given as an example. The 'E' and 'U' in brackets stand for Pakistani English newspapers and Pakistani Urdu respectively.
   'Fatima Jinnah (a female democrat, also the sister of the country's founder) more revered dead than alive challenging Gen. Ayub' (E)
   'Nineteen eighties (era of military rule in Pakistan) a fraught time to be a woman, woman lawyer, woman politician' (E)
   'No justice for women even under Gen. Zia' (E)
   'Gen. Zia era public hanging of rapists effective deterrence' (U)
   'Ex-army chief dances whereas Pashtoon female singers shot dead' (U)
   'Gen. Zia arrogantly crushed women rights' (U) (Zaidi, 2020, p 175).
2. Though in vivo codes are defined by Charmaz (2006, p 56) as 'symbolic markers of participants' speech and meanings', I redefined them as the language used in data as I did not have active participants as such – I was using extant literature. The focus on language (Grant, 2018) helped me spot intertextual references between the two language articles. I could trace similarities/differences with the help of the references used. In my case (evident in the example given), mention of the army acts as the overarching reference that holds all the discourses together.

Point by point the model can be explained as follows:

- D/doxa cuts across linguistic boundaries, as in the case of my data, which consist of two languages (English and Urdu) being used side by side in Pakistan. The

codes in point 1 of the list just given, refer to women's issues and they all use the military as a reference point.
- D/doxa stands for a single value (discursive) representation of an entire discourse/text, which means that the reference point for all discourses is the same but that opinions may vary. D/doxa is a product of interdoxtuality – referring to the structural–content interdependence that a text possesses. My meaning is conditional on the way the content is structured (one main discourse with minor discourses subscribing to it for their meaning).
- One doxa written with a capital 'D' subsumes all doxai under it. Doxai stem from one Doxa and become interconnected, deriving meaning from each other. To cite the example of my own PhD dissertation, there was one overarching discourse that provided the doxastic structure to all the minor doxai. Doxa with a small 'd' represents the different doxai present in a text.

The grounded theory that I generated using the model D/doxa is called ENGURD. I started my dissertation with the intention of reimagining the relationship between Urdu and English as they have a troubled relationship. English is considered as the ex-coloniser's legacy, and it is also a source of class division in society as it ensures a high status and better employment opportunities. Though the new entity ENGURD negates the supremacy of English as used in the given data, it still has the English part first and Urdu after, as I am an English scholar myself. ENGURD does not mean that every writer had the same views, but that they all would refer to one main discourse (the army's) whether they wrote on women, corruption, US–Pakistani relations, or education. This bound them in a doxastic relationship.

If my work is explained through the paradigm of ontology (what is there in the world?) and epistemology (what is acceptable knowledge?), there are two entities that I generate in my dissertation – ENGURD and D/doxa – which means that each has its own onto-epistemological value. I intertwine onto-epistemology (Barad, 2007) as I agree with Barad that they cannot be separated. Starting with ENGURD, in the beginning they were two different languages with (apparently) different discourses, Urdu and English, but, at the end, the two entities merged into one in the sense that they formed one unified discourse in the context of the 300 newspaper articles. The onto-epistemology of ENGURD is inseparable as the existence of the entity was dependent on the examiners' approval of it as acceptable knowledge. The reality (my substantive theory) constructed cannot be separated from what is deemed as acceptable by the committee. As for D/doxa, a GTM already existed when I started. A new model was devised that has an existence, but its existence was dependent on the defence committee's approval, as it came into being as a result of an academic exercise.

Besides the tangible results of my creative endeavour, I feel that this PhD dissertation has made me more confident in my abilities to re/imagine possibilities, especially in a developing country like mine. Moreover, humbly disagreeing with those who thought that a PhD was not the place to be creative, I feel that my dissertation demonstrates that it is the right place and time to start one's creative journey. One is fresh and thus mentally mouldable – an essential quality when being creative.

Some possible uses of D/doxa

D/doxa was conceived in a language study, but the make-up of the model is such that it can be used to study the formation of discourses in a single language or languages that do not exist in a dominant–dominated equation. Future usages of D/doxa need not be from sociolinguistics and can pertain to any field and across any time span. A possible weakness may be that it will need to be adapted for a study that has more than one overarching Doxa. My model can also be used in a non-grounded theory study, in which case the outcome may not be a substantive theory. The hierarchical make-up of D/doxa can be used to explain and even to deconstruct a hegemonic hierarchy, even if it is not purely discursive.

Conclusion

I had two objectives to meet in this chapter: to map my creative research journey and present my model D/doxa. As for creativity in academia, it is dependent on the environment and the perseverance of the scholar. My interactions with Global North scholars gave credence to my work. The hegemony of English is dismantled but the superior values of my Western scholars stay with me, and I acknowledge them in my thesis. Both D/doxa and ENGURD are interventions that encourage others to take a similar path.

Notes

1 Half of the entire country (erstwhile East Pakistan and present-day Bangladesh) became disillusioned because Urdu was imposed onto them and their local language, Bengali, was not accorded an official status (Jabeen et al, 2010). Linguistic controversy was one of the reasons that Bangladesh separated from Pakistan.

2 https://na.gov.pk/uploads/documents/1549886415_632.pdf.

References

Amossy, R. (2002) 'Introduction to the study of doxa', *Poetics Today*, 23(3): 369–94.

Barad, K. (2007) *Meeting the Universe Halfway: Quantum Physics and the Entanglement of Matter and Meaning*. Durham, NC: Duke University Press.

Bengston, E. (2017) 'The concept of doxa in the European invention of rhetoric', in N. Gutenberg and R. Fiordo (eds), *Rhetoric in Europe: Philosophical Issues*, Berlin: Frank & Timme, pp 79–88.

Blommaert, J. (2018) 'Durkheim and the internet: on sociolinguistics and the sociological imagination', *Working Papers in Urban Language and Literacies*, 204: 1–63. Available from: https://www.academia.edu/30540603/Durkheim_and_the_internet_On_sociolinguistics_and_the_sociological_imagination_

Bowen, G.A. (2006) 'Grounded theory and sensitizing concepts', *International Journal of Qualitative Methodology*, 5 (3): 12–23. Available from: https://journals.sagepub.com/doi/pdf/10.1177/160940690600500304

Charmaz, K. (2006) *Constructing Grounded Theory: A Practical Guide through Qualitative Analysis*, Los Angeles: SAGE.

Coupland, N. (1998) 'What is sociolinguistic theory?' *Journal of Sociolinguistics*, 2 (1): 110–17.

Glaser, B. and Strauss, A. (1967) *The Discovery of Grounded Theory Strategies for Qualitative Research*, New Brunswick, NJ: Aldine Transactions.

Grant, A. (2018) *Doing Excellent Social Research with Documents: Practical Examples and Guidance for Qualitative Researchers*. London: Routledge.

Hudson, R.A. (1996) *Sociolinguistics* (2nd edn), Cambridge: Cambridge University Press.

Jabeen, M., Chandio, A.A., and Qasim, Z. (2010) 'Language controversy: impacts on national politics and secession of East Pakistan', *South Asian Studies: A Research Journal of South Asian Studies*, 25(1): 99–124.

Jawad, S., Ali, B., Assad, M., and Sohail, M. (2021) 'Urdu as an official language: a constitutional mandate compliance, challenges prospective', *Review of Applied Management and Social Sciences*, 4(1): 261–70. Available from: https://papers.ssrn.com/sol3/papers.cfm?abstract_id=3833757

Kara, H. (2020) *Creative Research Methods in the Social Sciences: A Practical Guide* (2nd edn), Bristol: Policy Press.

Khan, A.J. (2006) 'Mesopotamian roots and language classification', in A.J. Khan, *Urdu/Hindi: An Artificial Divide*, New York: Algora Publishing, pp 13–32.

Lawson, A.E. (2015) 'Hypothetico-deductive method', in R. Gunstone (ed), *Encyclopedia of Science Education*, New York: Springer, pp 471–2.

Lubart, T. (1998) 'Creativity across cultures', in R. Sternberg (ed), *Handbook of Creativity*, Cambridge: Cambridge University Press, pp 339–50.

Ministry of Federal Education & Professional Training (2018) *Pakistan – National Education Policy Framework November 2018*. Islamabad: Ministry of Federal Education & Professional Training. Available from: https://aserpakistan.org/document/2018/National_Eductaion_Policy_Framework_2018_Final.pdf

Patton, M.Q. (2002) *Qualitative Research and Evaluation Methods* (3rd edn), Los Angeles: SAGE.

Rahman, T. (2002) 'Language, power and ideology', *Economic and Political Weekly*, 37(44/45): 4556–60.

Soulliere, D., Britt, D.W., and Maines, D.R. (2001) 'Conceptual modeling as a toolbox for grounded theorists', *The Sociological Quarterly*, 42(2): 253–69.

Suddaby, R. (2006) 'What grounded theory is not?' *The Academy of Management Journal*, 49(4): 633–42.

Tight, M. (2019). *Documentary Research in the Social Sciences*, Los Angeles: SAGE.

Tow, G.T. (2022) 'Defining creativity', in J.A. Plucker (ed), *Creativity and Innovation: Theory, Research and Practice* (2nd edn), New York: Routledge, pp 5–22.

Zaidi, S.B. (2020) 'Doxastic fault lines in a linguistically fractured society: a grounded theory study', PhD thesis, National University of Modern Languages, Islamabad, Pakistan.

Zaidi, S.B. (2022) 'Situating sensitizing concepts in the constructivist-critical grounded theory method', *International Journal of Qualitative Methods*, 21: 1–6. Available from: https://journals.sagepub.com/doi/pdf/10.1177/16094069211061957

23

Visual grounded theory: a new way of seeing, knowing, and constructing theories grounded in data

Jacquie Ridge

Eat the elephant: constructing a book chapter
About to write a book chapter,
Don't know where to start.
Listening to the radio, synchronicity of discussion: chatting all things procrastination,
Honestly, humbly, confronting fear … a message to the heart.
'Eat the elephant, one bite at a time' is the solution posed,
Not sure I'd like elephant, 'tastes like chicken?' not so sure.
Analogy draws memory to the fore,
A different location, time, and space; situation same,
Constructing something new, unorthodox, challenging; out of my comfort zone:
A substantive visual grounded theory.
Driven to create, share insights grown, taking steps hardly known.
Exploring new ways of seeing and thinking, joyfully recognising new possibilities.
So here we go, ready?

Introduction

If my poem has drawn you in, welcome. The purpose of this chapter is to share insights of using visual grounded theory, a modified version of constructivist grounded theory, incorporating visual data collection and analysis (Ridge, 2020), which I used for doctoral study to investigate professional identity in adult nursing in the UK. Firstly, though an explanation for the poem, which was written as an example of the courage required when trying something new or different, and the opportunities, such as using an unorthodox methodology or being invited to write a book chapter, that present if one is persistent. I am not saying it is easy to create something new, different, or novel, as change is often uncomfortable, but how do we challenge ourselves without it?

With that said, my intention is to provide insight into my experiences of constructing a visual grounded theory. A context to the study, its purpose, and application is provided, alongside justification for decisions made methodologically. A distillation of visual grounded theory's philosophical, theoretical, and methodological origins is provided for insight and application of this little-known methodology, with worked examples of 'the interconnectivity of tools for analysis' (Ridge, 2017) from my doctoral study to aid understanding, concluding with a critical reflection of using this methodology. At this point you will have reached the end of this chapter, hopefully with a sense of enlightenment of what this creative method has to offer; with the elephant well and truly eaten.

The trigger for this investigation began while working as a nurse academic in higher education. I witnessed a disconnect among nursing students between their aspirations and the perceived reality of what it means to be an adult nurse. Repeatedly hearing from those withdrawing from their programme of studies, that this 'wasn't what nursing is …', caused me to wonder, 'so, what is it that they think nursing is?' This led me to embark on a doctoral study investigating what it means to construct and sustain a professional identity in adult nursing, aspects of which are shared here.

With ethics approval granted from the Research Ethics Committee, University of Northampton, a purposive sample of 11 undergraduate pre-registration adult nursing students were recruited. The study had two phases of data collection, the first from 2012 to 2014, in which all 11 nursing students collaborated in three photo-elicitation interviews each over 18 months (33 in total, each lasting an hour approximately), reflecting different stages of progression through their programme of studies. It became a longitudinal research study with a second phase revisiting participants, now all registered nurses, to ensure rigour of findings. With re-approval of ethics, one additional photo-elicitation interview with 9 of the original 11 collaborators was conducted between 2018 and 2019, the other two declining participation.

In phase one, participants were invited to source five images of whatever type they wanted, representing 'who do you think you are?', and revisited repeatedly during data collection. As a nurse researcher investigating nurses' identity, I wanted to 'see differently' (Mannay, 2016), creating a 'language bridge' (Collier, 1957, p 858) to elicit nurses' thoughts of crafting of their identity through imagery. I originally thought that asking my participants to bring imagery would be a novel way in which to open discussions, provide insight to collaborations, and offer a different way of knowing their personal and professional selves among a population whose voice is often unheard. I was unprepared for just how much this collaboration would change the study, with participants owning data transforming the study from data collection to data production, causing a methodological shift in emphasis. I learnt to acknowledge that these images were representations of the participants themselves, with the student nurses initially being driven by engagement with the study to owning their engagement as collaborators. As one of the participant's reflected, *'I sort of didn't just flick off the internet and get a few off. I picked on the pictures that were in my mind, head. I wanted, about how I saw a nurse …'* (Participant 1: Interview 1 Data Production) (Ridge, 2020). And so data production and analysis of imagery transcended the original methodology, that of constructive grounded theory with visual imagery *simply* incorporated, to become a visual grounded theory. With participants as image makers, creating a subjective knowing of self in a visual format, this enhanced the rigour of the research. This collaborative working resulted in a substantive theory that proposes that an authentic aspiration of what it means to be an adult nurse, combined with critical resilience of being a nurse, results in a legitimate professional identity (Ridge, 2020).

Visual grounded theory: philosophical, theoretical, and methodological origins distilled

The philosophy underpinning this study is symbolic interactionism, a theoretical perspective that offers 'both a theory of human behaviour and an approach to enquiry about human conduct and group behaviour' (Goulding, 2002, p 39). Symbolic interactionism seeks to 'construct the realities of the participants (own) lives using their symbols and life language as the cornerstone of theory generation' (Williams and Keady, 2012, p 221). It sees meaning as social products, 'formed in and through the defining activities of people as they interact' (Blumer, 1969, p 5). This is learnt through an active, internalised social process of 'self-interaction', giving rise to the interpretation of meanings that negotiates, transforms, and reconstructs reaction (Blumer, 1969, p 5), As a researcher, an understanding of the participant's social acts, interaction, and interpretation is needed to investigate how this behaviour discerns meaning for the individual (Sandstrom et al, 2001). This way of interpreting reality aligns to a theoretical perspective known as social constructivism; an interpretive process in which individuals seek to understand the world in which they live and work and which requires close involvement or collaboration with participants in their setting or context for information gathering, with shared understanding of experience and background co-constructing multiple realities (Crotty, 1998).

The method of investigation used for this study – constructivist grounded theory – derives from the need to understand 'self-interaction' in order to capture the subjective knowing and meaning of participants, in this case when constructing and sustaining their professional identity. It places emphasis on the phenomena of study, seeing 'data collection and analytical process as being influenced by how the researcher and research participant interpret the meanings and actions of a situation/event' (Anaebere *et al*, 2013, p 118). With data collection negotiated, the researcher needs to constantly check how their perspective shapes interpretation of participants' explanations and categories through reflexivity (Charmaz, 2006). This means that the researcher's voice needs not to 'transcend experience but re-envis[age] it … bring[ing] fragments of fieldwork time, context and mood together in a colloquy of the author's several selves – reflecting, witnessing, wondering, accepting all at once' (Charmaz and Mitchell, 1996, p 299). Known as memoing, it is an essential part of grounded theory and a reflexive way of working to explore thoughts arising from undertaking research;

however insignificant, succinct, or random memos may seem at the point of creation. It enlists the 'use of coding language active in its intent' (Mills *et al*, 2006, p 7), and as memoing is rich with raw data, it ensures the continued presence of the participant's voice and meaning through collaboration. To exemplify, early on in my studies I remember being asked by a supervisor, 'you are memoing, right?' and panicking as I was unsure if the copious, scribbled, and scruffy notes that I had scattered in numerous texts, journals, and research diaries, and created on NVivo, would count as 'memoing'. Seeking out methodological literature and peers with grounded theory expertise for clarity, I was relieved to find that the reflexive writing of all things related to my study was memoing: from the early preliminary memos informed by researcher reflexivity to modifiable and advanced memos, combining initial and focused coding for memo sorting and construction of theoretical categories from data analysis. To exemplify this process, see Figure 23.1 for an early or preliminary memo, in comparison to that of the modifiable moving towards advanced memoing shared in Figure 23.2.

Figure 23.1: Memo writing: skill and application

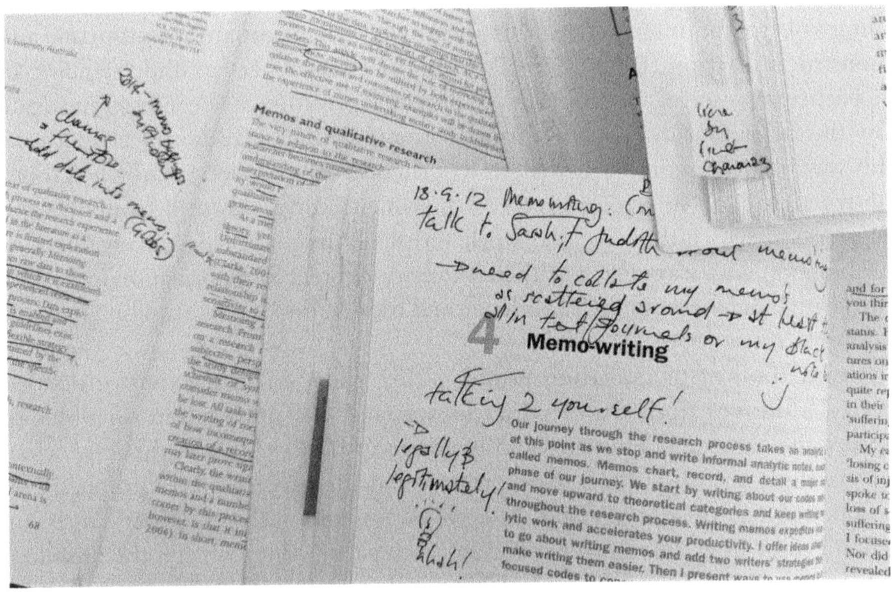

Source: Ridge (2020).

As noted, Figure 23.1 is a preliminary memo, whereas Figure 23.2, a modifiable memo, presents a montage of an image and its verbatim text shared by Participant 1 in her first interview. I then followed this up with my field notes, memoing dissonance of attributes felt by Participant 1, that triggered the need for theoretical sensitivity, another core component of grounded theory, requiring the researcher to recognise '*what is really going on*' with data to give it meaning. On revisiting

Figure 23.2: Hattie Jacques in *Carry on Nurse*

Note: Participant 1: Interview 1: Image 2 associated verbatim text and researcher reflexivity. Landmark Media/Alamy Stock Photo. Image identification: 2BC4HWC. Available from: https://alamy.com/. Licensed under CC BY-NC-ND 4.0.

Source: Ridge (2020, p 47).

this image some years later, our thoughts had remained constant; however, we shared the essential learning derived from this experience, with it contributing to our professional and research-driven development.

> Hattie Jacques, epitome of matron, at her absolute comic best, sense of humour very important to me, and I feel that she was the humour side, she was always in control, you know as that picture of the matron … sense of humour aside, when I see Hattie Jacques, in Carry On Nurse, I think 'yes'; we all sit there and say, 'oh, matron!', and its lovely, and it's funny, but at the same time she is the control, the centre, she is who everyone answers to, and I think that's quite good. That's why she's my, she's one of my pictures.
>
> Researcher reflexivity and memoing linked to Figure 23.2:
>
> Summer 2012: 'I dislike this imagery due to its unreal portrayal; conditional regard for self and others. Although of similar age to Participant 1, the point at which I undertook nurse education (late 1980s), images such as this and all they represent had been reality for me, with much of the control and none of the humour expressed in Participant 1's verbatim text. Interesting … need to link this to both Mannay's (2010) work on insider/outsider

perspectives, and consider the nature of humour, control, and historical reference influencing aspiration.'

Autumn 2018: 'These thoughts were revisited during our 4th interview; recognising the value that a difference of opinion can bring to simultaneously seeing the same image differently. I also reflected how her thoughts had made me appreciate the reality of collaboration core to this research and the methods used to "open up" data and its analysis for its multiplicity, such as in this example. This image has been reinterpreted in numerous ways informing theoretical sampling and as such acts as an aide memoire (re) informing my researcher positionality.'

(Ridge, 2020, p 47)

Visual grounded theory: constructing 'the interconnectivity of tools for analysis'

As noted, to enhance subjective meaning and its verification, this constructivist grounded theory approach has been modified, incorporating visual research methods, specifically photo-elicitation, in which 'the photo-elicitation interview seems like not simply an interview process that elicits more information but rather one that evokes a different kind of information' (Harper, 2002, p 13). Core to data collection is the collaboration between researcher and participants, co-creating meaning using imagery to 'figure out something together' (Harper, 2002, p 23). The use of visual imagery within grounded theory has been justified with the adage 'all is data' (Glaser 1998, p 8). Including visual images provided new ways of seeing, knowing, and constructing theories grounded in data (Ridge, 2020), shaping subsequent theory (Mey and Dietrich, 2017). This means that 'visual data is used to construct categories, describing properties, and generating/ constructing theoretical hypotheses, which account for the visual phenomena and visual sociological processes' (Konecki, 2011, p 133). Visual grounded theory offers a way to contextualise an image, its production, reception, and application within a sociological framework.

Mey and Dietrich (2017) have enhanced the idea of methodological fit, aligning the use of imagery to theoretical sampling and interpretation of data, key tenets of a grounded theory approach, and proposing 'we understand the image as a medium in need of explication, specifically with respect to its formal composition. We reject the transformation of an image into mere text and textual interpretation, encoded as latent forms of knowledge. Instead, reflection and coding elements are driven by "the image as such" (including the reciprocal relations between the image elements) and guided by the composition of the image' (Mey and Dietrich, 2017, p 283). Essentially, data production and analysis of imagery are used to substantiate this visual grounded theory, with participants as image makers creating a subjective knowing of self in a visual format (Ridge, 2020). This reaffirms their role as co-producers of data; subjective collaborators with creative freedom enhancing dedication to the process, alongside insight to construction and sustainability of their (professional) self, aspiration, and identity. It is this co-production that I would like to share next.

Visual grounded theory: a method shared

I am aware that the thread for this chapter is data analysis, and I reiterate that with grounded theory methods, data collection – in this case data production – informs analysis from start to finish to ensure a rigorous substantive theory. To provide clarity to this process I have detailed the nature of the imagery that my participants brought for elicitation, which informs the approach taken with my study and its analysis.

On commencing data collection, I recognised that the terminology I had used to inform participant data production to be used within the elicitation interviews was that of 'visual imagery', and I was initially asked by participants 'What do you want to see?' I had not specified photographic, which meant that a range of visual data and artefacts were brought that were a mixture of found and participant-produced materials. The majority of items brought by participants were found; that is, images that existed prior to, and outside of, research being undertaken (Rose, 2010). Other participants provided revised versions of established images to represent what they wanted known; these elements of participant production brought an authenticity and intimacy to what was shared. It was appreciating this notion of crafting identity via imagery of being 'made not taken' (Warren, 2002) that was of interest in this study; that is, imagery and its revelatory components, be these found materials, edited auteur, or participant production.

This variety of imagery meant that the 'fit' of the analytical tool created for using with photographic imagery would need adaptation to ensure consistency of coding with the range of visual data, documents, and artefacts produced. This led me to Saldana's (2016, p 65) guidance on visual, document, and artefact analysis, in which he proposes that researchers 'trust your intuitive, holistic impressions when analysing and writing about visual materials – employ a personal response methodology coined as "organic inquiry"'. This instinctive way of working brought another dimension to the work, with analytical memos created using the tenets of visual grounded theory of the image presented; beginning with first impressions of the image and what it meant to researcher and researched, followed by participant photo-elicitation. The image was then subject to interpretation by myself as the researcher, employing theoretical sensitivity of comparative experiences to recognise the implications of what was being said, and its significance to the individual and to the research as a whole. The final stage was to caption the image to provide synthesis of thoughts raised, informing memo writing and attributing the code to the data, with credibility confirmed by the participants.

A worked example of this process is provided alongside Figure 23.3. Given the nature of the image and its analysis, the implications of the disclosure were discussed. Practical actions and ethical considerations were applied for safeguarding of the participant within the remit of the study and their ongoing nurse education.

Figure 23.3: 'Square peg, round hole'

Note: Initial coding: Participant 3, picture 3.
Source: iStock by Getty Images Photo. Image identification: 157287760. Available from: https://www.istockphoto.com/. Licenced under CC BY-NC-ND 4.0. Ridge (2020, p 52).

Verbatim transcript (bold text identifies steps of analysis): 'felt like I didn't quite fit in … not quite fitting in … they had their team; they ran like clockwork … I felt a bit of a hindrance and not sure … where my place was' (Participant 3, Interview 1). **Visual analysis**: with **first impressions** being 'Childhood Game'; repurposed to represent Participant 3's subjective experience, **context**ualising placement orientation and identity. Its **interpretation** explained the maladaptation she experienced with every new placement experience was inextricably linked to self-worth and confidence (Jackson et al, 2011). Notably, she saw herself as the square peg being forced into the round hole. The **caption**, 'Enforced fit?' **Memo writing**, 'Are we looking for a perfect fit for nursing? Round peg, round hole, is there such a person? If so, what are we doing to our future selves? Virtuous or vicious circle? Force implies expectation to change through bullying' with **code** being acculturation theories.

In applying these methods, I initially reviewed the audio files and transcripts for accuracy, the value of this also being to refamiliarise myself with the audio files. These verbatim transcripts were read and reread in conjunction with the visual imagery that each participant had shared during the elicitation interviews. The action of engaging concurrently with the audio files, the textual transcripts, and the visual data made this an active kinaesthetic process and reaffirmed the point at which data analysis commenced (Charmaz, 2006). I chose to code the textual data electronically using NVivo software to manage the large volume of rich data that engaging with this research generated. The software functionality provided a platform in which to organise, store, and code the range of data that this research entails, which was accessible whenever and wherever I might require it (Saldaña, 2016). I did consider using NVivo for visual imagery, but instead preferred the immediacy of paper and pen that manual coding brings for this type of analysis; latterly I worked between analogue and electronic devices for analysis as the need arose.

Coding within the study followed a constructive grounded theory approach, with initial, focused, and theoretical coding being undertaken, albeit adapted as indicated for imagery. Initial textual coding was line by line, the purpose being to immerse myself in the data to seek out (in)significant occurrences warranting coding (Gibbs, 2013). The codes developed were purposively created in the form of gerunds to aid the process of 'seeing sequences and making connections' (Charmaz, 2006, p 42), fostering theoretical sensitivity through the selection of active words and processes to inform analysis. Initially, I felt overwhelmed by the volume of data collated, as I was new to grounded theory and was finding it difficult to comprehend its praxis of analysis; while seeking sense making theoretically to ensure rigour was troubling me. This concern must have been evident on my face while in a photo-elicitation interview with Participant 5. As she asked me to clarify the methods used for analysis, those adapted and adopted, she instinctively reached out and drew the following picture in my research diary, transforming my understanding through this shared learning and typifying my experience of coding, indeed of research, transmuting it to an active collaborative experience; hence the title of the image, 'In vivo artistry' (see Figure 23.4).

Initial coding of visual and textual data continued, with a conscious decision to move to focused coding at the point at which a number of codes appear to 'stand out and seemed to speak to the data' (Gibbs, 2013). I employed focused coding by using a significant code, or group of codes, to test against data, 'data and codes, codes, and codes' (Thornberg and Charmaz, 2014, p 10). This sees codes raised to tentative categories, refining what is going on using methods of constant comparison, memo writing, and reflexivity to inform concept development (Charmaz, 2006).

To exemplify, Figure 23.5 uses data derived from Participants 6, 1, and 10 to illustrate a series of images analysed for initial and focused coding to create a montage representing concept development: in this case 'role definition', specifically linked to the role of men in adult nursing. A long-established gender bias towards women in nursing is problematic, with negative stereotypes around the role and inclusion of men creating issues with the retention and recruitment of male nurses to the profession. This montage represents the analysis, including the initial and focused coding to create a subcategory. The first image is that from Participant 6, who particularly liked the message within this, that of an advert seeking to recruit men to nurse, its tagline being 'Are you man enough … to

Figure 23.4: 'Immersed in coding'

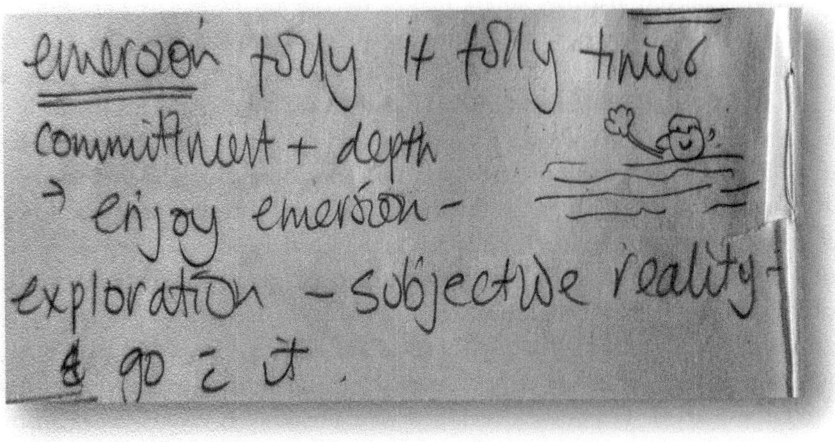

Note: Participant 5's in vivo artistry. Permission for use granted by Participant 5.
Source: Ridge (2020, p 51).

Figure 23.5: 'Battle ready'

Note: Montage of initial and focused coding demonstrating concept development showing Participant 6's 'Battle ready'. Permission to reuse first image provided courtesy of Oregon Centre for Nursing. Permission to reuse second image ('Participant 1 Culture Shift (Brother mine/Mona Lisa)') provided by Participant 1. Permission to reuse third image ('Role re-definition') provided by Participant 10, Painting Valley: attribution: https://paintingvalley.com/male-nurse-drawing. Image identification: https://paintingvalley.com/male-nurse-drawing#male-nurse-drawing-19.jpg. Available from: https://paintingvalley.com. Licenced under CC BY-NC-ND 4.0.
Source: Ridge (2020, p 57).

be a nurse?' It was coded as 'Battle ready' to represent the stance illustrated in the image of the courage required to address the antipathy of the stereotypical image of a nurse. Compare this to the second image, a photograph produced of Participant 1's brother, who retrained after leaving the armed forces to become a registered nurse, transferring his experiential learning to emergency care, coded as 'Cultural shift'; note his defiant stance and gaze ready for action, similar to that of the first image. The third, a found image, 'Role redefined', is that shared by Participant 10 and typifies his annoyance of having to defend his role, its practice, and of the label 'male nurse', as his verbatim text indicates, '"*You don't look like a nurse but by Christ you know what you are doing, and you love your job … it's obvious you love your job." But all I'm doing is giving the care as a patient would expect … So all I'm doing is what I think should be done, and that's what nursing is*' (Participant 10, Interview 4, 0:22:30). This montage represents the constant comparison used to ensure the fitness of the codes to inform the subcategory raised, as a component of the reality 'being a nurse' (Ridge, 2020, p 56).

The next step was to undertake theoretical coding, which required an exploration of the relationships between categories created to see 'how the substantive codes may relate to each other as hypotheses to be integrated into a theory … [and] weave the fractured story back together' (Glaser, 1978, p 55). Seeking clarity on how this worked, I discovered the concept of 'theoretical playfulness' (Thornberg, 2012, p 252), an approach that combines critical and creative thinking in the form of 'free-playing thinking' (Thornberg, 2012, p 252), thinking it would work perfectly with the data collated given the nature of its creativity. And it did, enabling me to open up the data to make innovative and unexpected connections, moving beyond 'descriptive cataloguing of data to theorizing imaginatively' (Thornberg, 2012, p 252), and providing the opportunity to 'refine, develop and revise the emerging theoretical structure' playfully to ensure participant meaning was generated (Chiovitti and Piran, 2003, p 431).

Figure 23.6 exemplifies my engagement with this theoretical playfulness. The photograph illustrates how analogue tools were used to picture the images that

Figure 23.6: Theoretical coding informed by 'theoretical playfulness'

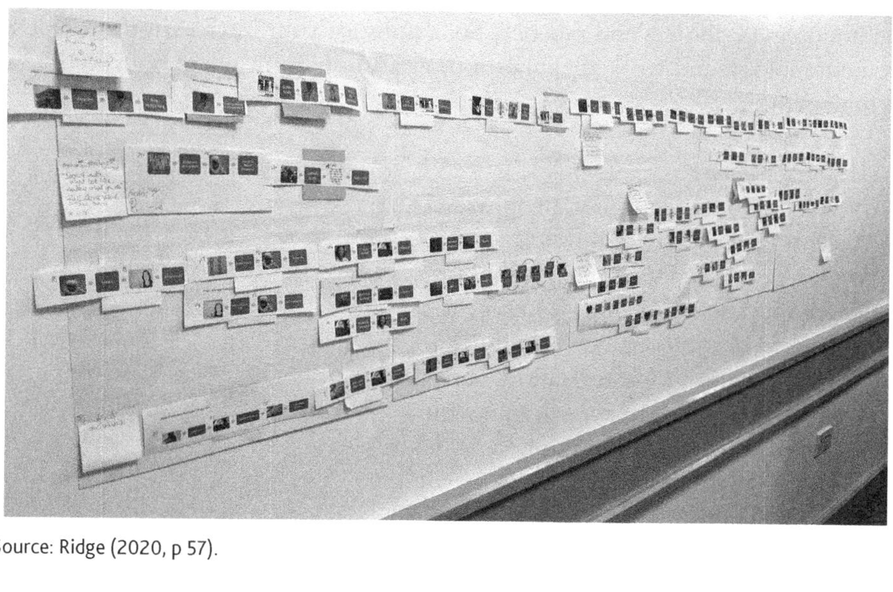

Source: Ridge (2020, p 57).

the co-producers shared with me for my research, and its coding and category construction. It was a wall in a conference room in a local hotel, where I was fortunate to spend a day on a research retreat. I went armed with paper, pen, sticky notes, tape, and scissors, which I used to playfully craft a tangible analysis and theoretically code my data into three core categories. This enabled the abstraction of my substantive theory in which a legitimate (nursing) identity is constructed by employing critical resilience and is sustained by authentic aspiration using the tenets of visual grounded theory (Ridge, 2020). It was so exciting to immerse myself in this way, realising that this was what 'real GT research' was all about, creatively crafting while reflexively thinking, what an opportunity! The only issue was when I had to remove my analysis from the wall, which is another story.

Conclusions

So, you've got to the end, insight gained? Elephant eaten? Still hungry for more?

Before you decide, take a moment to consider this succinct reflection, which informs the poem with which I intend to leave you. I started out having minimal understanding of grounded theory or visual research methods, but soon recognised the time, energy, and persistence required to develop understanding and application. This could be said of any research method chosen; however, I chose an innovative approach, for which neither resource nor expertise were readily available. While it has been exciting to create research using visual grounded theory, driven as it is by collaborative working, subjective data production, and analysis to construct a substantive visual grounded theory, I have also had to be prepared to face adversity, confronting traditional norms, culture, and attitudes that question new ways of thinking about and understanding research. In doing so, I have learnt to trust my intellectual curiosity, immersing myself in the creativity and rigour of the research process in order to make sense of the participant voice and challenge established methodological choices and practice. So, thanks for your interest in this chapter, its methodology, and worked application, vicariously from me to you. And now to my poem.

What do I know?
Non-traditional researcher, neophyte GT'er,
Accepting challenge of novelty.
Under-researched, complex, and obtuse.
Defending use; it fits don't you see?
Immersed in someone else's reality.
Image to the fore: joy, revelation, concern.
Voice owned; subjective reality known.
Co- production, collaboration,
Data emergent, not forced,
You can't control what you don't know.
Time, energy, persistence,

Be prepared for the unexpected,
Ready?
Validating a new way of seeing, knowing, and constructing research:
Visual grounded theory.
Want more?
Try it for size.

References

Anaebere, A.K., Maliski, S., Nyamathi, A., Koniak-Griffin, D., Hudson, A., and Ford, C. (2013) '"Getting to know": exploring how urban African American women conceptualize safer and risky sexual behaviours', *Sexuality & Culture*, 17(1): 113–31.

Blumer, H. (1969) *Symbolic Interactionism: Perspective and Method*, Berkeley: University of California Press.

Charmaz, K. (2006) *Constructing Grounded Theory: A Practical Guide through Qualitative Analysis*, London: SAGE Publications.

Charmaz, K. and Mitchell, R.G. (1996) 'The myth of silent authorship: self, substance, and style in ethnographic writing', *Symbolic Interaction*, 19(4): 285–302.

Chiovitti, R.F. and Piran, N. (2003) 'Rigour and grounded theory research', *Journal of Advanced Nursing*, 44(4): 427–35.

Collier, J. Jr (1957) 'Photography in anthropology: a report on two experiments', *American Anthropologist*, 59: 843–59.

Crotty, M. (1998) The Foundations of Social Research: Meaning and Perspective in the Research Process, London: SAGE Publications.

Gibbs, G,R. (2013) 'A discussion with Professor Kathy Charmaz on grounded theory', *Interview by Graham R Gibbs at the BPS Qualitative Social Psychology Conference, University of Huddersfield*, 14–16 *September* [online]. Available from: https://www.youtube.com/watch?reload=9&v=D5AHmHQS6WQ

Glaser, B.G. (1978) *Theoretical Sensitivity: Advances in the Methodology of Grounded Theory*, Berkeley: University of California Press.

Glaser, B.G. (1998) Doing Grounded Theory: Issues and Discussions (1st edn), Mill Valley: Sociology Press.

Goulding, C. (2002) *Grounded Theory: A Practical Guide for Management, Business, and Market Researchers*, London: SAGE Publications.

Harper, D. (2002) 'Talking about pictures: a case for photo elicitation', *Visual Studies*, 17(1): 13–26.

Jackson, D., Hutchinson, M., Everett, B., Mannix, J., Peters, K., Weaver, R. and Salamonson, Y. (2011) 'Struggling for legitimacy: nursing students' stories of organisational aggression, resilience, and resistance', *Nursing Inquiry*, 18(2): 102–10.

Konecki, K.T. (2011) 'Visual grounded theory: a methodological outline and examples from empirical work', *Revija za Sociologiju*, 2(3): 131–60.

Mannay, D. (2010) 'Making the familiar strange: can visual research methods render the familiar setting more perceptible?' Qualitative Research, 10(1): 91–111.

Mannay, D. (2016) *Visual, Narrative and Creative Research Methods: Application, Reflection, and Ethics*, London: Routledge.

Mey, G. and Dietrich, M. (2017) 'From text to image: shaping a visual grounded theory methodology', *Historical Social Research*, 42(4): 280–300.

Mills, J., Bonner, A., and Francis, K. (2006) 'The development of constructivist grounded theory', *International Journal of Qualitative Methods*, 5(1): 1–10.

Ridge, J.A. (2017) 'The interconnectivity of tools for analysis: combining textual and visual data to construct a substantive theory of professional identity of adult nursing', Paper presented at the *Royal College of Nursing International Nursing Research Conference*, 7 April, University of Oxford Examination School, Oxford.

Ridge, J.A. (2020) 'Authentic identity: a visual grounded theory of construction and sustainability of professional identity in adult nursing', PhD thesis, University of Northampton.

Rose, G. (2010) *Doing Family Photography: The Domestic, the Public, and the Politics of Sentiment*, London: Routledge.

Saldaña, J. (2016) *The Coding Manual for Qualitative Researchers* (3rd edn), London: SAGE Publications.

Sandstrom, K., Martin, D., and Fine, G. (2001) 'Symbolic interactionism at the end of the century', in G. Ritzer and B. Smart (eds), *Handbook of Social Theory*, London: SAGE Publications, pp. 217–31.

Thornberg, R. (2012) 'Informed grounded theory', *Scandinavian Journal of Educational Research*, 3(56): 243–59.

Thornberg, R. and Charmaz, K. (2014) 'Grounded theory and theoretical coding', in U. Flick (ed), *The SAGE Handbook of Qualitative Data Analysis*, London: SAGE Publications, pp 153–69.

Warren, S. (2002) '"Show me how it feels to work here": using photography to research organizational aesthetics', *Ephemera*, 2(3): 224–45.

Williams, S. and Keady, J. (2012) 'Centre stage diagrams: a new method to develop constructivist grounded theory: late-stage Parkinson's disease as a case exemplar', *Qualitative Research*, 12(2): 218–38.

24

But is it professional? Pairing creative practice and thematic analysis to illustrate organisational culture

Kyla Tully

Introduction

I am convinced that the PhD process is a treasure hunt to find a secret research handbook using a set of vague institutional guidelines rather than an explicit map. It seems once the handbook is discovered, successful adventurers are sworn to secrecy: my research supervisors have denied the existence of such a handbook, yet their research advice is often ambiguous in the manner of someone who knows the answer to a riddle but does not want to spoil the adventure. While such cautious guidance was helpful for many stages of my doctoral research, encouraging greater critical reflection on my research focus and data collection methods, I struggled to apply this casual mindset to select and justify a data analysis approach. With the secret handbook nowhere to be found, I decided I would retroactively establish my own analytical formula towards the end of my research process, a method that could be communicated through clear stages and compatible with a variety of research approaches. After all, by the last year of my doctoral research I would surely have a very clear understanding of how to conduct data analysis and talk about it. Surely.

This chapter is structured to provide a reflection of my experiences in designing and developing this analytical framework, using examples from a short but intensive period of my fieldwork to provide context to my decision-making processes as I established my research plan and subsequently watched it expand from a neatly structured project to an immersive experience. After providing some foundational information about the overall project and my original plan for conducting fieldwork and data analysis, I detail the different ways my research methods and focus deviated from this plan, the ethical and practical considerations I encountered, and the evolution of how I understand data analysis to relate to research practice. While I offer a framing of my approach as a formulaic analytical process that merges creative practice with thematic analysis that can be used as is or adapted to fit other preferred methods, my experiences in developing this approach demonstrate why a formulaic approach to data analysis is not, and should not be, everything.

Context

The examples I use to illustrate the development and use of my analytical framework are from my doctoral research project, which took place as an Applied Research Collaborative Studentship with Edinburgh's Queen Margaret University and the University of Glasgow from September 2020 through December 2023. With an aim to generate an understanding of the demands, expectations, and processes of professionalisation within a rural arts organisation, I was provided with a research residency at Alchemy Film and Arts (Alchemy), an experimental film organisation situated in the town of Hawick in the Scottish Borders. My initial research goal was to utilise participant observation to identify the management practices at Alchemy distinct to their rural contexts and distil them into a framework that could be replicated within other similarly situated cultural organisations. However, after three months of observing strategic planning and programming meetings I decided that this goal did not align with the operational patterns I was observing; I felt the organisation's management structures and practices were so informed by the personalities and values of the individual staff members and their interpersonal relationships that crafting a replicable framework devoid of these individualised aspects would be disingenuous. Therefore, I shifted my research focus, and overall methodological framework, from a case study that looked to isolate replicable processes and structures (Suryani, 2008) to an ethnography that framed Alchemy as a unique cultural phenomenon (O'Reilly, 2012; Barbour, 2014). The aim of my research then became to generate an account of Alchemy as a collective of people as well as a place of work, which I hoped would provide a detailed description of the unique contexts rural arts organisations operate within that could be useful for policy makers, funders, and other similarly situated cultural organisations.

My original timeline had involved a year of preparation before beginning fieldwork, but as the organisation had begun to implement several structural changes in response to the COVID-19 pandemic, I began virtual fieldwork via Zoom meetings in November 2020 after obtaining ethical approval for my study from the Queen Margaret University Research Ethics Panel. The end of the pandemic appeared imminent, with lockdown restrictions progressively easing since June 2020 (Scottish Parliament, 2022), and it seemed more efficient to document the changes as they were occurring rather than retroactively attempting to recall the context of the pandemic and its role in organisational decision making. This assumption was undeniably incorrect, and as a result of fluctuating lockdown restrictions my fieldwork took place within a wide variety of virtual and in-person contexts until I ended data collection in August 2022. However, while my fieldwork consisted of over 530 hours of participant observation spanning 22 months, my anecdotal points of reference for this chapter stem from a five-day period wherein I conducted 72 hours of in-person participant observation. I clarify this timescale not to encourage such intensive labour practices, but to provide context for the environment within which this data was collected, as the methods I used for documentation and sense making were curated in real time in response to the continual environmental, mental, and emotional fluctuations of the research field, research participants, and myself as the researcher.

The plan

Building on my interpretivist understanding of both reality and knowledge as socially crafted through the mediation of personal interpretations (Barbour, 2014), my initial research proposal looked to combine participant observation, interviews, discourse analysis, and practice-based research within a case-study framework. I planned to alternate between periods of data collection and data analysis: two separate processes that I would supplement with an intensive period of practice-based research to generate an embodied understanding (Candy, 2006) of what it was like to work within Alchemy. This shifted within the first five months to focus on participant observation, interviews, and thematic analysis, or the identification of theme and patterns of collected data (Coffey, 2018), within an ethnographic framework. This change was in response to both my previously described observations of the organisation and the realities of conducting fieldwork: the inclusion of discourse analysis meant I was examining the role of language within the organisation and required intensive transcriptions of verbal interactions (Barbour, 2014), but I realised I was more interested in the contexts around the conversations than the conversations themselves. Reframing my research as a cultural study provided space to follow these interests and establish methods of documenting my observations that came to me naturally, as opposed to forcing a framework in which I was not invested. I was able to put this new plan into action with my first round of in-person fieldwork in April 2021, observing the Alchemy Film and Moving Image Festival as it was streamed online by a small team in the Alchemy office. I had a clear vision for what my fieldwork would entail: spending my days in the office taking observational notes and documentational photos, then returning to my accommodation in the evenings to transcribe my field notes and categorise my observations by themes in a spreadsheet to revisit during my analysis stage.

The reality: creative deviations and complications

I arrived at the Alchemy office the day before the launch of their 11th film festival, the second virtual festival the organisation was conducting since the first pandemic lockdown in 2020. After officially meeting the staff in person, getting a tour of the building, and reviewing the schedules for the festival, I joined the team of four for dinner at a nearby restaurant and then returned to my accommodation to prepare for the week while the others returned to the office. While in the office the next morning, I realised that everyone had stayed on site working another four hours, which was notable from a series of inside jokes that this timeframe seemed to have generated. Panicked by having potentially missed foundational data, I abandoned my structured fieldwork plan and decided that I would stay in the office as long as any member of the team was still present. This was the beginning of a series of deviations from my planned approach.

Incorporating illustration

The first deviation from my initial data collection plans involved the incorporation of drawing as a substitute for documentational photography. Within my first hours of in-person fieldwork, I realised that although my research participants were accustomed to my observational presence in virtual meetings, being in the same room significantly changed the dynamics of our relationship. Without the restrictive framing of a video-call screen, my presence seemed more explicit and unnerving, especially as participants could now see where my focus was directed when documenting their movements and interactions. This became apparent as I attempted to take a photo of the team gathering for the first morning meeting, my raised phone causing everyone to freeze until it was lowered. Aware of the discomfort I was causing, I began sketching the participants and the objects they interacted with rather than taking photos (see Figure 24.1). As this was done alongside my field notes, it provided a bit of anonymity to both my focus and documentation practice because participants were not entirely sure if I was taking notes or drawing, and the subject of my interest was much less apparent in comparison to when I pointed a camera lens at a person, object, or area.

While my initial incentive to using drawing as a method of documentation was a reaction to the perceived discomfort of the participants, I encountered several benefits to this approach that kept me returning to it. While there were intensive bursts of conversation and interaction to take notes on, there were also significant periods of silence while everyone concentrated on their work. Drawing was a helpful way to document these quieter moments, as I could focus on physicality and non-verbal communication such as hand gestures and shifting feet (see Figure 24.2). Additionally, drawing became a way to help me remain engaged in the field itself, not only during quieter moments but also as exhaustion set in – by the third day of the festival with staff working 14 to 16 hours a day, we all struggled to string together coherent sentences. While this made for entertaining interactions to document, on a practical level I began to forget people's names and basic spelling, which increased the usefulness of drawing as a form of documentation.

Poetic transcription

In focusing on providing rich descriptions of how the observed staff interacted with each other and their shared space, I incorporated a variety of informal

Figure 24.1: Staff meeting

Note: Created in 2021 for Alchemy Film and Arts, Hawick, Scotland. Credit: Kyla Tully.

Figure 24.2: Feet

Note: Created in 2021 for Alchemy Film and Arts, Hawick, Scotland. Credit: Kyla Tully.

transcriptions from short conversations and offhand comments within my field notes. As the four members of staff that I was observing were from four different regions of England and Scotland, there was a notable variety of accents, vocabulary, and speech patterns within the room that I felt reflected the intersection of personalities and backgrounds within the organisation. In attempting to capture these aspects of speech, I initially utilised a combination of slashes and underlines to illustrate the rhythm of conversations; however, as I struggled to consistently use these symbols while notating fast-paced interactions verbatim, I shifted to moving my handwritten text down one line within my field journal every time there was change in speech such as a pause, interruption, or emphasis. As I crafted more and more of these short transcriptions, I noticed that my adjusted notetaking approach had unintentionally formatted the documented phrases and interactions as poems.

> **2:30PM (01/05/2021)**
> Underprepare
> The threat of failure
> Will propel the conversation to succeed

> **8:45PM (02/05/2021)**
> You just relax for a moment
> Just relax
>   Let me see my spreadsheet
> Step away from the spreadsheet

Reviewing this initial collection of poems embedded within my field notes, I felt that this approach was effective in representing the intersections of the participants' interpersonal relationships, communication styles, and patterns of work, such as the accidental humour of repeated conversations and the playfulness within staff interactions. At this stage of my research, I was unfamiliar with the established use of poetry within research (see Bhattacharya, 2008; Kara, 2015; Faulkner, 2016), but I recognised the potential of these poetic transcriptions as a form of data, as well as a possible research output, and I decided to shift my transcription approach to prioritise the generation of poems. This was ultimately a change in observational focus rather than writing practice, as while my process of poem generation remained consistent, the types of interactions and statements that caught my interest became increasingly based on the rhythms of speech or conversation over content. While not every poem was one that I would feel confident about offering for public viewing, this approach to notating the spoken word placed value on seemingly inconsequential interactions that had lasted for mere seconds. On an individual basis, each poem framed a moment of the workday as something significant and deserving of attention, while as a collection they provided an illustrative timeline of how participants engaged with each other and their work.

11AM (30/04/2021)
They've forgotten the name of one of the films
Can you send a list
        Must be nerves
                They've forgotten the name of one of the films
        Yeah
        I've sent them a list
                Must be nerves

6:30PM (01/05/2021)
It's raining
        Oh
        It's raining
What am I
A
Ghost
I
Literally
Just said that
        Did you
Yes
        In this room
Yes

Thematic analysis

Throughout the duration of this fieldwork, I saw my use of drawing and poetry solely as methods of data collection that I would process alongside my field notes utilising thematic analysis. I had come across thematic analysis while restructuring my research approach as a way to explain the intuitive nature of analysis within ethnography (Coffey, 2018). While this approach to analysis was not the explicit framework I hoped to apply to my own research, reading about thematic analysis was the first time I saw data analysis framed as 'an art, not a science' (Braun and Clarke, 2021, p 9), providing a theoretical justification for the development of my own strategy for categorising data (Coffey, 2018). While I had formally incorporated this guiding framework before investing in creative methods of data collection, its applicability to both visual and text-based forms of data was compatible with my incorporation of images and poems within my field notes.

The initial development of my thematic coding process unintentionally occurred while I transcribed my field notes. Due to the sheer quantity of text-based and visual data that I had generated within this period of fieldwork, I had decided to save my drawings, poems, and typed field notes in separate documents and folders. In the process of dividing these different forms of documentation,

I noticed that each creative product could be curated into thematic collections related to their content and context: poems could be grouped by explicit practicalities such as the date, time of day, who was speaking, or the present job roles, but also by themes based on keywords or the context of the poem such as time, scheduling, or reassurance. The drawings could be categorised by similar practicalities such as the time of day, or more abstractly by the nature of the composition, such as a time lapse or a still life, or the context of activity such as a morning meeting or working at computer. While I had initially begun assembling these collections for purely administrative reasons, the process of establishing how best to store the different outputs so they could be easily retrieved served as an initial coding process.

Once I had established an organisational format for these different outputs, I went back through the poems and drawings separately to experiment with different curatorial themes. Although I intended to keep the two creative formats separate, as I collated groups of poems based on patterns of conversational focus I had noticed within my initial administrative organising of data, I realised that different combinations of poems reminded me of specific drawings I had generated. Shortly after making this observation, two of the research participants contacted me with requests to see examples from my fieldwork during my next observational visit. Conscious of not wanting to provide access to my field journal itself, I decided to craft a small curation of my poems and drawings that linked the two formats and could be presented to both existing participants and potential new ones to provide context about my research. The resulting product, a document similar to a digital zine (see Figures 24.3 and 24.4), contained seven small collections of poems presented alongside one to three illustrations.

While this zine could be compartmentalised as a standalone research output resulting from a linear process of data collection and analysis, there are many aspects of its production and eventual presentation that intersect with processes of data collection and analysis. The timeframe within which this zine was created was a significant influence: with both the curation and presentation of the zine taking place during the first year of research, I was aware that I would be conducting fieldwork among most of the staff depicted for another full year at minimum. Therefore, I needed the included images and poems to be unthreatening but also recognisable to participants so as not to negatively impact my established relationship with the organisation or repel potential participants. Focused on avoiding possible negative reactions from continuing, new, and prospective research participants, I had not considered that the zine would act as a point of reflective conversation that would offer further insights for both my fieldwork and analysis processes. If this output had been offered at the end of my fieldwork, it might have been contained to a singular role where participant responses could be limited or less impactful. However, due to its place within the wider timeframe of my research, the zine became a conduit for mutual exchange between myself and participants. It was an output, a source of data, and an approach to analysis. This influential role was not one that I had intended

Figure 24.3: Excerpt from 'These times they will return': coordinating view

Note: Created in 2021 by Kyla Tully.

or predicted, but it helped grow my understanding about the integrated nature of research.

Clarifications, caveats, and considerations

The entanglement of collection and analysis

I have presented my approach to data analysis as a process separate from and subsequent to data collection, with drawing and poetic transcription categorised as forms of documentation to be revised within a thematic analytical framework. Yet, these methods, as I utilised them, were deeply enmeshed within both collection and analysis processes. I initially presented them as part of distinct stages because that is how I understood them to be during this fieldwork timeframe. However, following this period, as I began to communicate why I was utilising these approaches to my supervisors, research participants, and evaluators, I found it difficult to maintain these categorisations as separate and distinct. How could I talk about a theme I had noticed in reviewing my collection of poems and sketches

Figure 24.4: Excerpt from 'These times they will return': reassurance and regulation

REASSURANCE AND REGULATION

You've got
Six minutes to get it
Together

Don't worry
Just
Say the three names
You got it

You're doing fine
You know what you're doing
And you won't need to do this bit
Anyway
We can start 5 minutes late
If we need to
It's not a problem

Say something
That makes it sound like
We have everything
Under control
Like it's all going
According to plan

Note: Created in 2021 by Kyla Tully.

outside the field without discussing the observations that informed what I was documenting in the moment?

This realisation of the complex relationship between processes of data collection and analysis is not due to a lack of literature-based guidance or theoretical framing, as the texts that guided my research design continually reiterate the entangled nature of these stages within ethnographic fieldwork. For example, Karen O'Reilly (2012, p 186) describes research phases of data collection, analysis, and writing as 'inextricably linked', while James Spradley (1979, p 53) points to an overlap of data gathering and analysis due to a required 'constant analysis of utterances' within fieldwork. However, when I first read these texts, I could not see how such concepts could possibly apply to my work, as I was attempting to justify the inclusion of both textual and visual creative data and these reassurances centred on text-based outputs. As my fieldwork continued outside of this timeframe and I looked to texts regarding visual data collection (see Bhattacharya, 2008; Leavy, 2020; Ward and Shortt, 2020), I began to recognise not only how applicable this guidance had been all along, but just how interconnected the different stages of my research were.

I have come to understand that both forms of creative practice that I utilised to document my direct observations were embedded with an analytical process that Gannon and Fauchon (2021, p 67) refer to as 'ad-hoc decision-making', related to the composition of my drawings as well as my selection of phrases and snippets of conversation to document. This type of decision making is instinctive yet curatorial, difficult to explain afterwards but natural in the moment. These editing choices are present within the generated poems, each poem representing an interaction or offhand comment that I deemed worthy of documentation; after all, I did not create an epic poem containing every word spoken by my research participants, but rather jotted down things that interested me. However, this embedded analysis is more explicitly demonstrated within my drawings, my points of interest within the compositions a result of in-the-moment determinations of what to include or exclude. For example, during an observed staff meeting I decided to document that most of the staff were holding drinking receptacles (Figure 24.5). A series of choices were made between this initial decision and the final product, such as the utilisation of colour and the exclusion of context in relation to the full bodies in the space; choices that required a rapid determination of what was important to convey.

The enmeshed relationship between data collection and analysis within my approach to fieldwork is further complicated by the inclusion of thematic analysis and the resulting curated collections, as this frames the creative outputs from my fieldwork both as products of analysis and products to analyse. However, the additional layer of review and revision of these creative outputs

Figure 24.5: Staff meeting coffee

Note: Created in 2021 for Alchemy Film and Arts, Hawick, Scotland. Credit: Kyla Tully.

was crucial for making connections between themes and observations as it forced me to look at my accumulated data differently. Although the generation of the drawings and poems were hugely analytical in practice, they required my complete immersion within the field and meant that I was intensely focused on details within my fieldwork. The process of compiling and pairing select poems and drawings pulled me out of these specificities and allowed me to identify the broader patterns that had not only informed my decision making within generating my drawings, poems, and field notes, but were newly present within my selective curations. One of the key observations made from this process was the incorporation of care practices within the organisation's identity and professional practice; such practices were present to varying degrees within the generated poems and illustrations, such as the bags of sweets that sat between computer monitors for the staff to share, the collection of coffee mugs the management team washed every morning, and the pre-emptive verbal reassurance when something did not go to plan. I had documented these aspects instinctively because they caught my interest in the moment, but it was not until I looked at them as a broad collective that I noticed how they all related to a practice of care embedded within the organisation as both a natural occurrence and implicit expectation.

Navigating creative approaches in academic systems

In discussing the development and application of creative practice within my research I feel it is important to acknowledge the difficult or adversarial aspects of my research process and experience that resulted from external expectations and my own inexperience. While these were not the only challenges I encountered during my research, I emphasise the following points because of their direct link to my inclusion of creative practice and their prevalence within much of my research timeframe. I am confident that if the practices within my presented approach were not intrinsic to how I process information, and if I did not have the enthusiastic support from a selection of supervisors, examiners, and peers, I could not have maintained the inclusion of creative practice within my research.

Many of the challenges I encountered during my research resulted from a lack of prominence of creative methods within academic practice and research; although the utilisation of such methods can generate or illustrate data in a manner more approachable to non-academic audiences (Kara, 2015), I discovered academic systems and settings found it difficult to support or even tolerate such methods. Over the course of my research project, I navigated an absence of academic support for my approach that ranged from explicit discouragement regarding creative practice from institutional leaders to a general lack of guidance on methodology and dissemination. While I received encouragement from practice-based researchers and sociologists at Queen Margaret University, the formal support systems within both my universities, such as researcher forums, skills workshops, and thesis-writing guidelines, were not prepared to assist with creative methods of research. This was further compounded by the particular and exacting nature of keywords within academic search engines, and it took me months to discover the phrasing that linked me to not only a wider network of researchers invested in creative research practices but also the texts that became foundational to my work, namely visual ethnography (Pink, 2007), poetic inquiry (Faulkner, 2016), and illustration research (Gannon and Fauchon, 2021). The time it took to secure these theoretical justifications additionally unnerved my research participants, who were enthused about my creative outputs during my initial fieldwork sessions but soon began to express concern that my research approach would not be seen as valid or 'real' research either within or outside academia. I now have the research experience to challenge perceptions of validity and to convey my aims, justifications, and learning, but as a new student engaging with both doctoral research and creative research practice for the first time, this was a significantly isolating experience.

This inexperience with research additionally contributed to my negligence in considering my own needs and well-being within my research practice, exemplified by the intensive manner within which I conducted the fieldwork that forms the basis of this chapter. While I made this choice out of fear of missing valuable data, I did not consider how taxing being an active observer can be, not to mention over such an exhaustive timeframe. This consideration is not unique to creative practice within fieldwork, but my utilisation of drawing to both navigate data collection while exhausted and to circumvent my own lack of self-preservation was particular to this practice. This resulted from my own inexperience with field research outside the incorporation of creative practice, which I attempted to justify through sheer quantity of fieldwork hours. Although a significant aspect to my need to validate this approach stemmed from a lack of tolerance within academic settings for non-traditional formats of data collection, analysis, and dissemination, I equally failed to prioritise my own needs over the research process.

Conclusion

In presenting this approach as a singular analytical process that merges creative practice with thematic analysis, this framework can be simplified into a formula of (creative practice + creative practice) + thematic analysis = findings. Although this is a massive oversimplification, it underpins the variety of possibilities that come with incorporating creative practice into research, in that the processes of documenting, generating, and analysing data can be built from a researcher's specific creative skillset. Within my own research beyond the presented timeframe my creative practice continually evolved to incorporate other elements, such as making maps of office movements, illustrative tracking of focus in meetings, and still life photography, adapting to meet the intersection of my creative and academic strengths, interests, and abilities.

At the beginning of my research journey, I was desperate for a clear checklist I could work through to complete my research, and when I could not find one I wanted to be able to provide that framework for others. Unfortunately, what I have learned within the process of conducting research and attempting to relay the involved processes to others is that offering such a checklist would be disingenuous; it would not match my experience as a researcher. What I *have* created is something I unknowingly needed at the beginning of my research: a description of the research process that details the relationship between fieldwork and sense making and illustrates why a concise yet all-encompassing framework for most, if not all, research approaches cannot exist. That is, until someone finds that secret research handbook.

References

Barbour, R. (2014) *Introducing* Qualitative Research*: A* Student's Guide (2nd edn), London: SAGE Publications.

Bhattacharya, K. (2008) 'Ethnographic poetry', in M. Cahnmann-Taylor and R. Siegesmund (eds), *Arts-Based Research in Education*, Abingdon: Routledge, pp 72–8.

Braun, V. and Clarke, V. (2021) 'Conceptual and design thinking for thematic analysis', *Qualitative Psychology*, 9(1): 3–26.

Candy, L. (2006) *Practice Based Research: A Guide.* Creativity & Cognition Studios Report, Sydney: University of Technology.

Coffey, A. (2018) *Doing Ethnography* (2nd edn), London: SAGE Publications.

Faulkner, S.L. (2016) *Poetry as Method: Reporting Research through Verse* (2nd edn), Abingdon: Routledge.

Gannon, R. and Fauchon, M. (2021) *Illustration Research Methods*, London: Bloomsbury Publishing.

Kara, H. (2015) *Creative Research Methods in the Social Sciences*, Bristol: Policy Press.

Leavy, P. (2020) *Method Meets Art: Arts-Based Research Practice* (3rd edn), New York: Guilford Press.

O'Reilly, K. (2012) *Ethnographic Methods* (2nd edn), Oxford: Routledge.

Pink, S. (2007) *Doing Visual Ethnography* (2nd edn), London: SAGE Publications.

Scottish Parliament (2022) 'Timeline of coronavirus (COVID-19) in Scotland', Scottish Parliament Information Centre [online]. Available from: www.spice-spotlight.scot/2022/05/27/timeline-of-coronavirus-covid-19-in-scotland

Spradley, J. (1979) *The Ethnographic Interview*, New York: Holt, Rinehart, & Winston.

Suryani, A. (2008) 'Comparing case study and ethnography as qualitative research approaches', *Jurnal Ilmu Komunikasi*, 5(1): 117–27.

Ward, J. and Shortt, H. (eds) (2020) *Using Arts-Based Methods of Research: A Critical Introduction to the Development of Arts-Based Research Methods*, Cham: Springer International.

Section VIII

Analysis with participants

25

Using patient and public involvement, engagement and co-production to enrich the analysis, interpretation, and utilisation of sexual health research

Mar Estupiñán Fdez. de Mesa, Melvina Woode Owusu, Makeda Gerressu, India Henry, Chantel Sealey, Gwenda Hughes, and Catherine H. Mercer[1]

Introduction

Over an 18-month period, our research group developed and implemented a creative, pragmatic, and collaborative data analysis and interpretation framework to illuminate initial sexual health research findings. The basis of our methodology was the use of patient and public involvement and engagement (PPIE) and co-production to analyse, interpret, and contextualise research findings, and to facilitate knowledge translation. Using this approach allowed stakeholders to share ideas, unconstrained by conventional academic approaches (eg papers and statistical outputs). Further, this approach helped us develop a more comprehensive and richer understanding of the research findings, which fed into informing and co-producing novel research outputs. Creative data analysis interlinked with idea generation to help inform a dissemination strategy that prompted ownership among stakeholders. As a result, stakeholders championed dissemination and adoption of the recommendations, accelerating the process of translating research evidence into national guidance and practice.

This chapter includes six sections describing our creative approach from a methodological and narrative perspective, including supporting quotes and reflections from co-producers/co-authors. Firstly, we present the national guidance context followed by the sexual health context in the UK and the need for research to inform policy and practice in this area. Then, we describe how PPIE and co-production underpinned our knowledge translation framework, and explain how these frameworks informed our analytic process. Finally, we reflect on the strengths and limitations of our analytic approach, and share recommendations to help researchers explore, adapt, and ultimately improve ways of implementing this novel approach.

Context

Globally, governments are under increasing pressure to develop, implement, and evaluate evidence-based health-care policies and interventions that are tailored to the needs of research beneficiaries (WHO, 2021; Françoise et al, 2022). Using evidence-based policy approaches gives policy makers and providers a systematic framework to identify specific policy elements that are most likely to be effective for improving people's health and well-being outcomes and reducing inequalities (Cairney and Oliver, 2017). Evidence-based policy requires policy makers and providers to be agile in responding to rapidly emerging scientific research, and to the needs and priorities of those with lived experience (Oliver et al, 2014). To support policy makers and providers, academics are increasingly required to conduct robust research that addresses complex societal problems, to facilitate timely dissemination of research findings in an accessible way, and to influence policy agendas (Cairney and Oliver, 2017).

However, developing evidence-based policy is a complex process, further complicated by conflicting political agendas and priorities, variable understanding of interdependent policy and delivery systems and processes, and resource constraints impacting the prompt production and implementation of relevant evidence (Haynes et al, 2020). In the context of sexual health care, developing evidence-based policy is challenged by the stigma associated with sexually transmitted infections (STIs), inequalities in outcomes, and a complex provider and delivery partners network.

To bridge this research–policy gap and embed research-informed recommendations into policy and practice, creative, pragmatic, persuasive, and evidence-based approaches must be embraced (Kneale et al, 2019). One option is knowledge translation – 'a dynamic and iterative process that includes synthesis, dissemination, exchange, and ethically sound application of knowledge to improve the health of [residents], provide more effective health services and products and strengthen the healthcare system' (CIHR, 2012). Furthermore, the process of knowledge translation can accelerate the transfer of research findings into actionable recommendations, thus providing more effective services and programmes to individuals, and strengthening health-care systems (Straus et al, 2009; Esmail et al, 2020).

There is increasing recognition that PPIE and co-production enhance the quality of the research process and outputs, increase ownership, and empower community members to assume responsibility about how research findings might be interpreted and used (Adhikari et al, 2020; Tembo et al, 2021). PPIE in research has also received considerable attention in the past ten years, and, increasingly, is an essential requirement for research funders (UKRI, 2021). PPIE refers to conducting research *with* and *by* members of the public (involvement); and engaging, listening, and gathering insights from community members (engagement) (NIHR, 2021). Co-production involves researchers working collaboratively *with* providers and end-users to achieve a *collective* outcome through shared decision making (INVOLVE, 2023). PPIE and co-production can facilitate knowledge translation, help to create culturally competent workforces, and contribute to the development of policies, programmes, and services that are tailored and culturally appropriate (sensitive to people's cultural identity or heritage) (CQC, 2022; NIHR, 2022).

Sexual health

Sexual health and well-being are important for individuals and society and are 'determined by the quality and safety of people's relationships (with themselves and other individuals, and the society in which we live, including the gender norms that shape our experiences)' (WHO, 2022a). Achieving good sexual health and well-being relies on a wide range of factors, including access to good-quality information about sex and sexuality; accurate knowledge about behaviours including risky behaviours associated with poor sexual health; the availability of prevention and treatment of STIs; the ability to access sexual health services when needed; and living in environments that affirm and promote good sexual health and well-being (WHO, 2022b).

In the UK, despite the availability of comprehensive and high-quality sexual health services, some subgroups of the population persistently experience poorer sexual health outcomes. For instance, in 2022 in England the STI diagnosis rates remained highest among people of Black ethnicity, with diagnosis rates of gonorrhoea and trichomoniasis highest among people of Black Caribbean ethnicity (Migchelsen et al, 2023). These groups have experienced the greatest burden of STI since the early 2000s.

To better understand the behaviours, attitudes, and factors that increase STI risk among people of Black Caribbean ethnicity, the National Institute for Health and Care Research Health Protection Research Unit (NHIR HPRU) in Blood Borne and Sexually Transmitted Infections (BBSTI) (of which some of the authors are members) undertook a mixed-methods research programme between 2015 and 2021(Wayal et al, 2018, 2020).[2] Based on the initial mixed-methods research findings (Aicken et al, 2019, 2020), the HPRU proposed three policy and practice priorities to help address STI inequalities among people of Black Caribbean heritage:

1. Empower and equip women with safer-sex negotiation skills.
2. Improve knowledge of behaviours that increase STI risk.
3. Encourage a respectful approach to sexual relationships.

The authors then aimed to translate this research evidence to inform recommendations and outputs for targeted interventions and policy and practice that were culturally appropriate for people of Black Caribbean heritage. To achieve this, we designed a knowledge translation framework underpinned by PPIE and co-production, which included people from Black Caribbean communities and representatives, and public and sexual health (PHSH) professionals.[3] Following a systematic, iterative, and creative approach, we invited stakeholders to collaborate with the research team in further analysing and interpreting the initial research findings and to explore options to translate these findings to inform policy and practice (Aicken et al, 2019, 2020). This meant that the creative data analysis and interpretation of research findings were co-produced to maximise their relevance, impact, and reach, and ultimately to yield significant and timely individual and public health outcomes. We planned for this phase to take one year. However, due to the COVID-19 pandemic and related challenges and restrictions, this phase lasted 18 months.

Theoretical framework

Our approach was guided by the Transformative Paradigm. Transformative research addresses social issues including empowerment, inequality, and racism; aims to provide a voice for groups often overlooked in research; and focuses on influencing and advancing an agenda for change (Mertens, 2021). To this end, we took the epistemological stance that *all* voices would have *equal* value. The subsection 'Phase 1: stakeholders' engagement and involvement' gives details of how this was achieved in practice. We recognised the plurality of individuals' identities, roles, and expertise. This meant recognising that some stakeholders were simultaneously community members and professionals and, as such, their dual contribution added a unique value to the creative analytic process.

> 'As a person from the community that this research was aimed at, I felt both my lived and professional experiences would add validity and also provide insight from a different perspective.' (Chantel Sealey, co-producer and co-author)

Further, a core aim of the research was to influence policy and practice. This meant producing outputs that prioritised being relevant, rigorous, and actionable for PHSH professionals and beneficial for communities in contrast to conventional academic outputs (eg journal articles and conference presentations). Bridging this gap required us to work differently and creatively, as Chantel Sealey explained:

> 'The brainstorming activities and group work allowed us to work in a collaborative way and explore some areas from a different perspective. Co-producing the resources meant that our time, efforts were being recognised, would also benefit the wider community on a large scale, and more importantly support professionals and influence service delivery.' (Chantel Sealey, co-producer and co-author)

Ontologically, we assumed that reality is socially constructed, and recognised that people are influenced by multiple and unique realities that shape their sexual health experiences (Mertens, 2012). We therefore sought to understand the communities' perceptions about sexual health relative to their sociohistorical and cultural contexts. This included a focus on how culture influences sexual behaviours and how structural racism was perceived and influenced community members' help-seeking behaviours and the provision of sexual health services. We also explored the challenges encountered by professionals to engage and deliver sexual health services *for* and *with* people of Black Caribbean heritage.

Building on evidence-based knowledge-transfer frameworks (Lavis et al, 2003), transformative theoretical approaches (Mertens, 2021), and guidelines for PPIE and co-production in research (NIHR, 2022), we needed to develop a working definition that reflected a system-wide approach to knowledge translation, PPIE, and co-production, specifically: 'collaborative working of multi-disciplinary researchers, professionals, and community stakeholders to translate research findings, and to conceptualise, design, deliver, and evaluate ideas, solutions, and outputs that inform evidence-based policy and practice' (see Figure 25.1).

We consider that a system-wide approach to knowledge translation, PPIE, and co-production has the advantage of convening stakeholders across systems to work collaboratively to enhance the impact of research, to reduce the policy–research gap, and to improve outcomes for research beneficiaries, not all of which would be achieved using conventional research approaches. For instance, all the stakeholders involved assigned unique meanings to the research findings, which brought value to the creative process. Involving stakeholders throughout ensured varied perspectives were represented and needs considered.

We designed our approach around five pragmatic, evidence-based questions central to knowledge translation (Lavis et al, 2003):

1. What aspects of research-generated knowledge should be transferred?
2. To whom should research knowledge be transferred?
3. By whom should research knowledge be transferred?
4. How should research knowledge be transferred?
5. To what effect should research knowledge be transferred?

These questions guided our decisions to identify, involve, and engage stakeholders, and the process of analysing and interpreting data and translating findings into impactful research outputs. However, while a priori, these questions seemed simple to address, questions 3 and 4 in particular generated a relevant debate. Nevertheless, this approach provided flexibility to prioritise stakeholders' views and ideas as they emerged throughout the project. More importantly, it allowed us to gain an enriched understanding of our analytical process and to explore more nuanced answers to our research questions. This approach is explained in the next section.

Methods

Phase 1: stakeholders' engagement and involvement

Our approach to creative data analysis, interpretation, and translation of our research findings involved the research team collaborating with stakeholders,[4] as described at the beginning of this chapter. This encompassed Black Caribbean community members, their representatives, community-based organisations,

Figure 25.1: Outline of our approach

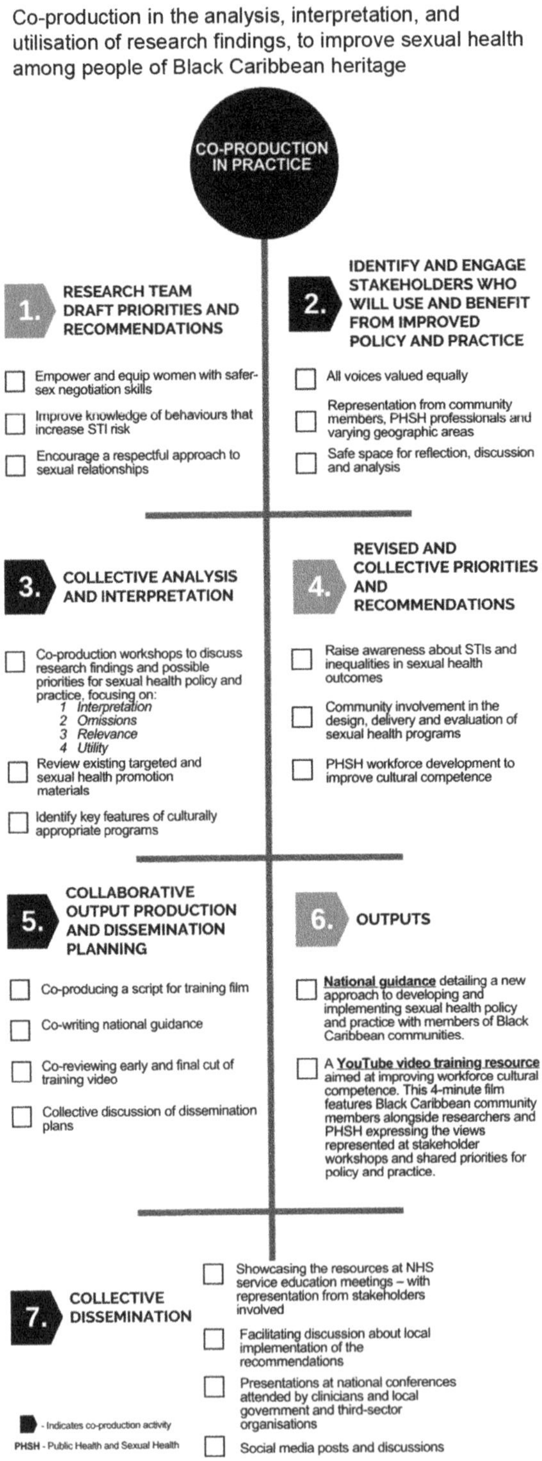

PHSH professionals, and researchers from various geographical areas across England. Stakeholders were invited to one of two workshops to analyse and interpret findings, co-produce recommendations, and revise priorities for policy and practice.

The research team facilitated two face-to-face workshops: Workshop 1 for community members and their representatives (n = 8 in November 2019) and Workshop 2 for PHSH professionals (n = 15 in January 2020). All travel expenses were paid, and community members were offered a £30 voucher as a token of appreciation.

We facilitated two distinct workshops to ensure all attendees felt comfortable and able to contribute fully. The aim was to enhance the community's assertiveness by creating a conducive setting where attendees felt comfortable sharing their lived experiences, perceptions, and creative ideas without feeling intimidated by the presence of subject-matter experts or other figures of power. Further, to increase engagement, we ensured settings were appropriate and trusted; for example, a community centre was used for the community workshop and a university setting for the PHSH professionals' workshop. and a university setting for the PHSH professionals' workshop. The community centre was a suitable venue to host workshops often used by our research team because it had convenient transport connections, particularly for attendees travelling from outside of London, and it had plenty of natural light and outdoor space. Moreover, our previous experience organising workshops where community engagement had been affected by venues representing institutions of power guided the decision to use the community centre. This decision translated in a better experience for our community members, as one attendee explained:

> 'The workshop allowed us to network and share our ideas and be transparent in a safe space. The brainstorming activities and group work allowed us to work in a collaborative way and explore some areas from a different perspective. Co-producing the resources meant that our time – efforts – were being recognised, [and] would also benefit the wider community on a large scale, and more importantly support professionals and influence service delivery.' (Chantel Sealey, co-producer and co-author)

Although this approach worked for our community stakeholders, we would recommend researchers seek community members' views when identifying a venue for community engagement. This avoids the often unconscious assumption that community centres are the most appropriate spaces for community members when this may not be the case and it could risk disempowering and reinforcing access barriers to already unheard communities. Ultimately, researchers should focus on creating safe and welcoming spaces that are informed by the community.

One reason for organising two distinct workshops was to mitigate unbalanced power dynamics between groups. However, it is important to consider other challenges that can arise within groups. For instance, despite all our efforts to convene a well-balanced PHSH professional workshop, it was challenging to identify senior decision makers of Black Caribbean heritage. This reflected an existing power inequity within the health-care structures. To mitigate the impact of this under-representation, we approached community and charity leaders who agreed to give a voice to communities often unheard and with whom they had close and professional working relations.

During these workshops, the research team presented phase one's initial findings and priority areas to stakeholders. We subsequently provided the following framework to enable a creative and free space for stakeholders to critically reflect on, and further analyse, the phase one findings and feedback on:

1. how *they* interpreted the research findings given their knowledge 'on the ground' and lived experience;
2. what *they* thought was missing (ie what additional analyses were needed to 'unpack' and better understand the initial findings);
3. what was most significant and relevant to *them*; and
4. what *they* saw as the implications for policy and practice.

Together with this framework, and to ensure attendees truly contributed towards data analysis and interpretation during Workshop 1, we used creative approaches that made research findings accessible to all individuals regardless of their knowledge and background. This meant converting specialist and/or technical terms into lay-friendly terms and using videos, photographic images, and visual cues to illustrate the concepts of health promotion and health inequalities. Workshop attendees reviewed generalised and targeted STI, HIV, and condom campaign materials, and we captured attendees' reflections, quotes, and feedback using flipchart pads. Subsequently, we divided attendees into three groups and asked them to create a new sexual health promotion campaign/initiative based on their shared understanding of the research findings. Using diagrams, illustrations, and notes, attendees pitched their interventions to the other groups, who then provided feedback about acceptability, barriers and facilitators, limitations, and ways to improve the initiative. This creative and free-flowing process enabled idea generation and helped attendees to co-produce a series of recommendations for targeted and culturally sensitive messaging resulting directly from their analysis and interpretation of the evidence provided. These recommendations were shared during the PHSH professional workshop, alongside illustrative (anonymised) quotes to inform the discussion. In the same vein as Workshop 1, PHSH professionals were asked to reflect on the research findings, attendees' recommendations, and their implementation feasibility. This enabled open reflection among PHSH professionals about the wider determinants of poorer sexual health outcomes, and fostered a renewed sensitivity towards, and understanding of, the collective experiences of Black Caribbean communities.

> 'Co-production makes a difference because it exposes you to real people and real challenges, for example in HIV prevention – how easy would it be to assume people are 'just' unaware or don't use preventative measures? When you hold co-production workshops you get to hear and understand every detail straight from the person. There's no assuming, you simply have to listen to know.' (India Henry, co-producer and co-author)

Notably, both groups emphasised the need to contextualise the research within the lived experiences of people of Black Caribbean heritage, using a multifaceted response to address the relationship between wider determinants of health, systemic racism, and poor health outcomes. Furthermore, PHSH stakeholders highlighted the importance of equipping their colleagues with knowledge about inequalities and training in cultural competence to inform and support collaborative programme design, delivery, and evaluation.

Priorities for policy and practice developments that could enhance research impact were discussed, with the following agreed as key recommendations:

1. Raise awareness among Black Caribbean communities about STI prevention, transmission risk, and diagnosis.
2. Raise awareness among the workforce about inequalities in sexual health outcomes that disproportionately affect people of Black Caribbean heritage.
3. Encourage collaborations with local partners and involve the community.

Phase 2: co-production of outputs

The workshops were also used to discuss the best approaches and vehicles for conveying the proposed recommendations, and for influencing and transforming policy and practice. Collectively, it was agreed that the initial resources should focus on producing policy guidelines and digital resources for sexual health commissioners and third-sector organisations. Stakeholders were invited to co-produce these resources with the research team in a two-stage process: firstly, we worked with PHSH stakeholders to produce the policy guidance, and subsequently we worked with community members and their representatives and PHSH stakeholders to co-produce a video resource.

To co-produce the policy guidance, we followed an iterative process whereby the research team used insights gained from the workshops to produce a first draft of the guidance, which was subsequently circulated to all PHSH workshop attendees and their extended networks for further input. The guidance required a series of revisions whereby stakeholders had the opportunity to advise, comment, and co-write this output via email, phone, or Zoom meetings. This involved general reviews of the structure and utility of the document and technical tasks such as checking whether the description of workshop insights was congruent with discussions on the day and advising on a need for more nuanced wording that better reflected the heterogeneity of Black Caribbean communities:

> 'I'd recommend that the document advises professionals acknowledge place of birth and socialisation when working with people of Black Caribbean ethnicity, as those born and socialised in the UK may have different sexual health attitudes and needs to those who have emigrated to the UK, but [were] born and initially socialised in the Caribbean.' (Uzochi Nwaosu, Sexual Health Adviser and co-producer)

Another co-producer, a Senior Public Health Specialist, advised that while a top-down approach should be acknowledged, the guidance should also highlight the dyadic nature of information flow and knowledge transfer.

Co-producing with stakeholders improved the quality of the guidance, as well as responsibility for and ownership of disseminating and actioning the recommendations. The process also facilitated the co-production of a tailored policy guidance that set acceptable and implementable recommendations.

For the video resource, the research team and stakeholders worked in collaboration with a professional producer. Importantly, during the PHSH workshop, stakeholders advised that involving PHSH professionals who also identify as being of Black Caribbean heritage was essential. On their request and advice, two self-identifying Black Caribbean PHSH professionals refined the video script alongside a member of the research team. This process helped to assess and adapt the tone and language so that it was accessible and acceptable to PHSH professionals, as well as to lay audiences, especially those from the communities involved in the research and affected by any resulting changes in policy or practice. The video features five cast members, including two Black Caribbean professional actors (delivering verbatim community-workshop quotes), a high-profile Black Caribbean professor and public health professional, a professor from the research team, and a research translation consultant. We piloted the initial film, which was very well received, with the exception of the relatability and strength of delivery of a community member quote. As all opinions carried equal weight in the co-production process, the actor was recast and replaced with another whom the PHSH and community stakeholders advised was more relatable and better able to convey the community co-producers' quote more credibly. This demonstrated that ensuring stakeholders are at the centre of decision making enables co-production with improved and relevant outputs that speak to those who will benefit from them and are more likely to be acceptable to the target audiences.

As a result, and through an iterative 18-month process of creative analysis, interpretation, and reflection, the research team, in collaboration with stakeholders, co-produced, published, and actively disseminated:

- Public Health England guidance detailing a new approach to developing and implementing sexual health policy and practice with members of the Black Caribbean community (PHE, 2021).
- A YouTube video training resource aimed at improving workforce cultural competence. This four-minute film features researchers alongside two

representatives from the Black Caribbean community expressing the views represented at the stakeholder workshops and shared priorities for policy and practice.[5]

Once the outputs were completed, the research team, in collaboration with stakeholders, discussed dissemination plans. Subsequently, stakeholders helped to co-disseminate the outputs, showcasing the resources at NHS service education meetings, presentations at national conferences, social media platforms, and more.

> 'Congratulations! This is great work. Wonderful to see it all come together. Showed a snippet of the video at a health outreach event this weekend – it was really well received.' (Dr Vanessa Apea, Consultant Physician in Sexual Health and HIV and co-producer)

Strengths and limitations

This evidence-based and creative PPIE, co-production, and knowledge translation project suggests that it is possible to accelerate the translation of research findings into national guidance and practice when academics, policy makers, providers, and communities work collaboratively. Furthermore, our example shows how PPIE and co-production have the potential to inform and improve the quality of current and future research processes; to use more creative, interactive approaches when analysing and interpreting data; and to enhance ownership and shared responsibility when seeking to increase the impact of research. For instance, here the creative approach led to the refinement of the research findings; the regeneration of policy and practice priorities initially proposed by the research team; and collaborative development of tailored, culturally appropriate, and actionable recommendations informed by a shared interpretation of the research evidence (Table 25.1).

Our co-production method of creative data analysis and interpretation was informed by existing theoretical frameworks to provide a robust and systematic approach to knowledge translation, while enabling opportunities to capitalise on stakeholders' creativity and unique lived experiences and expertise. Embedding PPIE and co-production in the knowledge translation framework was instrumental to refining our analyses and interpretation of the research findings. Together, the resulting policy and practice priorities and recommendations are likely to be more meaningful and actionable for PHSH professionals and beneficial to the community they are seeking to support.

> 'It is important for researchers to consider creative ways of working because we have to be inclusive and accessible to all. The thought of research findings only being accessible to a few, or academics[,] widens the inequality gap further. If we truly want to work towards equity then creative ways to interpret research findings isn't an add on, it's a way forward.' (India Henry, co-producer and co-author)

Table 25.1: Priorities for sexual health policy and practice

Initial priorities identified by the research team
Empower and equip women with safer-sex negotiation skills
Improve knowledge of behaviours that increase STI risk
Encourage a respectful approach to sexual relationships
Refined and co-produced recommendations
Raise awareness among Black Caribbean communities about STI prevention, transmission risk, and diagnosis
Raise awareness among the workforce about inequalities in sexual health outcomes that disproportionately affect people of Black Caribbean heritage
Encourage collaborations with local partners and involve the community

Furthermore, all stakeholders were invited to complete an anonymous survey to share their feedback on the approach we had taken. This unveiled beneficial indirect health impacts of the PPIE, co-production, and knowledge translation exercise for our stakeholders, including:

1. Raising awareness of STI inequalities and services among community members.
2. Empowering community members and representatives to champion and advocate for sexual health in their communities.
3. Equipping PHSH stakeholders with knowledge to better support Black Caribbean communities.
4. Establishing PHSH champions who organised video screenings and facilitated group discussions about the PPIE, co-production, and knowledge translation approach with local teams, establishing both ownership among the champions and buy-in among their colleagues.

While our approach is adaptable to different research questions, it is resource-intensive and requires strategic planning from the outset (Woode Owusu et al, 2023). Retaining a focus on the Transformative Paradigm, providing an equal voice for groups often overlooked in research, and taking steps to influence and advance an agenda for change (Mertens, 2021) were at times challenged by time and financial constraints. In practice, maintaining the core principles of co-production (diversity, equality, accessibility, and reciprocity) meant, for example, although we organised events where the research team was based, we booked and reimbursed travelling expenses to ensure that representatives from a range of geographical areas across the country could join us in person. It has to be recognised, however, that this approach still poses some challenges to stakeholders, as one attendee described:

> The main and only challenge for me was that the event was held in London; working in Manchester this meant committing to a full day being out of office and travelling. As my team is small and we have quarterly targets depending on the programme deliverables of that day, it could have been difficult for me to attend. (India Henry, co-producer and co-author)

Recommendations

We recommend planning and budgeting for PPIE, co-production, and knowledge translation from the inception of the research project. This will ensure the appropriate resources and budget are obtained for and committed to this purpose. Furthermore, we recommend embedding PPIE and co-production throughout the research cycle to enhance the quality and appropriateness of the research and its outputs. Additionally, integral to the research cycle is the incorporation of an evaluation of the knowledge translation process to examine the extent to which the implementation of the research findings achieved the intended effects.

When planning co-production activities, it is essential to manage and balance group dynamics and power inequities. We recommend that group work, such as workshops and focus groups, are carefully and sensitively designed with a focus on equality, diversity, inclusion, accessibility, and reciprocity.

While remote meeting options using platforms such as Zoom and Microsoft Teams are now more commonly used to enable people who are geographically dispersed to come together, events such as those we held are likely to benefit from in-person or at least hybrid opportunities, for example, in terms of the length of the event and level of interaction required.

> The workshop allowed us to be in a room full of bright ideas and lived experience which sparked further discussions and other lived experiences. It was authentic and rich; you couldn't achieve the same outcome just online. (India Henry, co-producer and co-author)

Finally, convening multidisciplinary and multicultural research teams facilitates the establishment of trust, helps with relationship building, and provides a natural basis for rapport building among stakeholders. In turn, this enhances opportunities to spark more creative, engaging, and reflective discussions for data analysis and interpretation.

> Researchers should be more creative in their way of working as one size does not fit all. Translating findings into resources is a great way to INFORM, EDUCATE, and EMPOWER people. Making it accessible to ALL (protected characteristics) in different formats, languages, platforms and tailored for faith communities (cultural differences, religious beliefs) allows for a wider reach in terms of promotion. (Chantel Sealey, co-producer and co-author)

Conclusion

Throughout this research, we used creative approaches to enable co-produced analysis and interpretation of research evidence. We achieved this by creating and conducting an evidence-based yet pragmatic knowledge translation framework, in which each step was informed by feedback collected in previous activities. This flexible approach allowed space to accommodate and respond fluidly to stakeholders' needs and to foster idea generation among stakeholders.

Adopting a system-wide approach and shifting from only *generating* findings to strategically including defined activities to guide and accelerate the process of *translating* research findings allowed us to collaborate with stakeholders to co-produce evidence-based recommendations for policy and practice. Our approach helped amplify and accelerate the dissemination of research through innovative channels such as sharing an informative video resource on social media, sexual health clinics, and conferences, thereby increasing the research impact.

In conclusion, we consider that using creative ways of analysing and interpreting research findings, while following robust, transparent, and systematic research methods, will help researchers to contextualise and gain a deeper and enriched understanding of research findings, and will increase research impact that would have not been possible using conventional research methods.

Notes

1 We would like to acknowledge the contribution of Hamish Mohammed and Adam Winter, who were part of the research team and contributed meaningfully to the development of this research translation project, as well as the input received from all our community and public health and sexual health sector colleagues and co-producers.

The authors acknowledge members of the NIHR HPRU in the BBSTI Steering Committee: Professor Caroline Sabin (HPRU Director), Dr John Saunders (UK Health Security Agency (UKHSA) Lead), Professor Catherine Mercer, Dr Hamish Mohammed (previously Professor Gwenda Hughes), Professor Greta Rait, Dr Ruth Simmons, Professor William Rosenberg, Dr Tamyo Mbisa, Professor Rosalind Raine, Dr Sema Mandal, Dr Rosamund Yu, Dr Samreen Ijaz, Dr Fabiana Lorencatto, Dr Rachel Hunter, Dr Kirsty Foster, and Dr Mamoona Tahir.

2 The views expressed are those of the authors and not necessarily those of the NIHR, the Department of Health and Social Care, or UKHSA (previously Public Health England).

3 In line with guidance from University College London (UCL) and NIHR, ethical approval was not required for this PPIE, co-production, and knowledge translation project. However, we conducted the work in line with available PPIE guidance at the time (https://www.nihr.ac.uk/documents/briefing-notes-for-researchers-public-involvement-in-nhs-health-and-social-care-research/27371). For the research conducted within the NIHR HPRU on BBSTIs, ethical approval was granted by the National Research Ethics Service Committee South Central – Oxford C, ref.: 15/SC/0223.

4 The research team is ethnically diverse, including members who identify and present as Black Caribbean or Black African, have experience living and working with Black communities, and have multisector professional experience across national surveillance, sexual health services, local authority public health teams, and community-based organisations. The diverse personal and professional networks among the team supported the identification of a professionally diverse group of stakeholders and those who have experience of living within Black communities.

5 https://youtu.be/h0O5sTjQWAk?si=v2eT72cXmqtcbGgC.

References

Adhikari, B., Pell, C. and Cheah, P.Y. (2020) 'Community engagement and ethical global health research', *Global Bioethics*, 31(1): 1–12. Available from: https://doi.org/10.1080/11287462.2019.1703504

Aicken, C.R.H., Wayal, S., Blomquist, P.B., Fabiane, S.M., Gerressu, M., Hughes, G., and Mercer, C.H. (2019) 'Pathways to, and use of, sexual healthcare among Black Caribbean sexual health clinic attendees in England: evidence from cross-sectional bio-behavioural surveys', *BMC Health Services Research*, 19(1): 668. Available from: https://doi.org/10.1186/s12913-019-4396-3

Aicken, C.R.H., Wayal, S., Blomquist, P.B., Fabiane, S., Gerressu, M., Hughes, G., and Mercer, C.H. (2020) 'Ethnic variations in sexual partnerships and mixing, and their association with STI diagnosis: findings from a cross-sectional biobehavioural survey of attendees of sexual health clinics across England', *Sexually Transmitted Infections*, 96(4): 283–92. Available from: https://doi.org/10.1136/sextrans-2018-053739

Cairney, P. and Oliver, K. (2017) 'Evidence-based policymaking is not like evidence-based medicine, so how far should you go to bridge the divide between evidence and policy?' *Health Research Policy and Systems*, 15(1): 35. Available from: https://doi.org/10.1186/s12961-017-0192-x

CIHR (Canadian Institutes of Health Research) (2012) 'Guide to knowledge translation planning at CIHR: integrated and end-of-grant approaches' [online]. Available from: https://cihr-irsc.gc.ca/e/documents/kt_lm_ktplan-en.pdf

CQC (Care Quality Commission) (2022) 'Culturally appropriate care' [online]. Available from: https://www.cqc.org.uk/guidance-providers/adult-social-care/culturally-appropriate-care

Esmail, R., Hanson, H.M., Holroyd-Leduc, J., Brown, S., Strifler, L., Straus, S.E. et al (2020) 'A scoping review of full-spectrum knowledge translation theories, models, and frameworks', *Implementation Science*, 15(1): 11. Available from: https://doi.org/10.1186/s13012-020-0964-5

Françoise, M., Frambourt, C., Goodwin, P., Haggerty, F., Jacques, M., Lama, M.-L. et al (2022) 'Evidence based policy making during times of uncertainty through the lens of future policy makers: four recommendations to harmonise and guide health policy making in the future', *Archives of Public Health*, 80(1): 140. Available from: https://doi.org/10.1186/s13690-022-00898-z

Haynes, A., Rychetnik, L., Finegood, D., Irving, M., Freebairn, L., and Hawe, P. (2020) 'Applying systems thinking to knowledge mobilisation in public health', *Health Research Policy and Systems*, 18(1): 134. Available from: https://doi.org/10.1186/s12961-020-00600-1

INVOLVE (2023) 'Co-production' [online]. Available from: https://involve.org.uk/resource/co-production

Kneale, D., Rojas-García, A., and Thomas, J. (2019) 'Obstacles and opportunities to using research evidence in local public health decision-making in England', *Health Research Policy and Systems*, 17(1): 61. Available from: https://doi.org/10.1186/s12961-019-0446-x

Lavis, J.N., Robertson, D., Woodside, J.M., McLeod, C.B., and Abelson, J. (2003) 'How can research organizations more effectively transfer research knowledge to decision makers?' *Milbank Quarterly*, 81(2): 221–48, 171–2. Available from: https://doi.org/10.1111/1468-0009.t01-1-00052

Mertens, D.M. (2012) 'Transformative mixed methods: addressing inequities', *The American Behavioral Scientist (Beverly Hills)*, 56: 802–13. Available from: https://doi.org/10.1177/0002764211433797

Mertens, D.M. (2021) 'Transformative research methods to increase social impact for vulnerable groups and cultural minorities', *International Journal of Qualitative Methods*, 20: 16094069211051563. Available from: https://doi.org/10.1177/16094069211051563

Migchelsen, S.J., Enayat, Q., Harb, A.K., Daahir, U., Slater, L., Anderson, A., et al (2023) *Sexually Transmitted Infections and Screening for Chlamydia in England, 2022.* London: UK Health Security Agency. Available from: https://www.gov.uk/government/statistics/sexually-transmitted-infections-stis-annual-data-tables/sexually-transmitted-infections-and-screening-for-chlamydia-in-england-2022-report

NIHR (National Institute for Health and Care Research) (2021) 'Briefing notes for researchers: public involvement in NHS, health and social care research' [online]. Available from: https://www.nihr.ac.uk/documents/briefing-notes-for-researchers-public-involvement-in-nhs-health-and-social-care-research/27371

NIHR (National Institute for Health and Care Research) (2022) 'Resources for co-producing research' [online]. Available from: https://arc-w.nihr.ac.uk/patient-and-public-involvement/co-producing-research-resources-and-toolkit-list/

Oliver, K., Lorenc, T., and Innvær, S. (2014) 'New directions in evidence-based policy research: a critical analysis of the literature', *Health Research Policy and Systems*, 12: 34. Available from: https://doi.org/10.1186/1478-4505-12-34

Straus, S.E., Tetroe, J., and Graham, I. (2009) 'Defining knowledge translation', *Canadian Medical Association Journal*, 181(3–4): 165–8. Available from: https://doi.org/10.1503/cmaj.081229

Tembo, D., Hickey, G., Montenegro, C., Chandler, D., Nelson, E., Porter, K. et al (2021) 'Effective engagement and involvement with community stakeholders in the co-production of global health research', *British Medical Journal*, 372: 178. Available from: https://doi.org/10.1136/bmj.n178

UKRI (UK Research and Innovation) (2021) 'Stakeholder engagement' [online]. Available from: https://www.ukri.org/councils/epsrc/guidance-for-applicants/what-to-include-in-your-proposal/health-technologies-impact-and-translation-toolkit/stakeholder-engagement/patients-and-the-public/

Wayal, S., Gerressu, M., Weatherburn, P., Gilbart, V., Hughes, G., and Mercer, C.H. (2020) 'A qualitative study of attitudes towards, typologies, and drivers of concurrent partnerships among people of Black Caribbean ethnicity in England and their implications for STI prevention', *BMC Public Health*, 20, 188. Available from: https://doi.org/10.1186/s12889-020-8168-0

Wayal, S., Reid, D., Blomquist, P.B., Weatherburn, P., Mercer, C.H., and Hughes, G. (2018) 'The acceptability and feasibility of implementing a bio-behavioral enhanced surveillance tool for sexually transmitted infections in England: mixed-methods study', *JMIR Public Health and Surveillance*, 4(2): e52. Available from: https://doi.org/10.2196/publichealth.9010

WHO (World Health Organization) (2021) 'Evidence, policy, impact: WHO guide for evidence-informed decision-making' [online]. Available from: https://www.who.int/publications/i/item/9789240039872

WHO (World Health Organization) (2022a) 'Redefining sexual health for benefits throughout life' [online]. Available from: https://www.who.int/news/item/11-02-2022-redefining-sexual-health-for-benefits-throughout-life

WHO (World Health Organization) (2022b) 'Sexual health' [online]. Available from: https://www.who.int/health-topics/sexual-health#tab=tab_1

Woode Owusu, M., Estupiñán Fdez. de Mesa, M., Mohammed, H., et al (2023) 'Race to address sexual health inequalities among people of Black Caribbean heritage: could co-production lead to more culturally appropriate guidance and practice?', *Sexually Transmitted Infections*, 99: 293–5. Available from: https://sti.bmj.com/content/sextrans/early/2023/05/02/sextrans-2023-055798.full.pdf

26

Creative collaborative data analysis: co-constructing and co-analysing the data together

Jess Mannion and the R&S (Relationships and Sexuality) Research Team[1]

Introduction

This chapter documents a cooperative inquiry with co-researchers with intellectual disabilities, exploring their experiences of relationships and sexuality. The method was devised for adults with intellectual disabilities who were new to data analysis. This project combined existing methodological techniques in the literature and then developed them to make them more accessible to the group. Analysis workshops began with the most common words from the group's previous meeting transcripts. Words were written on a card and provided on large sheets of paper and art materials. Co-researchers were asked to choose the word that was most important to them and creatively express their thoughts in a poster. Posters were discussed by the group for sense making and to see if anything was misunderstood or missing. Once the data was generated, we then dived into the analysis to see what would happen, which resulted in us co-creating a collaborative approach to analysis.

The chapter begins by introducing our project, and the epistemological and ontological positions that underpinned this way of working and engendered a creative method of collaborative data analysis. This is a six-step systematic method that includes visual analysis, member checking, content analysis, arts-based analysis, embodied data analysis, and creative collaborative reflections. The chapter evaluates this method from the group's perspective, considering the limitations of the method we encountered and reflecting on the related ethical considerations.

Background to the study

This project stemmed from my doctoral study in the School of Nursing and Midwifery in Trinity College Dublin, the University of Dublin, in Ireland (Mannion, 2024). I worked with a team of co-inquirers with intellectual disabilities, called the R&S (Relationships and Sexuality) Research Team. This research team formed and participated in a cooperative inquiry group, where its members acted as co-researchers and co-subjects, who were actively involved in all the decisions in the study and co-created the research findings on topics relating to relationships and sexuality, based on their own life experiences. The study was granted ethical approval from the university's faculty ethics committee.

As it was a collaborative project, it was important that the co-inquirers had the opportunity to be meaningfully and authentically engaged in all aspects if they so wished. The co-inquirers were involved in choosing research topics. We explored a range of topics and the four main themes that they selected from these were, firstly, sexual health education, to include LGBTQ+; secondly, online safety, to include dating apps, social media platforms, and the impact of Grand Theft Auto on our perceptions of sex work; and, thirdly, rights, to include marriage and children, parental and staff perceptions, group homes, and the law on disability rights. We also felt the theme of disability ran throughout all the themes.

For each fortnightly meeting over the year-long study, the co-inquirers chose a different topic that arose from our group conversations to explore and decided upon one visual and creative method to aid them to do this. Initially the co-inquirers chose the visual and creative methods from a selection of alternatives I suggested. As the study went on, the co-inquirers selected visual and creative methods themselves through group conversations we had. These methods were art and drawing, participant poetry and rap, creative performance and practice, joint sandboxing, LEGO® SERIOUS PLAY®, and third objects. Art and drawing helped us to make images during each data analysis phase and when we were exploring the topic of marriage and children. When utilising poetic form, we wrote short poems individually on our experiences of sexual health education, and together as a group reflection at the end of the study. We wrote and performed raps in groups to consider our thoughts on the topics of online safety and how the video game Grand Theft Auto influenced our perceptions of sex work. We used a drama activity to examine the topic of marriage and children for people with intellectual disabilities and made a short film to disseminate some of our research findings on this topic. The final three creative methods were used to facilitate our reflection on the project at the end. Questions we devised to respond to included what we learnt about ourselves, others, disability, and relationships and sexuality on the project; what the biggest thing we were taking away from the study was; and what our hopes and dreams were for the future regarding relationships and sexuality for people with intellectual disabilities. For joint sandboxing (Mannay, 2019), we created scenes with sand, water, figures, and objects in response to reflective prompts. Through LEGO® SERIOUS PLAY® (Gauntlett, 2007), we built metaphors based on reflective prompts. Third objects are where an individual uses an object as a point of reference to focus on (Winnicott, 1968), which can help frame the topic and lead the conversation (Kara et al, 2021). We each brought a made or found object to the research meeting as a point of reference to a discussion prompt. After creating these visuals, we shared them with the group and engaged in a discussion to help draw out a shared understanding of these topics. We identified that our favourite visual and creative research methods were LEGO® SERIOUS PLAY® and sandboxing because these methods were new to us, the sandboxing was relaxing, we liked how the sand felt in our hands, and they were both fun methods. Bruce (one of the co-inquirers) did not want to use the words LEGO® SERIOUS PLAY® at a conference we presented at about the study, because this did not reflect how he felt about it; he changed the name to LEGO® FUN PLAY® and we all agreed.

Together we decided and planned for actions arising from the study. One of these actions pertained to the dissemination of some aspects of the research findings to create awareness on these topics with the aim of change; this is because we identified other people's judgements as a barrier. To facilitate dissemination of the key findings some members successfully applied for funding, which enabled us to co-produce a presentation (Mannion et al, 2022a). The presentation was shared at an international action research colloquium and in a co-authored academic blog (Mannion et al, 2022b), and we created a short film, devised from some of the research transcripts, to form part of an awareness event we held at Trinity College Dublin (Mannion et al, 2022c).

The key focus of this chapter is how we worked together to co-create the data analysis method and co-analyse the data. To make this accessible and interesting to inexperienced data analysts, we used a range of creative methods to help us in this process, which are outlined in the following sections.

Epistemological and ontological positions

The decision to work collaboratively in research, including co-constructing the data analysis method and co-analysing the data, was born out of my epistemological and ontological positions (Mannion, 2024). My epistemological stance is constructivism. I reflected on the paradigms underpinning the research by undertaking a first-person inquiry (Marshall, 2016), displaying who I was (Bosher, 2002) and writing myself into the study. I wanted to be aware of how I impacted on the study's construction, participation, and outcomes (Coghlan and Brydon-Miller, 2014). Through this I identified past experiences of lacking voice and agency due to my own disability and classed intersectional identities. However, as a doctoral researcher I had the privilege of being able to change my situation. This was the driving force that led me to doing research in a manner that gave others the opportunity to speak, and to be heard, where together we could act and possibly make positive changes. Through this exploration I could see that the study was influenced by my personal values of social justice, equity, rights, access, and participation that arose from my lived experiences. Consequently, the most authentic approach to take was cooperative inquiry. Researching collaboratively enabled us to collectively construct an understanding of our experiences, while breaking down power barriers and creating a space to have a voice.

My ontological position was concerned with social constructivism. Social constructivism is where jointly constructed perceptions of the world are created, which result in shared assumptions that can impact on the concept of identity (Vygotsky, 1978). It is a very useful concept when analysing disability as it helps the researcher to look at influences that shape the meaning of disability such as culture, sociopolitical factors, societal attitudes, language, and power relationships (Żółkowska, 2016). Understanding how disability is constructed, interpreted, and experienced helps us to increase social awareness and to understand the reality of labelling and classification for people with intellectual disabilities (Żółkowska, 2016). Although I did not use this language to discuss this with the group, as it is inaccessible to some, and because I did not want to influence their thoughts, this came up in our conversations. For example, when the group identified experiences of other people's negative judgements, this led to the group discussing labelling and perceptions on disability and the negative impact this had on them.

Identifying the impact of labelling and perceptions on disability brought social justice and social constructivism together. We used this research to try to change some of these perceptions. When people with intellectual disabilities are discussed in research, for example, in ethics applications, they are labelled as 'vulnerable' and 'at risk'. The R&S Research Team demonstrated they were researchers, public speakers, writers, and authors. The team demonstrated skills such as data analysis, applying for funding, presentation skills, and writing for publication. We educated others about relationships and sexuality for people with intellectual disabilities through our dissemination activities. It is hoped that we contributed to changing people's attitudes through all these activities, challenging and moving beyond dominant discourses of vulnerability, risk, and othering.

Creative collaborative data analysis

Creative data analysis 'can refer to the analysis of data gathered using creative methods … or it can refer to creative methods of analysing data gathered using creative or conventional methods (or a combination of the two)' (Kara et al, 2021, p 83). This study used creative ways to analyse the multimodal data, using an embodied and collaborative approach (Mannion, 2024). If a study is fully inclusive then there should be the option to collaborate in the data analysis, and if co-inquirers have been fully involved and feel the project is theirs, and that the method is accessible, then they may want to engage in this stage (Kara, 2015). All members of the R&S Research Team did want to engage in the data analysis stage of the research.

It was important that the method was systematic as it needed to yield findings that were embedded in the data, and that helped with the next stages of writing, dissemination, and application (Kara, 2015). Mixed-methods analysis was used in this study, which uses two or more different methods of analysis and allows for the creation of a richer understanding of the phenomena (Kara et al, 2021). I brought existing ideas and methods from the literature together to devise a systematic method of creative collaborative data analysis that would be accessible and interesting, where we could work through the analysis together. These ideas came from visual analysis (Kara et al, 2021), member checking (Call-Cummins et al, 2019), content analysis (Grbich, 2013), and embodied data analysis (Thanem and Knights, 2019). All these ideas and methods are described in detail in the creative collaborative data analysis section below, where I outline the devised systematic method we used.

Initially, I did not consider co-creating this method, as data analysis was new to the group and I was concerned with overburdening or daunting them. Barriers to participation in collaborative data analysis may be that it is time consuming, tedious, and repetitive (Kara, 2015). These factors needed to be considered, to ensure it was accessible and engaging to the co-inquirers. As a result, most of the analytic work took place collaboratively in the research team meetings and different accessible methods were used to support the analysis. Examples of accessible methods included making video summaries of each research meeting instead of written summaries to undertake member checking of the data, and using arts-based methods to analyse the data.

However, as things transpired, this process resulted in a co-creation of the method in several ways. During the final process of data analysis, one of the co-inquirers contributed to the methods development, as explained in the embodied data analysis section. Also, after the main themes were decided and I had brought the leftover data to the group, they came up with the idea to use some of these for subthemes. The group also devised one of the reflective sessions as part of this process by designing a method and developing a research question. This co-creation made the method truly collaborative.

Creative collaborative data analysis method

1. Visual analysis:
   The group chose a different topic each research meeting, along with one creative activity to explore that topic. For the first hour we engaged in art and drawing, participant poetry and rap, creative performance and practice, joint sandboxing, LEGO® SERIOUS PLAY®, or by bringing in found or made third objects to share and aid discussion. We sat around a large table to work individually on our creations, except for when we wrote raps, a news story, and a poem in groups. In the initial stages I facilitated the group activities until the group felt comfortable to either facilitate individually themselves or to contribute to the facilitation when they wanted. As this was a collaborative project, I did not want to give this data my own interpretations. As we created, we often spoke, sometimes about the topic, sometimes socially. It was important that we had space for discussion on what we had created. Therefore, we dedicated the second hour to sitting together to share and discuss our creations and explore their meanings. An example of this is when the group decided on two topics to explore in one research meeting as these had come up in the previous research meeting briefly and we wanted to delve into them more. They were online safety and the impact of the video game, Grand Theft Auto, on our perceptions of sex work. The group decided they wanted to try rapping, so each member chose which of the two topics they

wanted to rap about. When the two groups had written and rehearsed their raps, we performed them one at a time. After each rap we sat together and discussed it further. All these discussions were recorded to create a transcript.

2. Member checking:
   Member checking helps to ensure the interpretations of the co-inquirers' perspectives are accurate. However, as Sandelowski (2002) points out, member checking should be viewed as an additional opportunity to discuss the findings as opposed to validating them, as perspectives can change over time. Having the space to discuss these topics again helped us to do this. I borrowed ideas from Call-Cummings et al (2019), who sent copies of transcripts to participants for feedbackand further reflection, to help in the analysis. However, I wanted to find a more accessible method to do this. I listened back to the research meeting recording and made a ten-minute video summarising the main interpretations of each meeting after they took place. This video was emailed to each co-inquirer with five questions to respond to. The individuals had the choice to email back or discuss the video and questions at the next research team meeting.
   Questions:
   1. Did I misinterpret any information? If I said anything wrong, please let me know what it was.
   2. Did I miss any important information in the video? If I did, please let me know what it is.
   3. What did you find most interesting?
   4. What did you find least interesting?
   5. What would you like to do or talk about next?
3. Content analysis:
   When there is a large amount of textual data, it can be systematically categorised or coded by finding trends or patterns in the words used. In content analysis, the repetition of words is presumed to signify their importance (Grbich, 2013). Instead of waiting till the end of the study when we would have an overwhelming amount of data after 20 meetings, I undertook content analysis twice during the research. The first time we decided to do this was after seven research meetings, as at this point we moved the study from online, due to COVID restrictions, to in person. We wanted to look at where we were in the study after discussing some topics at the start, and where we wanted to go next, and felt this would guide us. We were going to do the content analysis again after meeting 15, before we broke up for a month's summer break, but instead we decided to do it on our return to refresh us on where we were and to help us decide where we wanted to go for the remainder of the study. Here I undertook content analysis of the data from the next seven meetings. The approach used was where word frequency was identified by counting the main words used by the group and selecting the most common words from the transcripts (Grbich, 2013). After omitting words such as 'a', 'in', and 'the', the remaining key words were written on cards. This approach enabled us to work with data that was now manageable but still meaningful. However, this was

a preliminary activity as we wanted to work on the data together, to dig deeper, to include all our interpretations, and to provide a more complete picture.

4. Arts-based analysis:
   Arts-based analysis is when arts-based methods are used to analyse the data. The word cards from the content analysis were placed on a large table at two research meetings (after the first seven meetings and after the next seven meetings). For example, in the first data analysis session the words were 'sex', 'girlfriend', 'boyfriend', 'sex education', and 'nightclubs'. The group looked at the words, deciding whether to remove or add words, or to bring words together. Each member then chose a word most important to them and made a visual with drawings and words to express its meaning. Each co-inquirer then shared their visual and its interpretation.
   For example, in the first data analysis session Chloe, one of the co-inquirers, chose the word 'nightclubs' and drew an image of ladies on a dance floor (Figure 26.1).
   Chloe explained that she picked the word 'nightclubs' and drew people dancing because she would love to go out with friends, meet new friends, and possibly meet a boyfriend. She expressed she did not get the opportunity to do this and spent a lot of time alone. The group agreed they would like more opportunities to meet potential partners and felt this was an important topic. The sharing of our art and interpretations led to further discussions and through this process we decided the main themes at that point based on what was important to us.

Figure 26.1: Chloe's artwork on 'nightclubs'

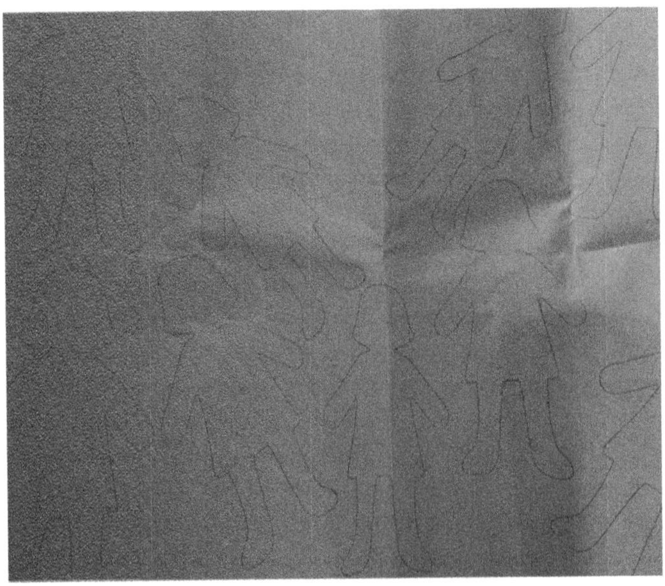

Source: From the first data analysis session (Mannion, 2024).

5. Embodied data analysis:
   Embodied data analysis helps to conceptualise the work and realise new perspectives of the data during the process (Kara et al, 2021). I was inspired by the work of Thanem and Knights (2019), who used hard copies of the data, printed field notes, and recordings of data to cut out key quotes or observations and, along with their creative artefacts from the study, spread them out on the floor and walked, crawled, and played with the data. The key quotes and observations can then be arranged into themes and subthemes. Photographs or images of the data can be taken during this process to keep track of the data. Thanem and Knight (2019) recommended one keep doing this till new learning takes place and the data is seen differently.
   Ellingson (2017) highlighted that embodiment can be done in accessible ways such as online. I adapted Thanem and Knights' approach to make it accessible to the group by having two sessions instead of one – so we were not working for more than two hours at a time and so nobody was rushed – and by having the material on the table instead of the floor so it was more easily accessible. Our artefacts included our artwork, poems, and the news reports, which were displayed on a large boardroom table (see Figure 26.2) along with the word cards, while the raps we wrote and performed played on a speaker, like an art installation. I asked the co-inquirers to walk around the room and play with the data, do what felt right.
   One co-inquirer began lifting artefacts and themes from the table, and suggested others do the same, advising them to choose what was important or stood out. When everyone sat down, we shared what we had chosen and why, and had a group discussion. We removed everything else off the table and the members placed their data in the middle. Here they stood around and discussed the data, moving one item when they felt it was not relevant on its own and could join another theme. In this process they decided on the four most relevant themes and subthemes in their study. The four main themes that they selected were: disability, relationships and sexuality education, online safety, and rights. I took photos as we developed themes. The exciting part was that a co-inquirer suggested that we should choose artefacts and themes of importance, and it worked well.
   In the following research meeting, I placed the themes in the centre of the table but put the removed data back onto the table so it was in front of them in case they wanted to add any items they had omitted. We went through each piece of removed data one at a time. The group decided to come up with subthemes, debated what main theme each subtheme should go under, and removed any repeated or unrelevant subthemes. The group were happy with the four themes they had agreed upon in the previous meeting and added important subthemes to the relevant themes. The themes and subthemes we had now were: firstly, relationships and sexuality education, to include LGBTQ+; secondly, online safety, to include dating apps, social media platforms, and the impact of Grand Theft Auto on our perceptions

Figure 26.2: Data analysis table

Source: Mannion (2024).

of sex work; thirdly, rights, to include marriage and children, parental and staff perceptions, group homes, and the law on disability rights; fourthly, the theme of disability, which ran throughout.

6. Creative collaborative reflection:
   We ended the study with three creative reflective sessions. One of these sessions was devised by the group. They stated they wanted to either make something in their own time or to bring something in that represented 'the biggest thing they are taking away from the group'. This approach is known as third objects, where an individual uses an object as a point of reference on which to focus (Winnicott, 1968), which can help frame the topic and lead the conversation (Kara et al, 2021). We followed this with a discussion on personal transformation, skills developed and acquired, and what we learnt about research. For example, Alice, one of the co-inquirers, shared her third-object creation on the biggest thing she was taking from the group (see Figure 26.3).
   Alice said the string represented life and that people were expected to be on one linear path, but life does not work like that, just like the string will not stay straight. She said the shells represented us and even though we may look different we are

Figure 26.3: Example of a third object

Source: Made by Alice (Mannion, 2024).

all on the same path, living our lives and wanting similar things. The group agreed that this reflected what they had brought up in many discussions throughout the study on their rights. They also discussed how people's perceptions of them affected how they wanted to live their lives, but they agreed with Alice that they wanted the same things as other people – relationships, marriage, children, and so on – and if they did not want those things that is okay too. What was important is that they made their own decisions, and their rights were not taken away.

As illustrated in Figures 26.4 and 26.5, using sandboxing (Mannay, 2019) we explored the learning that took place in relation to ourselves, others, disability, and relationships and sexuality during the research project.

Bruce's sand scene explored what he learnt about himself during the research (Figure 26.5).

Bruce explained that in his scene there was a person who used a wheelchair, that they could be good at sports and driving and, if they were not, they could find another way of getting around, such as catching a bus. This led to a discussion where we learnt we did not need to be held back on what we wanted to do in our lives solely based on our disabilities. Bruce further explained his sand scene

Figure 26.4: Sandboxing materials

Source: Mannion (2024).

Figure 26.5: Example of a sandboxing scene

Source: Made by Bruce (Mannion, 2024).

Figure 26.6: Example of a LEGO® SERIOUS PLAY® metaphor

Source: Made by Louise (Mannion, 2024).

> was a metaphor for what he learnt about himself in the study, which was that he can enjoy life more and have fun, especially with the group. This generated a discussion on the relationships we had built in the group during the study, and how we were more accepting of our disabilities from being in the group.
> LEGO® SERIOUS PLAY® (Figure 26.6) is an experiential form of communication and analysis (Kara et al, 2021). Using LEGO® SERIOUS PLAY®, the group explored successes, dissatisfactions, and hopes for the future from the study. Louise, one of the co-inquirers, shared her LEGO® SERIOUS PLAY® metaphor on the hopes she had for the future regarding relationships and sexuality for people with intellectual disabilities (Figure 26.6).

Louise noted her LEGO® construction was of a tower and bridge to show how people who use wheelchairs are affected. She said it also represented how we built a connection and that she learnt we should have rights. Louise voiced she hoped that people with intellectual disabilities could get a partner no matter what and shared her personal relationship hopes for the future. The group agreed and shared their hopes for the future.

The end of the data analysis

When we finished analysing the data, we stood over the table looking at our main themes. There was a sense of pride and achievement that we had got to the end of the year of the study. We could see all our hard work in front of us and we had brought it together and highlighted what was the most important to us.

Ellen: It felt good. It's our work, and we did it and I'm proud of that. I am glad I did it.
Jess: Ellen, you said let's pick what's important to us and that made us come to this point, so thank you. Your contribution has helped us develop a new method of data analysis.
Ellen: That makes me feel good.
Bruce: I am happy that the work is done.
Hermione: It was good looking at all the stuff we have done.
Alice: I felt overwhelmed at the amount of work. It's amazing when you see it all out. The amount of work that's gone into this, in just a year.
Jess: Well done to you all. You have worked for a year on this PhD project, that's not easy. It's amazing the work that you have put in and I could not have done this without you. It was important to work together as a team. So, thank you so much. It's not the end but we have finished analysing our data. I'm really excited. I will take this away and have lots of fun writing it up. So that's us done, five minutes early as well!
Everyone clapped with huge smiles on their faces. (Mannion, 2024)

Later we presented our data at an international action research colloquium (Mannion et al, 2022a). As part of this presentation we discussed how it felt to be involved in co-creating the data analysis method and in co-creating the data. We identified that one of the things we learnt about research was data analysis and we were interested in doing so. Although we identified that it can be hard to think or talk sometimes, we felt it facilitated our voices being heard and it was fun talking about these topics. We felt it was nice to see what was important to each other and it was exciting to see the research come together. We felt excited and confident going over the data, it went smoothly and seemed to fly (Mannion et al, 2022a).

I then took the list of themes and subthemes we agreed upon to my desk and printed all 20 research meeting transcripts. I colour coded each theme and subtheme, then went through each data transcript and highlighted these themes with different coloured highlighter pens. I was then ready to write our findings.

Limitations of the data analysis method

At the start of the study, we were online, and the group decided to work on creative activities in their own time. However, in practice this did not work for everyone as some members did not engage with the creative activities. It worked much better when we moved to in-person research meetings and integrated the creative activities into our meeting time. Also, during the LEGO® SERIOUS PLAY® it took quite some time for one person to grasp the idea of a metaphor. These sessions take time. Sometimes we ran out of time in the sessions to fully discuss everything we wanted to. When we met two to four weeks later, the momentum was gone for that topic/creative activity.

The member checking worked better in person rather than through email. People did not always watch the video summary at home, and some did not reply to the questions via email, so it was not as accessible as in person.

The content analysis was done exclusively by me, and where it did result in more manageable, accessible work and left the focus on the meaningful work for the group, it did negate the participatory element.

During the embodied data analysis stage, it could be argued that the co-inquirers just chose their own work. While there was an element of this, co-inquirers picked up others' work too. Even when they did pick their own work it was not necessarily a bad thing. Bruce chose his own poem and said he did so because it was the first poem he had ever written, he was proud of it and writing it helped him to think about being safe around sex. That poem detailed his past sexual health education, and it was a topic important to him and the group. We also spent two sessions looking at all this data together. However, Bruce noted near the end of a session that it was a bit tiring.

Finally, all this work took time; we worked together on this for a year, mostly meeting for two-hour fortnightly meetings. If a research project does not have this time, it may not work. This was a small group who had built up a strong working relationship, which added to the effectiveness. The co-inquirers had mild to moderate intellectual disabilities; this method may not work for groups with severe or profound disabilities.

Ethics in the data analysis stage

When using visual and creative data construction, the ethical considerations needed to be contemplated. As I mentioned earlier in the chapter, I was concerned with how I may impact upon the study, and one of these ways is how my bias may have affected the analysis phase (Kara et al, 2021). This is another reason why it was important that I engaged in first-person and inquiry and reflective practice.

Not everyone is comfortable with creative activities (Mannay, 2016). Barriers to creativity in adults are based around lack of confidence (Kearney and Hyle, 2004), fear of the unknown, and concerns about being judged (Kelley and Kelley, 2012). As a result, at the end of each meeting we chose, from a range of possibilities, what method might work next. I always reminded the group that it was not about what the end piece looked like, and they should only share if they felt comfortable to do so. Only non-identifiable objects were photographed for the thesis and other forms of dissemination, ensuring none of the co-inquirers were in view, and only with permission from each co-inquirer each time.

Working collaboratively can result in more ethical practice (Kara et al, 2021). I wanted to ensure that the work was truly participatory and not just tokenistic. There was an onus on me as the initiating researcher to ensure the analysis was conducted ethically and collaboratively, to be true to the cooperative inquiry. This way of working can widen perspectives (Sword et al, 2018) and is essential in fully participatory research (Cole, 2017). In the final analysis session (Mannion, 2024), I was concerned that something important would be left out because it seemed too easy for the co-inquirers to choose what was important for them out of so much data. I had to recognise this was their data, based on their experiences, and so the choices were up to them. Ethically, this would ensure that emphasis was not placed in areas unimportant to the co-inquirers and prevent my unconscious bias from affecting the data. At one point, I disagreed with the co-inquirers when they wanted to include a subtheme under a theme where I felt it did not belong, but they stood their ground and argued that it should be included. The co-inquirers took ownership of the data and the method, and had the confidence to stand their ground and prove me wrong. Without their inclusion, I would have omitted this important data.

Conclusion

One aim of this cooperative inquiry group was for the co-inquirers to be meaningfully and authentically engaged in all aspects of the research. This included the data analysis phase. This way of working originated from my constructivist and social constructivist positions, my own disability and class intersectional identity experiences, and values of social justice, equity, rights, access, and participation. I was interested in how intellectual disability, in the context of relationships and sexuality, was constructed, interpreted, and experienced. We undertook the research collaboratively, which resulted in having a space for our voices, while breaking down perceived power hierarchies enabled us to collectively construct an understanding of our experiences. Combining existing methods, making it creative and accessible, along with diving in and asking the group what they felt we should do, resulted in an in-depth analysis of the data to unearth what was important to us. This facilitated all voices being heard, and a sense of pride, achievement, and excitement in seeing a whole year's work come together. We started the research together and we ended it together, creatively and collaboratively analysing our data and sharing the findings generated in this co-analysis process.

Note

1 We would like to acknowledge Dr Fintan Sheerin and Dr Vivienne Brady, School of Nursing and Midwifery, Trinity College Dublin, the University of Dublin, as supervisors of this PhD.

References

Bosher, M.A. (2002) 'How can I as an educator and professional development manager working with teachers, support and enhance the learning and achievement of pupils in a whole school improvement process?' PhD thesis, University of Bath.

Call-Cummings, M., Hauber-Özer, M., Byers, C., and Mancuso, G.P. (2019) 'The power of/in photovoice', *International Journal of Research & Method in Education*, 42(4): 399–413.

Coghlan, D. and Brydon-Miller, M. (eds) (2014) *The SAGE Encyclopaedia of Action Research*, vols 1–2. Thousand Oaks: SAGE Publications.

Cole, P. (2017) 'An indigenous research narrative: ethics and protocols over time and space', *Qualitative Inquiry*, 23(5), 341–51.

Ellingson. L.L. (2017) *Embodiment in Qualitative Research* (1st edn), New York: Routledge.

Gauntlett, D. (2007) *Creative Explorations: New Approaches to Identities and Audiences*, London: Routledge.

Grbich, C. (2013) *Qualitative Data Analysis: An Introduction*, London: SAGE Publications.

Kara, H. (2015) *Creative Research Methods in the Social Sciences: A Practical Guide*, Bristol: Policy Press.

Kara, H., Lemon, N., Mannay, D., and McPherson, M. (2021) *Creative Research Methods in Education: Principles and Practices*, Bristol: Policy Press.

Kearney, K.S. and Hyle, A.E. (2004) 'Drawing out emotions: the use of participant-produced drawings in qualitative inquiry', *Qualitative Research*, 4(3): 361–82.

Kelley, T. and Kelley, D. (2012) 'Reclaim your creative confidence', *Harvard Business Review*, 90(12): 115–18.

Mannay, D. (2016) *Visual, Narrative and Creative Research Methods: Application, Reflection and Ethics*, Abingdon: Routledge.

Mannay, D. (2019) 'Artefacts, third objects, sandboxing and figurines in the doll's house', in L. Pauwels and D. Mannay (eds), *The SAGE Handbook of Visual Research Methods* (2nd edn), London: SAGE Publications, pp 322–31.

Mannion, J. (2024) 'Establishing a co-operative inquiry group for people with intellectual disabilities to explore relationships and sexuality', PhD thesis, Trinity College Dublin, the University of Dublin. Available from: http://www.tara.tcd.ie/handle/2262/104363

Mannion, J., Blee, P., Gallagher, R., Gallagher, T., Higgins, B., McHugh, M., and Mulligan, J. (2022a) 'The R&S (Relationships and Sexuality) Research Team: a collaborative inquiry group: sharing our journey and learning, of inclusive and accessible research', Paper presented at the 10th International Action Research Colloquium, 'Action Research for Collaboration and Conversation in Today's World', TU Dublin, 24 June.

Mannion, J., Blee, P., Gallagher, R., Gallagher, T., Higgins, B., McHugh, M., and Mulligan, J. (2022b) 'We are proud to be researchers with disabilities', Methodspace: SAGE. [Online blog.]

Mannion, J., Blee, P., Gallagher, R., Gallagher, T., Higgins, B., McHugh, M., and Mulligan, J. (2022c) 'We have the right to get married and have children', Paper presented at the School of Nursing and Midwifery, Trinity College Dublin Public Lecture, Trinity College Dublin, the University of Dublin, 23 September.

Marshall, J. (2016) *First Person Action Research: Living Life as Inquiry*, London: SAGE Publications.

Sandelowski, M. (2002) 'Reembodying qualitative inquiry', *Qualitative Health Research*, 12: 104–15.

Sword, H., Blumenstein, M., Kwan, A., Shen, L., and Trofimova, E. (2018) 'Seven ways at looking at a data set', *Qualitative Inquiry*, 24(7): 499–508.

Thanem, T. and Knights, D. (2019) *Embodied Research Methods*, London: SAGE.

Vygotsky, L.S. (1978) *Mind in Society: The Development of Higher Psychological Processes*, London: Harvard University Press.

Winnicott, D.W. (1968) 'On the use of an object and relating through identification', in D.W. Winnicott (ed), *Playing and Reality*, London: Tavistock, pp 86–94.

Żółkowska, T. (2016) 'Construction of intellectual disability: (de)construction of the social role of intellectually disabled persons', *International Journal of Developmental Disabilities*, 62(4): 213–23.

27

Perspective Narrative Analysis: analysing data creatively within a participatory research group

Chloe East, Feydakeen Smith, Iris, Zaynab Charafi, Zoe Fordham, Annie Champion, and Carys Jones

Introduction

This chapter describes how a participatory research group – named Fizzacc after the initials of our names – approached analysing our data.[1] Fizzacc is a neurodiverse group that comprised six autistic girls between the ages of 13 and 18 and a postgraduate researcher with 20 years' experience as a secondary school teacher.[2] It was created to explore three research questions:

- What do autistic girls feel are the barriers to successful transition from primary to secondary school?
- How can creative data collection and analysis methods promote the voices of autistic girls?
- What are the challenges and benefits of including autistic girls in participatory research?

The focus here is on how we used creative data analysis methods within a participatory framework to explore autistic girls' transition experiences. The process involved developing stories describing events from two different perspectives: a first-person narrative to reflect the lived experiences of the group and a third-person account of the same event. A series of questions and talking points were also included to further explore what the data meant to us.

Challenges of data analysis within a participatory research framework

The creative analysis process in the research came about as a way to address the challenges of including data analysis as part of a participatory research project. Participatory research refers to the practice of meaningfully including the community affected by the research in the research process (Fletcher-Watson et al, 2019). It aims to address the imbalance of power between researchers and the researched, and to draw on the wisdom of those with lived experiences to strengthen research findings (Pellicano, 2020). Although participatory research is increasing in popularity in some areas of study, there remains a disparity in how much those with lived experience are involved in the research beyond the data collection phase. This is particularly apparent in how group members are involved in the process of analysing data.

Analysis can be particularly challenging when working within a research team that includes members without training or experience in research practices. This is common in participatory research, which often includes researchers from marginalised or seldom-heard communities. For Fizzacc, an additional challenge was analysing data – often a very time-consuming process – with young people who had limited time to devote to the project as it was being conducted alongside schoolwork and exams. To support meaningful participation in data analysis for all members of the team, it was therefore necessary for us to be creative with our approach.

Challenges of data analysis within a neurodiverse research group

To explain how our data analysis process worked this chapter will use Fizzacc's research to provide a worked example. This will show just one possibility of a process that is flexible enough to support the sharing of a wide range of voices.

We know that the transition between primary and secondary school can be challenging for autistic children (Nuske et al, 2019). We also know that there are differences in the way autism is perceived in girls, which can lead to a later diagnosis (Whitlock et al, 2020) or lack of support (Wood-Downie et al, 2021). What we wanted to do with our research was to capture the power of what the girls were sharing about their transition experiences and analyse the stories they told in a way that didn't diminish them. We collected our data using body mapping (Solomon, 2007), which is a creative method that investigates the embodied experiences of participants through a series of exercises based around the creation of a life-sized representation of the body (see Figure 27.1).

When exploring how to analyse the body-mapping data, I recognised that as a non-autistic researcher I needed to be very mindful of how my participation in the analysis process would impact on the authenticity of it. I was particularly mindful of the impact of the double empathy paradigm (Milton, 2012), which refers to the differences in social interaction between autistic and non-autistic individuals leading to barriers to understanding. It is a 'double' problem because it is experienced by both sides, and it should not be incumbent on the autistic individual to adapt their behaviour to fit in with the perceived norm. Ableist assumptions about autism led to theories arguing that autistic people lack theory of mind, but these only apply if autistic people are being asked to assume the mental states of non-autistic people; it has been shown that non-autistic people find it equally difficult to assume the mental states of autistic people (Mitchell et al, 2021). For research to address the double empathy problem it is important to have autistic voices involved in all stages, and to avoid neurotypical researchers making assumptions based on perceptions of best practice drawn from conjecture. If this is not addressed, the ability to achieve a 'true knowledge exchange' will be

Figure 27.1: Body map

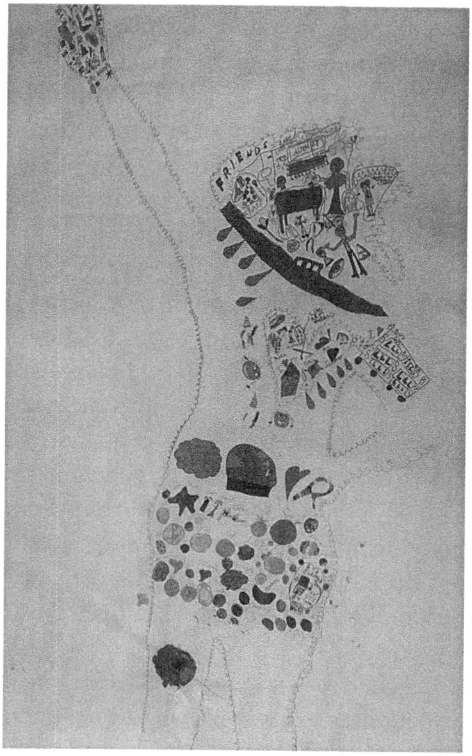

impossible (Fletcher-Watson et al, 2019, p 950). Therefore, for Fizzacc's research to achieve participatory aims, it was not only necessary to include autistic young people in a participatory way, but to also be continually mindful of how my understanding of their perspective is shaped by the double empathy problem.

We therefore needed to seek out a way of approaching the data analysis that would enhance the participatory nature of the research and promote the voices of the co-researchers. This search led me to Narrative Inquiry.

Drawing on the methods of Narrative Inquiry

Narrative Inquiry is a broad approach that encompasses a range of methods for data collection and analysis, centred around using stories as a way to understand human experience.

Barone (2007, p 466) states that:

> Our aim as researcher-storytellers is not to seek certainty about correct perspectives on educational phenomena but to raise significant questions about prevailing policy and practice that enrich an ongoing conversation.

Caine et al (2017) defined three purposes for using stories to explore data: firstly, to protect the identities of participants; secondly, to harness the imagination to enrich understanding; and, thirdly, to create distance between self and experience. By sharing thoughts and feelings within fictionalised accounts, experiences are presented authentically but without fear of individuals or establishments being identified. Fictionalisation also allows for an amalgamation of experiences to be presented when working collectively. As Fizzacc was using creative methods of data collection, continuing along this imaginative trajectory to create stories based on real experiences seemed like a natural progression.

Polkinghorne (1995 p 12) talks of two different types of narrative inquiry. The first he calls 'analysis of narratives' – a process whereby stories are gathered together, and where analysis is carried out on those stories to allow the development of themes. This is akin to using stories as data and then applying a thematic analysis to this data. The second, 'narrative analysis', involves the researcher gathering data that they then organise into story form. 'Narrative analysis' is, therefore, a process of synthesising data rather than of separation and categorisation. Our interest in the process was very much focused on Polkinghorne's second interpretation of narrative inquiry, as body mapping had provided a depth and richness of data for analysis. Our focus was on ensuring the process of data analysis would enhance knowledge and understanding, rather than diminish the data by unnecessary division and categorisation.

Bignold and Su (2013) argued for two forms of narrative inquiry: epistemological, which explores lived experiences as a way of considering the nature of knowledge; and ontological, which removes the researcher from the process to delve deeper into the truth as experienced by those in the narrative. The epistemological form was more aligned to our work in the way we sought to use narrative inquiry to explore the double empathy problem and what that means for our understanding of lived experiences. It was important that the presentation of the data not only supported the sharing of autistic voices, but that I considered how my expertise as a researcher and teacher could add to the way the knowledge and understanding of the data could be explored. Moreover, by involving all group members in data analysis we could think about how our stories developed from the lived experiences of one individual into something that represented the whole group.

Perspective Narrative Analysis

What kind of data can this be used for?

Any qualitative data can be analysed using Perspective Narrative Analysis. What is important is that the meaning of the data is defined by the person who contributes it.

For our research it was important that analysis was based on the individual's own descriptions of their body maps, as, if we were to interpret what other group members had drawn, it would have become an analysis of our perceptions of their drawings rather than of the ideas, feelings, and memories that the individuals who created the maps chose to represent. This is also true of other arts-based data such as photovoice.

Whose voice is being represented?

A key element of this analysis process is the consideration of an event from different perspectives. This promotes deeper thinking about how the events or ideas being analysed are experienced. Fizzacc members reflected on how they experienced a lack of understanding by peers and staff members in school, so we chose to consider the impact of this in the telling of stories. The introduction of different perspectives also allowed us to address the double empathy problem in the way stories are created.

To give a coherence to our narratives in a way that allows confidentiality and represents the experiences of the whole group, stories were told from the perspective of an imagined character – Maggie – who was the central character in the transition resources created by Fizzacc in another part of our research. Maggie's perspective was juxtaposed with another narrative, which tells the same story but from an alternative perspective. The first story is told in the first person to emphasise that Maggie is an autistic girl and represents the collated voices of the autistic members of Fizzacc; the second narrative was written in the third person to accentuate that this is not the voice of someone who experiences the situation first-hand, but who is providing an imagined perspective that provides contrast in order to provoke reflection. While this alternative perspective is not based on real events, unlike the stories voiced by Maggie, they are based on my experience as a teacher as well as the co-researchers' experiences with teachers and therefore offer an authentic perspective on the school experience. With the voice of Maggie being provided by those with lived experience of autism and discussed as a group to agree on how effectively her story portrays shared experiences, I believe we go a long way to achieving authenticity. By adding the narrator's voice, an added level of depth is included that allows my lived experience as a teacher to enhance the analysis and presentation of the data (see Figure 27.2).

> *'Being invited to parties and having friends were rare.'*
> *'I think I got invited to a party once, but the person who the party was for didn't want to invite me but their parent felt bad for me.'*
>
> The teacher, Mr Parsons, watches as the gate opens and the children rush into the playground. They are trailed by parents, laden with book bags, coats and PE kits. Mr Parsons takes a deep breath. The start of a new school year. Time to establish rules and expectations. One of the children, Ruby, must already be turning 9. She is bustling around handing out purple envelopes. Mr Parsons overhears her excitedly share:

'It's a pamper party! We are going to have face masks and get our nails done!'

This is not the best start to a new term. The girls will all be beside themselves with excitement and the chances of them focusing on this morning's numeracy activity are slim. Mr Parsons notices Ruby's mother pushing her towards Maggie and her grandmother. Ruby silently hands Maggie an envelope then runs off to chat to her friends. Maggie stands perfectly still with the invitation in her hand. Mr Parsons smiles. It will be nice for Maggie to be included.

By choosing these quotes and creating a story in which the teacher is oblivious to the subtext of the invitation offered by Ruby, I wanted the co-researchers to understand how the stories could be structured to represent the misunderstandings and miscommunication they had experienced.

Figure 27.2: Stages of the perspective narrative process

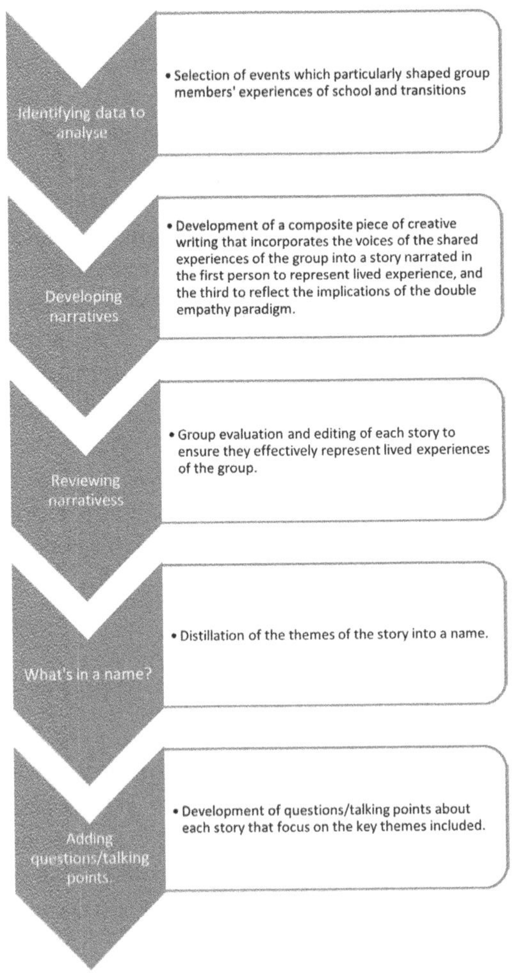

From the discussion of their body maps and the example narrative, images were chosen that would form the basis of a narrative that would reflect their experiences and the shared experiences of the group, and an agreement was reached about a basic plot to follow.

Developing narratives

The method for developing the narratives needs to be devised to suit the needs of the individual and the group. Some Fizzacc members felt they would like to write their first-person narrative first, others preferred the third-person story to be created first, giving them a structure on which to base their story. Some chose to narrate their stories aloud while I transcribed them. For these stories I edited the transcriptions for sense and structure before returning them to the co-researchers for approval. It could be that this is a process conducted by the whole group or taken on by volunteers from within the group. What is important is that the development stage is only the first element of analysis and therefore the narratives will be subject to further scrutiny and development.

Reviewing narratives

Once agreed on individually, the narratives should be shared with the group and edited to ensure they represent the views and experiences of the group as a whole. This stage of the process could be conducted in a group meeting or by sending the text of the narrative to group members for written comment. Again, it is important to adapt the process to best support the participation of all members of the group.

Any amendments should be agreed by the group. If there are any differences of opinion the casting vote should be with the group member whose data initiated the narrative.

The order of the stories was considered both in terms of which narrative should come first for each story as well as the overall order in which to present them. It was decided that each individual story should be considered separately, with some starting with the first-person narrative and some with the third. This was to ensure that the narratives were clear and that the message that each story was conveying was given maximum impact.

Fizzacc members were keen to discuss the narratives during one of our group meetings. We considered how effective the stories were at representing shared incidents as well as telling individual stories. All group members felt they were able to identify situations they had encountered that were similar and therefore the stories were felt to be representative of Fizzacc members' school experiences.

> The story is so relatable. Someone stole my diary and started looking through it! (Feydakeen – Fizzacc member)

> I've literally heard my teachers say the exact line about a uniform not being the height of fashion or something. (Zoe – Fizzacc member)

What's in a name?

It is important to consider carefully how to name the narratives, just as naming themes in thematic analysis is important. Braun and Clarke (2022, p 111) argue that a good theme name will be 'informative, concise and catchy'. The name could be a line taken from the narrative or phrase that exemplifies the message that is being conveyed. The reflexive process that takes place when naming a narrative is an important part of the analysis as it involves a distillation of ideas and careful consideration of what meaning the narrative is created to convey.

Adding questions/talking points

This process involved rereading the stories and creating a series of questions and talking points for each one. The decision to do this was based on a conversation we had in one group meeting in which one of the co-researchers expressed confusion as to how the themes were expressed clearly in the story:

> I guess I'm a bit confused on how, like, people would understand it as much. (Zoe – Fizzacc member)

Another group member agreed, saying:

> Yeah, kind of the same because there's no context. (Carys – Fizzacc member)

From this confusion it was suggested that a question or statement could be added at the end of the story to emphasise the point the story was trying to make. This encouraged a focus on the pertinent issues for each story and facilitated discussion between us as a group, as well as providing a medium for discussion for readers of the narratives. While this process was used as part of the data analysis process, we recognise that it offers potential for these stories to be used as a training tool (see Figure 27.3 for a body mapping image with associated story and talking points).

Figure 27.3: Body-mapping image showing the wall and the gate

Maggie and the gate

Maggie is 8 years old. She is tall for her age with long, dark brown hair and golden skin with a dusting of freckles. She looks very like her mother, Sally, who walks her to school every morning. Sally is smartly dressed and has an efficient air about her. It is Maggie's grandmother who collects her at the end of the day.

Maggie always speeds up when she passes the last house before the school boundary. She moves closer to her mother, tightens her grip on her hand and looks intently at the toy she is holding. It looks like she is in a hurry to get to school but [she] doesn't look around to greet her friends like many of the other children are doing.

When they get into the playground, Sally passes Maggie's hand to Miss Bird as if she was presenting her with a precious object.

'Any problems this morning?' Miss Bird asks Sally.

'No. We've had a good morning haven't we Maggie?'

Maggie brings her toy closer to her face.

'Come on then Maggie. We are going to go into the classroom, hang up your things, then have a look at a book about Sierra Sloth that I found in the library!'

As Maggie walks into the classroom with Miss Bird, she glances briefly back at the school gate where her mother is hurrying off to work.

It's time for school and my mum is calling me downstairs. I hold her hand as she opens the door and we set off down the road. As we get nearer, I start to worry about if I'll be alone at break time again today and having no friends. We arrive at school, and I see the wall and gate which isn't tall enough and I get even more anxious about someone breaking in[to] the school. I hold my mum's hand even tighter and try not to make eye contact with the wall and gate. I go over to Miss Bird with my mum because I have no friends to talk to. Miss Bird speaks with my mum, and I hide myself behind my elephant toy. After their conversation, Miss Bird tells me about a book which I'm excited about; my mum hands me over to my teacher and I look down because I don't like looking at people's eyes or mouths. We walk into school, and I look back at the dangerous wall and gate and I think 'that is the reason why I don't feel safe in school'.

Questions/talking points

- What kind of an impact does her worry about friendships have on Maggie?
- Feeling safe at school is really important. Were there any signs in the story that Maggie didn't feel safe?
- What assumptions might we make about Maggie from the narrative that we might change when reading the first-person account?[3]

Advantages and limitations of Perspective Narrative Analysis

As with any method of analysis there are advantages and limitations to using perspective narratives. These are both procedural and philosophical.

- *Balancing power.* Perspective Narrative Analysis supports members of a participatory research group – who may lack knowledge and experience of conducting research – to have power and influence over the analysis process. It is a way to bridge the gap between experienced researchers and group members with lived experience. This can support the promotion of the voices of rarely heard communities within the process of data analysis, which can be difficult to achieve in a way that is not tokenistic.
- *Promoting voice.* When looking for ways of analysing the body-mapping data produced by Fizzacc, we were looking for methods that would allow us to capture the power of the stories the girls were telling and analyse these stories in a way that didn't diminish them. The process of creating narratives that represented the participants' experiences of education as a group of autistic girls was an empowering one that produced results that Fizzacc members were keen

to share: 'To be a part of something that I know can help even more people is just a really lovely to do' (Zoe – Fizzacc member).

- *Impact on practice.* Fizzacc's narratives were created to analyse our body-mapping data in a way that allowed all group members to be involved in the process in a meaningful way. They also prompted a depth to the analysis that would not have been possible with other methods. In addition to this, however, as a participatory research group working to raise awareness of the challenges of school transitions for autistic girls, it was important for us to consider how our analysis could make a difference to practice and therefore to consider who our potential audience for the narratives might be. Group members expressed the importance of thinking about 'what we want people to take away from it' and about how the stories could be used to change people's perceptions and understanding. Poulsen et al (2022 p 3) said that 'meaningful change can be driven by those who have a true and deep understanding of the strengths and needs from their own experiential expertise'. The potential for narratives to be used as a training tool or stimulus for discussion is one that deserves exploration.
- *Rejecting positivist critique.* As with many creative qualitative methods of data collection and analysis within research, there are those who will criticise the robustness of Perspective Narrative Analysis. In considering the strength of perspective narratives as a method of analysis, however, it is necessary to move away from the positivistic notions of reliability and validity in favour of factors such as credibility, persuasiveness, and authenticity. For Fizzacc, this involved being conscious of ontological and epistemological presumptions, which we addressed by being reflexive in the awareness that knowledge and reality for my autistic co-researchers was different to mine, and therefore that viewing our data through a lens of 'double-empathy' was vital.
- *Cherry-picking the data.* Perspective Narrative Analysis is not a systematic procedure that results in a synthesis of an entire data set. It does not attempt to offer an overview of a research field. It is not an objective or accurate representation of a phenomenon. Instead it allows closer analysis of minutiae within data and provides a different way of investigating phenomena. For any one of our narratives, a multitude of perspectives could be written and each one could teach us more about the experiences of autistic girls in schools.

The following quote from Clough (2002, p 8) encapsulates what Fizzacc was hoping to achieve with the Perspective Narrative Analysis:

> The fictionalisation of educational experience offers researchers the opportunity to import fragments of data from various real events in order to speak to the heart of social consciousness – thus providing the protection of anonymity to the research participants without stripping away the rawness of real happenings.

Further development opportunities

The development of perspective narratives for analysing data is about drilling down into the impact of the data rather than trying to provide an overview of the problem. It could sit alongside other methods of analysis that offer a more holistic view of the data, or could be incorporated as a way of reflecting on data in a different way. For example, a narrative could be generated for each theme developed as part of a thematic analysis of data. This could help to ensure that the implications of each theme are thoroughly investigated.

Fizzacc developed the use of Perspective Narrative Analysis partly as a response to the double empathy paradigm, in recognition that the experiences of autistic people are often misunderstood and misrepresented in research and practice. There are many communities whose voices could be represented more accurately with better consideration of how their stories are influenced by their situation.

In conclusion, a quote from a Fizzacc member about how she felt about participating in participatory research, including the development of perspective narratives, follows:

> More people now know what it is and what it feels like, so we changed the world. (Iris – Fizzacc member)

Notes

1 This work was undertaken at the University of Southampton and was supported by the Economic and Social Research Council.

2 Our group is neurodiverse because it includes both neurotypical and neurodivergent members.

3 More information about Fizzacc, including the rest of the narrative, can be found at: https://acorns-soton.org.uk/2022/03/fizzacc-information/.

References

Barone, T. (2007) 'A return to the gold standard? Questioning the future of narrative construction as educational research', *Qualitative Inquiry*, 13(4), 454–70.

Bignold, W. and Su, F. (2013) 'The role of the narrator in narrative inquiry in education: construction and co-construction in two case studies', *International Journal of Research & Method in Education*, 36(4), 400–14. Available from: https://doi.org/10.1080/1743727X.2013.773508

Braun, V. and Clarke, V. (2022) Thematic *Analysis. A Practical Guide*, London: Sage.

Caine, V., Murphy, M.S., Estefan, A., Clandinin, D.J., Steeves, P., and Huber, J. (2017) 'Exploring the purposes of fictionalization in narrative inquiry', *Qualitative Inquiry*, 23(3), 215–21. Available from: https://doi.org/10.1177/1077800416643997

Clough, P. (2002) *Narratives and Fictions in Educational Research*, Maidenhead: Open University Press.

Fletcher-Watson, S., Adams, J., Brook, K., Charman, T., Crane, L., Cusack, J. et al (2019) 'Making the future together: shaping autism research through meaningful participation', *Autism*, 23(4), 943–53. Available from: https://doi.org/10.1177/1362361318786721

Milton, D.E. (2012) 'On the ontological status of autism: the "double empathy problem"', *Disability & Society*, 27(6), 883–7. Available from: https://doi.org/10.1080/09687599.2012.710008

Mitchell, P., Sheppard, E., and Cassidy, S. (2021) 'Autism and the double empathy problem: implications for development and mental health', *British Journal of Developmental Psychology*, 39(1), 1–18. Available from: https://doi.org/10.1111/bjdp.12350

Nuske, H.J., McGhee Hassrick, E., Bronstein, B., Hauptman, L., Aponte, C., Levato, L. et al (2019) 'Broken bridges – new school transitions for students with autism spectrum disorder: a systematic review on difficulties and strategies for success', *Autism*, 23(2), 306–25. Available from: https://doi.org/10.1177/1362361318754529

Pellicano, E. (2020) 'Commentary: broadening the research remit of participatory methods in autism science – a commentary on Happé and Frith (2020)', *Journal of Child Psychology and Psychiatry*, 61(3), 233–5. Available from: https://doi.org/10.1111/jcpp.13212

Polkinghorne, D.E. (1995) 'Narrative configuration as qualitative analysis', in J.A. Hatch and R. Wisniewski (eds), *Life History and Narrative*, London: Falmer Press, pp 5–25.

Poulsen, R., Brownlow, C., Lawson, W., and Pellicano, E. (2022) 'Meaningful research for autistic people? Ask autistics!' *Autism*, 26(1), 3–5. Available from: https://doi.org/10.1177/13623613211064421

Solomon, J. (2007). *'Living with X': A Body Mapping Journey in Time of HIV and AIDS. Facilitator's Guide*, Johannesburg: REPSSI.

Whitlock, A., Fulton, K., Lai, M. C., Pellicano, E., and Mandy, W. (2020) 'Recognition of girls on the autism spectrum by primary school educators: an experimental study', *Autism Research*, 13(8), 1358–72. Available from: https://doi.org/10.1002/aur.2316

Wood-Downie, H., Wong, B., Kovshoff, H., Mandy, W., Hull, L., and Hadwin, J.A. (2021) 'Sex/gender differences in camouflaging in children and adolescents with autism', *Journal of Autism and Developmental Disorders*, 51(4), 1353–64. Available from: https://doi.org/10.1007/s10803-020-04615-z

Section IX

Pushing the boundaries

28

Analysing the unspoken: finding the richness created in dialogue with people who cannot speak

Katherine Broomfield[1]

Introduction

The Unspoken Voices Project is concerned with understanding the experiences of people who rely on augmentative and alternative communication (AAC), also known as communication aids, because they cannot speak clearly.[2] People who use AAC are frequently excluded from being involved, or they are misrepresented, in research because they cannot provide the 'rich narrative' often demanded by qualitative analytic methods. The project was inspired by clinical practice as I am a speech and language therapist with experience of working with people who use AAC. I have frequently had cause to wonder at interactions between people who use AAC and their familiar communication partners and have marvelled at the nature of the mutuality that exists beyond words. These observations led me to search for, but not find, analytic methods that would enable me to explore and authentically represent people who use AAC and their experience of communication.

A dialogic theoretical lens provided the conceptual tools to extend my understanding of communication and voice, and to develop a creative data analysis method incorporating my embodied experience as a speech and language therapist and researcher. I will draw on data from the Unspoken Voices Project, a research project concerned with understanding more about the experiences of communicating using AAC, to elucidate and illuminate my application of this method and the impact that it had on my research. This method synthesises multimodal data sources through attending to the complexity and nuance of dialogue with this population.

AAC and speech and language therapy

Speech and language therapy (SLT) is a profession that supports people who have speech, language, or communication needs. Clinical practice is influenced by the International Classification Functioning, Disability and Health (ICF) (WHO, 2001), within which the impact of disability is categorised into areas of impairments of function, activity or execution of a task, and participation in a life situation. SLTs identify the nature of the communication disability and the impact it has on activity and participation. They might suggest or provide strategies to augment communication if people cannot speak clearly because of conditions acquired either in birth or early childhood, such as cerebral palsy, or in adulthood, such as a stroke, motor neurone disease, or cancer. AAC strategies include gestures and sign, for example Makaton™,[3] paper-based systems to which people can point,[4] and computer-based systems that have synthetic voice output of messages entered in or pre-stored within them. Communication using AAC is slow and frequently sparse because of the time and effort taken to construct and generate a message. People who use AAC rely on highly skilled and/or familiar communication partners for the joint establishment (co-construction) of meaning. Communication partners recognise acknowledge, and frequently interpret, a range of multimodal communication signals during interactions with people who use AAC (Beukelman and Light, 2020). The mechanisms by which meaning making occurs are often implicit and reflect hidden connections between the context of the individual communicating and the relationship they have with the person with whom they are interacting (Broomfield et al, 2022a).

Understanding the lived experience of people who use AAC is vital to being able to provide support and services to meet their needs. This is of particular interest in the professional field of SLT, but the inclusion of seldom-heard voices in qualitative research is also important to academia and the wider health, education, and social care systems. Traditional qualitative data analysis methods rely on the collection of narratives that represent rich descriptions of personal experience, collected via interviews, focus groups, or questionnaires (Denzin and Lincoln, 2018). Data are analysed by coding words and sentences, then deriving themes from clusters of codes based on the semantic connections between words and concepts (Miles et al, 2020). This dependence on linguistic narratives can overlook the unspoken richness in dialogue; the moments that point towards hope, create intrigue, and shape a sense of meaning informed by embodied responses within interactions. However, to gain an understanding of the experiences of people who cannot easily provide narrative descriptions of their experience requires an analysis method that values multiple forms of knowing and expressions of knowledge. Consequently, for the Unspoken Voices Project, it was essential to identify and employ a conceptualisation of communication that encompasses more than speech and language. Engaging with a theory of communication that better reflects the multifaceted and dynamic nature of interaction can enable a deeper, conceptual interpretation of the meaning generated during human communication exchanges.

Theoretical framing: dialogism

Dialogic philosophy recognises that communication occurs in the context of social relationships and is influenced by cultural, historic, and situational factors. This theoretical perspective attributes the generation of shared understanding to the dynamic interaction between people within a communicative dyad and the constituent linguistic, non-linguistic, contextual, and embodied messaging as it occurs (Broomfield et al, 2022a). It provides a perspective that encourages qualitative researchers to employ methods that take account of the physical and metaphysical nature of understanding that occurs in-between the linguistic features of communication (Phillips, 2009). John Shotter describes this approach in terms of social constructionism, which situates knowledge brokering between people in dialogue within a shared context that shapes and moulds meaning making dynamically, drawing strands of interrelatedness together into a cohesive whole (Shotter, 2010).

A dialogic theory of communication and interaction was developed for this project based on the work of Shotter (2010, 2016) and extended to incorporate AAC-mediated communication (Broomfield et al, 2022a). This theory enabled me to include all aspects of the multimodal toolkit of expressive methods employed by people who use AAC during data collection, and to incorporate my clinical skills and experiences as an SLT. By attending to the embodied understanding that manifests within and between interviewer and participant (and, when present, the communication partner of the participant), I created a set of conditions and applied a theoretical lens that allowed me to be drawn to key moments during data collection and analysis – key moments that pointed towards mutuality and the confluence of ideas.

Study design and data collection

The Unspoken Voices Project followed five participants on their journey through NHS specialist AAC assessment to the provision of AAC devices, and for 12 months after receipt of equipment. The aim of this longitudinal study was to understand more about people's hopes and expectations from AAC equipment and services to inform the development of a patient-reported outcome measure.

Collaborative method development

An advisory group of people with lived experience of communication difficulties and using AAC supported the development of this method (Broomfield et al, 2021). The group consisted of people who use AAC and their family and carers. They helped to create accessible and acceptable recruitment materials and interview topic guides. Written information in recruitment materials was minimised and pictures were included to support the participant in understanding the nature of their involvement in the project. The topic guides for each interview stage included questions that intended to provoke responses that added to existing knowledge about the experiences of communicating using AAC (Broomfield et al, 2022b). The advisory group were also crucial to the integrity of the analytic method, providing check and challenge to the application of the method and to the interpretations generated through it.

Ethics review and approval

This study was approved by the Health Research Authority in the UK (IRAS ID: 227722/REC reference 18/YH/0001).

Recruitment

Participants were recruited from a regional specialist assessment service for AAC in the south-west of the UK. People were identified by the assessment service from their referrals list and were approached to be involved in the project if they met the following inclusion criteria: were aged 12 or over, have a communication disability, have the potential to benefit from AAC, are able to consent to participate, and are able to participate in an interview. Those who consented to having their information shared with the research team were contacted by me to arrange the initial interview and to confirm participants had capacity to consent to involvement in the project. Communication partners were invited to be part of the interview where that was the preference of the participant. Consent for their involvement was gathered where they contributed to the interview. For a summary of participant demographics see Table 28.1.

Interviews

A series of four interviews took place with each participant: (1) prior to their assessment; (2) after their assessment; (3) after they received AAC equipment; and (4) after six months. I carried out semi-structured interviews with the participants. A copy of the topic guide was sent to the participant in advance so that they had the option of preparing responses. Questions included topics such as: their current communication, hopes from AAC, expectations of the services supporting AAC, and thoughts about future opportunities for communication that AAC could enable. I facilitated the involvement of participants in several ways: adapting the questions by using simpler language, clarifying responses with closed questions, and using gestures and images to support understanding and facilitate expression. Interviews were video- and audio-recorded, and detailed field notes were written consisting of basic contextual information and reflections on the interview process.

Data preparation

The focus of this inquiry was phenomenological. It aimed to understand more about the lived experience of using AAC to communicate and to understand the outcomes from using AAC that are important to people. In collaboration with the wider research team, I explored several phenomenological research approaches to create the analytic method. The data preparation phase built on the process developed for interpretative phenomenological analysis (IPA)

Table 28.1: Participant demographics

Participant	Age	Diagnosis	Communication mode	Communication partner	Interview 1	Interview 2	Communication aid issued	Interview 3	Interview 4
1*	68	Stroke	Speech (short phrases); alphabet spell chart	0	Nursing home	Nursing home	0	0	0
2*	67	Stroke	Speech (single syllable); gesture; intonation	Daughter	Home	Home	0	0	0
3	64	Multi-systems atrophy	Speech (whisper, interviews 1–3); AAC only (Interview 4)	0	Home	Home	Tellus Vibe; Me 4	Home	Home
4	26	Cerebral palsy	AAC; facial expression for yes/no	Mother; key worker	Home	Home	Tobii Dynavox i16; Communicator 5	MS Teams®	MS Teams®
5	17	Learning difficulties	AAC; Makaton	Mother; Teacher	School	#	iPad mini in a ChatWrap; Super Core 50 on Grid for iPad	MS Teams®	MS Teams®
6	50	Down Syndrome	Speech (short phrases); Makaton	SLT; Father	Health centre	MS Teams®	iPad; Grid 3; Symbol Talker	MS Teams®	MS Teams®
7	44	Cerebral palsy	Written sentences (computer); facial expression for yes/no	0	Home	Home	Grid Pad; Grid 3	Email	Home

Note: *Participants 1 and 2 withdrew after Interview 2; # Interview 2 with Participant 5 did not take place due to the COVID-19 pandemic; Participants 1, 3 and 7 did not have a communication partner present during the interviews.

(Smith et al, 2009). Data was prepared with reference to the initial stages of IPA: familiarisation, initial noting, and developing emergent themes relating to outcomes from AAC.

Interview audio recordings were transcribed verbatim and video recordings were reviewed. I noted important aspects of gestural or non-verbal communication in the transcripts. Details added to the transcripts included gaps and pauses in the dialogue (often when messages were being constructed on the AAC), gestures used to support communication, additional non-verbal communication such as facial expressions, and observations pertaining to the context, for example, if people used cues or objects within the room to support their communication. In the transcripts natural speech was denoted in sentence case and synthetic speech in upper case letters. Drawings and images used to facilitate data collection were collated, reviewed, and stored alongside transcripts in the data management software (NVivo for Mac, Version 11), along with any additional data provided prior to or after the interview took place, such as email correspondence. NVivo was used initially as a data storage tool, and then to organise ideas and themes from the data during analysis. The interview topic guides and impressions of the physical environment noted in field notes provided contextual data. The field notes written after the interviews, along with reflective notes made during analysis, provided data about the role and reactions of all those involved in the interviews and offered insights into the embodied responses of researcher, participant, and communication partner.

Data was organised into tables that were generated in a MS Word® document to allow for the range of data sources to be considered simultaneously (see Table 28.2). Questions from the topic guide were added to the first column of the table and sections of the transcript that responded to the questions were arranged in the second column. Descriptive comments were added to the third column to provide any additional contextual detail from the video recordings that would enhance interpretation of the transcript section presented in the second column. A linguistic comment was added to the fourth column, representing an explicit, explanatory interpretation of the information in the first three columns. Reflective comments from the field notes were added to the fifth column and further reflections made during data preparation, also reflecting initial analytic comments, were added to the sixth. The final column was used to document a conceptual interpretation – a word or comment that encapsulated the accumulated, iteratively derived meaning generated from the first five columns (see Figure 28.1 for diagrammatic representation of initial stages of data preparation and analysis).

Analysis

Analysis began during data collection and continued into the data preparation stages just described. A dialogic theoretical lens was applied to the data to tease out novel interpretations across the data set.

Table 28.2: Example of a data table

Topic guide questions *Prompts in italics*	**Data: video transcript**	**Descriptive comments**	**Linguistic/ non-linguistic comments**	**Data: field notes**	**Reflection**	**Conceptual comments** *Conceptual interpretation*
'Tell me about your current communication.' *What do you use to support you to communicate?* *How does it help?* *In what ways does it not help?* *In which situations?* *With whom?* *Why?*	**P1:** No (whispers). **K:** No. Do you use anything to support you to communicate? **P1:** (Pointing with left finger towards K. Whispers something unintelligible.) **K:** Are you saying a letterboard? **P1:** Laptop (whispers). **K:** A laptop, ok. Not this one the one we were looking at earlier? (Shows P1 the iPad.) Do you ever use this? **P1:** (Mouth movement. *Unintelligible speech.*) **K:** Yes. An iPad. And you mentioned a laptop. Do you have a laptop as well? **P1:** (Pointing to iPad.)	I look for cues within the room for an existing means to support communication during the interview and find only an uncharged iPad.	P1 currently uses single words and short phrases to communicate.	The interviewee is bedbound, side-lying, and largely non-verbal. Some very dysarthric single-word and short-phrase utterances. His iPad lay uncharged and unused on the bedside table and his communication book was only located at the end of the interview; the care centre manager insisting that he refused to use any communication aids.	P1 is minimally communicative, relying on single words and short phrases. He requires a lot of prompting from the interviewer to be able to participate in the interview.	Current communication is ineffective/ insufficient. The role of the communication partner is critical to generating meaning. ***Communication dependent on connection***

Note: This data table was used to organise a range of data sources and reflections on the analysis process.

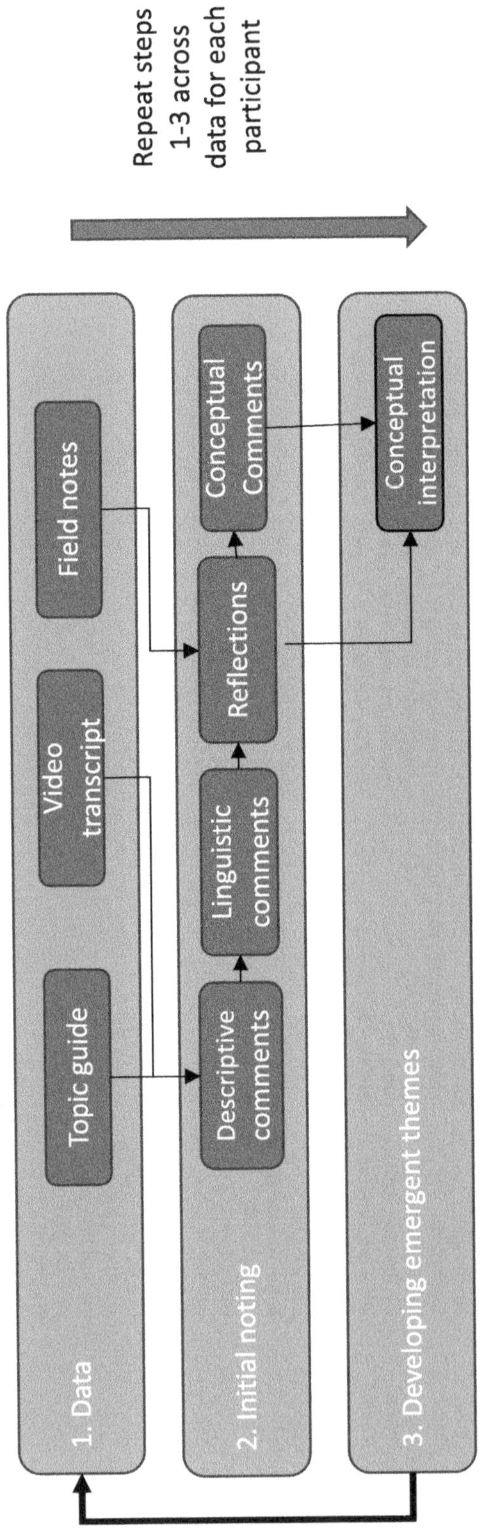

Figure 28.1: Diagrammatic representation of the analytic process

Developing emergent themes

During phases of familiarisation and initial note taking, some moments in the data were 'arresting' (Sullivan, 2012); that is, they made me *feel* like I needed to take notice. These moments frequently signified novelty, change, or salience. I immersed myself in the data; drawing on the arresting moments, working abductively between participant data seeking dissonance and resonance, and stepping back to look across the data set to find wider explanations. This required deep, critical, and active reflexivity – noticing what was happening in the data, but also reflecting on how meaning was emerging during the analytic process. I then challenged my own observations and sought critique from the wider team – both the advisory group and clinical academic supervisors on the project.

I coded key moments in NVivo and, by noticing conceptual similarities and differences rather than semantic ones, iteratively clustered concepts and ideas together into themes. This stage of analysis was driven by the data; by observing a range of signals within the data and being alert to my own embodied reactions when engaging with the data during analysis.

Researcher positionality and embodied analysis

As an SLT researcher, engaging with my biographical landscape was an important element of the data collection and the analytic processes. The early stages of data preparation were informed by IPA. However, in considering my embodied response to the data, the analysis diverted from IPA and became more creative and interpretative. The reactions observed in the actions of those involved in the interviews, as well as reactions elicited within me in response to the data during analysis, were incorporated into the interpretative process through the reflective notes and iterative phases of analysis. Through engaging with and noticing these embodied responses and reactions within and between the multimodal data, I was drawn towards moments of novelty, those that 'create the complexity intertwined influences at work in dialogical spaces' (Shotter, 2016, p 114).

Researcher reflexivity is a fundamental element of conducting excellent qualitative research as it speaks to the sincerity, transparency, and therefore authenticity of how the research is conducted (Tracy, 2010). For clinical researchers who have competing 'insider–outsider' perspectives informed by their practice experience, reflecting on arresting moments in the data can help to embed reflexivity in the analytic process (Latchem-Hastings, 2018). I was aware that there is a risk, when researching with people who use AAC, that the voices of others involved in supporting a data collection event can overshadow the meaning intended by the participant. When dealing with the data I kept this risk – and how it may influence both the dialogue in vivo during interviews and also the interpretations generated during analysis – in mind.

Rigour

Preparing the tables provided structure, or 'bureaucracy' (Sullivan, 2012), to the analytic process. Building up the content of the tables allowed me to develop conceptual interpretations iteratively as layers of data and reflections on that data were added. Key moments relating to the aims of the project were identified from the conceptual interpretations once tables were fully populated. These key moments were then compared and placed in contrast across participant-level data to generate multidimensional pictures of experiences and the nature of people's expectations from AAC devices. Themes were generated across the data set, from all the data collection stages, that represented a confluence of ideas reflected by the key moments identified. I looked for familiarity, similarity, and dissonance within the data to develop a deeper, richer, and integral representation of the experience of communicating using AAC. I was unable to check interpretations with participants due to time and resource constraints within the project, so instead the representations I generated from the analysis was checked for resonance and authenticity with the advisory group.

Check and challenge

The advisory group worked in close collaboration with me during the data analysis period, discussing my initial reactions to the data, emergent ideas, and conceptual interpretations. Through discussion of the analysis and exploration of the underpinning constructs of interest, the key moments and concepts identified during the analysis were mapped onto the areas of interest defined by the research aims.

Attention to rigour was pursued by drawing together a range of data sources and presenting them in the detailed tables, thus elucidating the iterative process of analysis and explicating connections between data and conceptual interpretations. The multiple data sources and perspectives of the advisory group and wider research team provided different lenses through which the analysis was considered; hence, the analysis identified areas of consensus and highlighted differences across and between data.

Impact

The results of using this creative analytic method enabled me to reconsider the nature of data in qualitative research with people who use AAC and how my embodied reactions as a clinical researcher informed the research process. Reconceptualising communication and the idea of voice and where it emerges in interactions with people who have communication difficulties enabled me to add depth to the analysis and to identify novel insights into people's interactions and experiences using AAC.

Multimodal data

The participant cohort in the longitudinal study included people who used natural speech, single sound utterances, Makaton™[ii] (gestural) signs, paper and computer-based AAC, or a combination of several of these communicative modalities. The theory-informed analysis supported the inclusion of all these forms of communication and incorporated the embodied responses of all those involved in the interview interaction.

Participant 2 had aphasia following a stroke and was only able to verbalise a single, two-phoneme utterance – 'me'. She used this utterance repeatedly and, in conjunction with elaborate intonation, facial expression, and gesture, was able to convey her message. However, she was heavily reliant on her communication partner; her daughter, who was present in each interview. Her daughter supported her by establishing the shared context for the interaction and providing simple yes/no questions to support her to assent or dissent from the interpretation offered (Box 28.1).

Box 28.1: Excerpt from video transcript: Interview 2, Participant 2

P2: Me me ah, me me, me uh (raises left hand in the air and looks up, as if trying to think of something), um me me, ah mm me me (uses frown/unsure facial expression while raising left finger in the air), me me me, me me, ah (rising following tone on 'ah', as in understanding something, raised eyebrows and holds up second finger on left hand) me me
KB: Ok
P2: Ah
KB: So having a bigger vocabulary maybe as well ...
P2: Uh me me (points assertively towards KB and nods, as in 'that's it')

I used the field notes to reflect on the negotiated nature of the interview: 'Communication was negotiated … I was able to glean new information from P2 about her communication, hopes and expectations from AAC' (field notes, Interview 2). During analysis, I reflected further on how interpretations were mediated by the relationship and contextual understanding of the communication partner: 'The interpretation of P2 tends to rely on the interpreter and their understanding of the context i.e. is it related to the question (KB) or is it related to communication (Daughter)?' (analytic comment, Interview 2). I was both intrigued and challenged by the almost automatically reactive responses of Participant 2's daughter and had to rely on an embodied sense of the coherence of their co-constructed meaning during the interview.

The iterative analytic process enabled me to explore the relational and contextual factors that mediate meaning making and explicitly connect these to my interpretation. In the example presented in Box 28.1, I attended to the nature of the mother–daughter/communication partner relationship, communicative context, and non-verbal messaging to help to illuminate the multi-perspectival nature of meaning making during interactions with Participant 2.

Communication

A conceptualisation of communication rooted in speech and language was an insufficient basis on which to understand the important outcomes from AAC. Dialogic theory enabled me to delve beyond semantic meaning during analysis and to connect with the deeper, more implicit messages that participants conveyed during interactions.

Participant 3 has a deteriorating medical condition, and her speech became progressively worse during the series of the interviews. She was able to use natural speech during Interview 1 but was entirely dependent on AAC by Interview 4. In Interview 1, she described that she liked chatting to people (Box 28.2). Chatting does not necessarily involve an expression of wants or a transfer of information in a traditional communicative sense but may be used to make another person feel at ease and to open the possibility of further conversation, or simply to form a human connection with another person.

Box 28.2: Excerpt from video transcript: Interview 1, Participant 3

P3: So for example if I'm in a taxi or in a shop ...
K: if you're in a taxi or a shop ...
P3: I can't chat to people ...
K: you can't just chat away
P3: Do you know what I mean?
K: Yea. So you used to enjoy a bit of chitchat with people ...
P3: Yes. Yes

Participant 3 also felt that her inability to chat to people represented her as somebody other than the person she wanted to be perceived as: 'Sometimes people think I have a learning disability. ... Drives me nuts. ... The other thing that drives me nuts, think I'm deaf' (P3, Interview 1). Once Participant 3 was given an AAC device, she used pre-stored phrases as a means of accessing 'chat', to connect with others rather than to transmit specific meaning (Box 28.3).

Box 28.3: Excerpt from video transcript: Interview 4, Participant 3

P3: YES. I USE THE SET PHRASES ...
KB: Have you got them there?

P3: (picks up device and turns it towards KB)

KB: Do you want me to have a look? – okay – (KB is holding and looking at the AAC device). Okay so you got set phrases like – (reads out set phrases from AAC device):

Yes; no; hello; goodbye; how are you; I'm fine thanks; I'm not so good today; I'm great; it's nice to talk to you; this is my voice it sounds different because I use a computer to talk; please ask me yes/no questions; that's good isn't it; I can't come to the phone

I was able to recognise what was important to Participant 3 about communication not through analysing what she explicitly told me, but by reconceptualising communication as an interactive process that expresses individuality and identity, and fosters the development of relationships rather than the expression of wants and needs. By engaging with the wider sense of who she was and what she had lost because of her deteriorating neurological condition, I could begin to see how AAC could help her to reconnect with important aspects of her identity and that for her this was an important outcome. This extended concept of communication can be used to inform the development of a patient-reported outcome measure that recognises the impact of AAC as being more than a replacement for disordered speech and language, and move towards an understanding of what communication provides people in terms of meaning and opportunity.

Voice

The use of dialogic theory informed the research team's conceptualisation of voice as emerging between all the people involved, shaped by the interview context and against the background of the research project (Broomfield et al, 2022a). As a research practitioner, I engaged actively and responsively with the participants, communication partners, and their context to generate interpretations during data collection and analysis.

I used reflective field notes to interact with my initial feelings and observations about the interview context and look at how I responded within this context. Box 28.4 provides an example of the situation when I greeted Participant 7 on my arrival at his house for Interview 4. I documented my assumptions and considered how they may have shaped a clinical interaction. I was in tune with my embodied responses as a clinician and I then used these reflections to inform how I interacted with the transcript data from this interview.

Box 28.4: Excerpt from reflective field notes: Interview 4, Participant 7

When I arrived, I was a little concerned because P7 was in his usual situation, in the office at the front of the house, in front of his regular desktop set-up. I didn't see any sign of a new communication aid anywhere. This made me assume that the communication aid wasn't working for him for some reason. I wonder why I was instantly so pessimistic about the AAC?

Had I already assumed that AAC wouldn't be useful for P7? How would this assumption have informed my clinical decision making if I had been his clinician?

Dialogism is crucial to the creativity of this method as it enables consideration of and interaction with the embodied responses observed during this study. Rather than translating all the images, pictures, sounds, and words into linguistic data and sorting these into semantic categories, I tuned into the embodied responses evoked within me and the feelings that led me to notice or engage with particular events and excerpts. Through engaging with these feelings, described as 'living', 'arresting', or 'poetic' moments by Shotter (2016), I became part of the voice that emerged during data collection and analysis. Creativity enabled a voice to emerge in the data that is metaphorical, transcendent, and polyphonic – being shaped and tuned by the contours of the dialogue and the actions and activity of those involved.

Depth

The creative approach to data analysis described in this chapter added depth and breadth to current understandings about the experience of using AAC to communicate. Being able to listen and attend to the experiences of people who have communication difficulties in the Unspoken Voices Project in a more expansive way enabled me to understand more about what communication means to individuals and what aspects of communication they value. I also learned about how the various aspects of their circumstances and life histories informed their experience of and their requirements from communication. The interdependencies between current context, communication need, and access to future opportunities for growth and self-actualisation were highlighted through this analysis. This complex and fulsome picture of communication offered insights into the outcomes that are important to people who use AAC.

In essence, this approach to analysis entailed noticing key moments in participant-level data, putting them in dialogue with moments from across the data set (Sullivan, 2012), and then synthesising these to generate interpretations that represent a confluence of ideas represented across participants' experiences. Applying a dialogic lens to qualitative data requires the researcher to attend to their own positionality rather than try to extract themselves from the analytic process; they must critically engage with their background and skills, and the influence that these will inevitably have on data collection and analysis.

Challenges and limitations

I struggled to reconcile developing this method with my clinical academic background. The use of a dynamic and relational theory of communication, despite feeling instinctive as a researcher, disrupted core features of my clinical training. Working interdisciplinarily with colleagues from research, health, and creative practices helped me to see beyond the status quo and to develop a method that suited my data rather than trying to fit my data into existing methods. Artistic practitioners enlightened me to the strengths of employing creativity in this research project and expanded my understanding about knowledge and the value of embodied experience. Leaning on the experience of academic supervisors and mentors enabled me to develop confidence in my critique of existing methods and to apply theory in developing a new method.

The wider research team involved in the Unspoken Voices Project consists of researchers with psychology, SLT, and clinical science backgrounds. This interdisciplinary team of fellow researchers, supervisors, and mentors helped me to develop the confidence and the vocabulary to add my clinical experience, and hence my voice, to the data; thereby challenging the status quo in qualitative praxis in communication disability research.

Wider applications

The use of theory enabled me to critique the concept of communication and consequently disrupt traditional assumptions concerning what constitutes qualitative data. It also enabled me to extend my consideration of the role that the researcher-practitioner plays in data collection and analysis, from reflexively responding to the participants to explicitly drawing on the practitioner's professional skills and experience during the creation of the data and the co-construction of meaning. Engaging in theory to develop a novel method for analysis enabled me to generate an analysis that extended beyond the spoken words and into the unspoken experience of participants.

This theoretically informed analytic method could be extended for use with a range of creatively collected data. The Unspoken Voices Project used a semi-structured interview format, in part to align to my existing clinical, communicative, and interactive skills. Although facilitated by non-verbal communication, the interviews were still rooted in words and language. The method could be applied to a range of artistic and creative methods of data collection as the focus of the analysis is on the dialogue that an interaction provokes; dialogue can occur in many forms of making, creating, and building. This creative data analysis method will be of value to any practitioners who have complex data that requires attention to how meaning and shared understanding are negotiated throughout data collection and analysis. The key lies in the researcher being connected to their biography while responding flexibly and dynamically during data collection and analysis: noticing, noting, and interpreting moments that point towards meaning, novelty, and mutuality.

Notes

1 This chapter was based on learning from my doctoral research project. I would like to acknowledge and thank my supervisory team for supporting me in carrying out the project, completing my doctoral training, and in preparing outputs, including this chapter. They include

Professor Karen Sage, Professor Deborah James, Professor Georgina Jones, and Dr Simon Judge. I would also like to acknowledge the invaluable role of members of the expert group in guiding me towards the method I have described in this chapter: John Hammil, Dr Ben Codling, Jamie Preece, Emma Sullivan, and Patrick Bates.

I was funded by Health Education England/National Institute for Health and Care Research (HEE/NIHR) for this research project. The views expressed in this publication are those of the author(s) and not necessarily those of the NIHR, HEE, NHS, or the UK Department of Health and Social Care.

2 Further details about this project can be found on the project blog: https://unspokenvoicesproject.wordpress.com/.

3 Makaton is a language system that uses symbols, signs, and language to enable people to communicate: https://makaton.org/TMC/AboutMakaton.aspx.

4 An example of picture- or paper-based AAC is Talking Mats™. Talking Mats™ uses symbols or pictures to facilitate conversation about a specific question or topic: https://www.talkingmats.com/.

References

Beukelman, D.R. and Light, J.C. (2020) *Augmentative and Alternative Communication: Supporting Children and Adults with Complex Communication Needs* (5th edn), Baltimore: Paul H. Brookes Publishing.

Broomfield, K., Craig, C., Smith, S., Jones, G.L., Judge, S., and Sages, K. (2021) 'Creativity in public involvement: supporting authentic collaboration and inclusive research with seldom heard voices', *Research Involvement and Engagement*, 7(1): 17. Available from: https://doi.org/10.1186/s40900-021-00260-7

Broomfield, K., Harrop, D., Jones, G.J., Sage, K., and Judge, S. (2022a) 'A qualitative evidence synthesis of the experiences and perspectives of communicating using augmentative and alternative communication (AAC)', *Disability and Rehabilitation: Assistive Technology*, 1–15. Available from: https://doi.org/10.1080/17483107.2022.2105961

Broomfield, K., Sage, K., Jones, G.L., Judge, S., and James, D. (2022b) 'The Unspoken Voice: applying John Shotter's dialogic lens to qualitative research from people who have communication difficulties', *Qualitative Health Research*, 33(1–2): 3–12. Available from: http://journals.sagepub.com/doi/10.1177/10497323221139803

Denzin, N.K. and Lincoln, Y.S. (eds) (2018) *The SAGE Handbook of Qualitative Research* (5th edn), Los Angeles: SAGE.

Latchem-Hastings, G. (2018) 'The emotion of "doing ethics" in healthcare research: a researcher's reflexive account', in T. Loughran and D. Mannay (eds), *Emotion and the Researcher: Sites, Subjectivities, and Relationships*, vol 16. Studies in Qualitative Methodology, Bingley: Emerald, pp 213–28.

Miles, M.B., Huberman, A.M., and Saldaña, J. (2020) *Qualitative Data Analysis: A Methods Sourcebook* (4th edn), Los Angeles: SAGE.

Phillips, L.J. (2009) 'Analysing the dialogic turn in the communication of research-based knowledge: an exploration of the tensions in collaborative research', *Public Understanding of Science*, 20(1): 80–100. Available from: https://doi.org/10.1177/0963662509340092

Shotter, J. (2010) *Social Construction on the Edge: Withness-Thinking and Embodiment*, Chagrin Falls: Taos Institute Publications.

Shotter, J. (2016) *Speaking, Actually: Towards a Fluid Common-Sense Understanding of Relational Becomings*, Farnhill Everything is Connected Press.

Smith, J.A., Flowers, P., and Larkin, M. (2009) *Interpretative Phenomenological Analysis: Theory, Method and Research*, Los Angeles: SAGE.

Sullivan, P. (2012) *Qualitative Data Analysis Using a Dialogical Approach*, London: SAGE.

Tracy, S.J. (2010) 'Qualitative quality: eight "big-tent" criteria for excellent qualitative research', *Qualitative Inquiry*, 16(10): 837–51. Available from: https://doi.org/10.1177/1077800410383121

WHO (World Health Organization) (2001) 'International Classification of Functioning, Disability and Health: ICF'. Available from: https://apps.who.int/iris/handle/10665/42407

29

Speculative a/r/tography

Kathryn Coleman, Sarah Healy, Abbey MacDonald, and Peter J. Cook

Introduction

We are a/r/tographers who work in pre- and in-service teacher education in Australian universities. We write this chapter for researchers interested in encountering and conducting speculative creative inquiry. Data analysis in speculative a/r/tography is an active doing coupled with quiet reflection in the space of 'not yet'. It is 'less about discovering something pre-existing and more about constructing knowledge through the process itself' (Reason, 2018, p 49). Doing data this way advances our understanding of small data via close-to-practice, aesthetically oriented, process-driven, human-scale analyses. This approach to data resists data imperialism and is decolonial, post-human, and activist in its politics – making way for a multiplicity of ontologies, knowledge systems, and methodologies. Our creative methods are informed by an interplay of methodology and pedagogy that contiguously respond to relational understandings of being an artist/researcher/theorist and the many creative (in-)betweens that being a practitioner affords. In this sense, we dance between theory and practice, whereby theory becomes data as much as the artefacts created in and through the a/r/tistic practice. As collaborators, we explore complex problems with similarly curious colleagues and students, bringing together multiple perspectives, sites, and spaces to assemble meaning (MacDonald et al, 2022). This is an important place to begin this chapter on creative data work because it positions where and how we do speculative a/r/tography and what the opportunities might be as you begin to play within creative possibilities.

A/r/tography has its roots in education (see LeBlanc and Irwin, 2019). However, our particular iteration of a/r/tography transcends the disciplinary boundaries of education, living and breathing across speculative inquiry, creative methods, and creative data analysis. A/r/tography renders its practices in ways that enable thinking with and doing through iterative and looping actions: contiguity, living inquiry, openings, metaphor/metonymy, reverberation and resonance, excess and absence, and gaze (Coleman, 2017). Our rendering of a/r/tography as speculative emphasises inquiry as being complex, imaginative, playful, and risky. Speculative a/r/tography enacts possibility thinking, 'designerly' ways of knowing and thinking, prototyping and failure in novel ways that other forms of a/r/tography might not. We shift away from the T:teacher towards T:theorist to reconfigure the artist–researcher–*teacher* triad in relationships as artist–researcher–*theorist*. Our 'take' on a/r/tography harks back to the early a/r/tographic work of Professor Rita Irwin, who says 'Theory as a/r/tography as métissage is a way for those of us living in the borderlands to creatively engage with self and others as we re-imagine our life histories in and through time' (Irwin, 2017, p 146). Speculative a/r/tography is highly evocative and located in lively sites of possibility as a form of knowledge mobilisation. It is a weaving and knotting of voices, a métissage of knowns and unknowns. 'Single authors/researchers use métissage as a theoretical construct and textual practice ... [and] weave the repressed languages and traditions of local cultures and vernaculars (particularly incorporating autobiographical material and local oral traditions and stories) with the dominant (often colonial) languages and traditions of literacy' (Chambers et al, 2007, p 142). It is practice-related creative research, bouncing playfully between making and theorising in a place of not-yetness (Ross et al, 2016).

For those of you who have worked with or read about a/r/tography, you know that it is inherently about the self as artist/researcher/teacher [or in our case artist/researcher/theorist] as we come to making and knowing *as* becoming. Yet it also provokes a radical togetherness through sociocultural and ethico-political thinking when collaborations and communities of a/r/tographers come together to engage in shared inquiries, act as critical friends, or present their collective provocative works to others (Irwin, 2006). A collaborative model or a/r/t triadic is employed to facilitate research that is relational, participatory, and reciprocal through knowledge sharing, activation, and mobilisation (Muff, 2014); if you have worked with an a/r/tographer, you know the effect of being tucked and folded into a collaboration.

In speculating ways for a/r/tographic data (both small and big) to be created, we pull and push at edges of metonymic possibility (Aoki, 2000). Thinking *with* data informs our pedagogic intentions. We draw attention to serendipity, accidents, and unlearning as they occur (McLeod et al, 2020), maintaining an acute awareness of how ethico-political, methodological, and pedagogical knots cannot be untangled. As researchers who also teach, we are particularly sensitised to the co-constitutive nature of pedagogy and methodology and think of this coupling as metho-pedagogy (MacDonald et al, 2022; PlayTank Collective et al, 2022). It is through a metho-pedagogical lens that we reimagine the *t* in a/r/tography as theory.

Positioning the speculative process

This chapter is deeply situated in creative practice and methods. As we write *as* method, our aim is to show our speculative a/r/tographic process as renderings (Irwin and de Cosson, 2004; MacDonald et al, 2022), both known and yet to be known, and how these can be organisers, informers, and speculations for collected, created, choreographed, and curated data. To achieve this, we propose four practices central to our creative method as collaborators:

- Proposition 1: speculative futuring as data analysis
- Proposition 2: one hundreds as data analysis

- Proposition 3: one hundreds otherwise as data analysis
- Proposition 4: walking as data analysis

Our examples illustrate how speculative a/r/tography translates into different types of interdisciplinary research designs, including those which are *not* located in the arts or education. In converging deep root systems/spaces of artistic ways of knowing, doing, and becoming, speculative a/r/tography involves the grafting of theories (educational, arts-based, creative, digital, and post-human) as data with emergent pedagogies. Speculative a/r/tographies fold analogue, digital, and virtual practices, creating opportunities to activate all senses, including the sense of movement, as they refold and unfold between art–research–theory. In attending to the root systems (rhizome) and loamy, liminal space around those (rhizosphere), this chapter articulates a process and possibility for data analysis 'in' speculative a/r/tography. In doing this, we create space for slippage and speculative discourse (Bauwens, 2021) of practice, process, product, and place making with/in a relational scholarship of creative research.

Proposition 1: speculative futuring as data analysis

Speculative methods are situated in a space between knowing and not knowing. Attending to speculation, we linger in a not knowing process of making meaning that can help reveal possibilities for what might be worth saying (Berlant and Stewart, 2019). In this way, speculative turnings of data slip between inferences of influence, contradiction, expansion, reflexivity, dissensus, speculation, and change (Kalin, 2014). Speculative methods are powerful because they enable us to manipulate the present through future casting (Rae and Coleman, 2021), which in turn entails casting ourselves into possible, probable, and preferable futures to look at the present *and make change* – what we like to think of as initiating little activisms or freedoms. As Claire G. Coleman reminds us in *Refugium*, 'most revolutions in history are seeded in story and song' (Rae and Coleman, 2021).

One example of speculative futuring can be found within the Master of Teaching Capstone data (2017) of the initial teacher education course for two of our collaboratory/authorship team. These workshops were designed to invite ways of seeing the self in relation to what 'might be' in a decolonisation of education. This example, from a Possible, Probable and Preferable Futures workshop led by Susan Wright with Kathryn Coleman, shows speculative a/r/tographic methods in action, with teacher candidates collaging and drawing collaborative and conspiratorial data before performing their research as an act of change making through activism.

Proposition 2: one hundreds as data analysis

We adopt Berlant and Stewart's (2019) hundreds as method at distinct stages in our speculative a/r/tographic process. This enabling constraint for ideation and languaging allows you to work ideas through generating 100 individual words

Figure 29.1: Speculative futuring Master of Teaching Capstone data

Note: Collaborative data work with teacher candidates in 2017.

in a reflexive process or a pairing of words, and/or in passages of 100 words as a métissage (a writing praxis). At the beginning of a project, it is a generative way to bring co-research teams together and start the process of collaborative speculation. When this is the intent, it works well to start the hundreds with a found provocation related to the shared inquiry. Once a project is underway, the hundreds become a generative way to codify and edit data by selecting data/artefacts/[d]artefacts to act as the 'provocation'. This coding is performative as you weave in and out of genres and blur what began. The hundreds are but one approach to writing as method – an approach that we find generative and collaborative; however, it may not always be the right fit for an individual, team, or project. What we find useful about what we refer to as the '100s' as writing method is that it provides a productive enabling constraint that creates 'openings' and a felt reverberation that prompts attention to contiguity, metaphor, metonymy, excess, and absence; all key a/r/tographic concepts.

Box 29.1: A webinar in lockdown (1)

Data generated during a walking, listening, writing webinar in lockdown (2021) as educators across the South-East Asia Pacific region met to collaborate and create community through embodied drawing

A pandemic hundred (Kate):

There is an intimacy online, particularly during COVID. At home, often without shoes, many of us in parts of our home not often used for work and study. This blurring of space and time can lead to more anxiety, more pressure to continue to work and live as the university seems to not notice what is happening. Life continues to move on as daily numbers rise and lockdowns continue. These things often drive our dialogue as we meet in Zoom. We say hi and discuss daily numbers and vaccination rates. Meanwhile each day is a little bit more Groundhog Day.

Box 29.2: A webinar in lockdown (2)

Data generated during a walking, listening, writing webinar in lockdown (2021) as educators across the South-East Asia Pacific region met to collaborate and create community through embodied drawing

See Anderson et al (2022). As Sarah shows in this '100' she weaves theory through her poetic writing as a collective métissage.

Cutting 100 (Sarah):

Cutting conjures action, redaction, blackout poetry. What are we going to reveal? Splicing together hybrid creations, reordering events, changing time. Relief printing, carving woodblocks, cutting out, cut outs, paste ups. Cut it out and paste it up. Resequencing through analogue and digital cuts. And agential cuts (Barad, 2007). Negative space and positive space, in and out. Cut hard, cut sharp, cut fearlessly, cut carefully. Don't cut your finger. You cut your finger. Careful what you cut. Can we splice it together again? Offcuts, woodcuts, hotcuts. Or is that supposed to be a hot take? A hot take on an offcut? Redact that.

Box 29.3: A webinar in lockdown (3)

Data generated during a walking, listening, writing webinar in lockdown (2021) as educators across the South-East Asia Pacific region met to collaborate and create community through embodied drawing

Redaction 100 (Kate doing a 100 otherwise):

There is an intimacy online, ████████████████ without shoes, ██████████ home ████████████████ blurring of space and time can lead to more anxiety, more pressure to continue to work and live as the university seems to not notice what is happening. Life continues ████

[redacted] numbers rise and lockdowns continue. [redacted]
we meet in Zoom. We say hi [redacted]
[redacted] each day is a little bit more Groundhog Day.

This process of collaborative list making and writing 100s comes from a 2019 workshop by Stacy Holman Jones and Kathryn Coleman, facilitated for Lisa Grocott's Wonder Lab at Monash University (for more on the lineage of this method, see Healy and Edwards, 2020). This process (adapted from PlayTank Collective et al, 2022), generally involves the following steps:

1. Identify a provocation, for example, image, object, film, sound, idea, question, data, [d]artefact …
2. Engage in exhaustive listing, collectively or individually. If a first-time exhaustive listener, aim for 100 items.
3. Identify affective affinities, for example, contiguity, openings, metaphor and metonymy, reverberation and resonance, excess and absence. You can do this by curating and/or annotating your list in a group or with a critical friend.
4. Write 100-word texts (eg take an idea for a walk or try a listicle). And yes, exactly 100.
5. Share your 100 (100s lend themselves to slam poetry and spoken word-style sharing).
6. Do it all again but do it otherwise (eg write or make your 100 slant).

Writing in this way is initially an experimental data-making process. It both enables and constrains the creation of precise 100-word blocks of texts, or multiple 100-word blocks. When collaborating, the writer of each 100-word block then works with specific encounters as 'incitement[s] to composition' (Berlant and Stewart, 2019, back cover). These incitements as experiments through w/riting from the creative encounter offer time to unfurl Berlant and Stewart's (2019) hundreds, and to stay with what opens or is opening or might open as speculative a/r/tographic analysis unfolds, refolds, folds.

Proposition 3: one hundreds otherwise as data analysis

For our example of Proposition 3, data was generated 'in' a Drawing in the Wild: Mapping Site through the Senses digitally mediated workshop. The workshop begins synchronously; participants are welcomed and countries from which we gather acknowledged. Then the workshop is positioned in relation to Rita Irwin's (2006) 'Walking to create an aesthetic and spiritual currere'.

With bodies and minds suitably primed for a period of drawing in the wild, participants consider how attending to their surroundings through non-visual senses creates different kinds of knowledge. A 60-minute pre-recorded audio guide leads them through a series of activities, prompting them to walk, listen, draw, rub, photograph, and record in their local surroundings. After the

Figure 29.2: Kate's pocket zine

Note: InSEA Hundreds Otherwise webinar.

Source: Healy and Coleman (2022).

drawing in the wild is complete, participants e-convene (a metonymic play on a 'reconvene' that slips between digital space) together on Zoom to critically reflect on the process through a see/think/wonder routine developed by Project Zero, a research centre at the Harvard Graduate School of Education. Creative data and wonderings are uploaded to Padlet (for an example of data and wonderings, see Figure 29.2) to create a collaborative, sensory map of the collective surroundings of participants - across time zones and countries. The Padlet simultaneously becomes a dataverse; a creative artefact and living exhibition space in which future participants engaging in a recorded version of the session can continue to post. The dataverse is both analysis *and* data. It invites further analysis, which we do via a 'warm-up' hundred (following steps 1–5 in Proposition 2), followed by the main event: one hundreds done otherwise. This involves everybody making a hundred that is a creative riff off the classic hundreds. In Figure 29.3 Rita Irwin does the hundreds as a hundred letters not words. The blackout poetry (Box 29.3) is also an example of the hundreds done otherwise. You could also try the hundreds backwards or perhaps using images, marks, sound, or even collage.

Proposition 4: walking as data analysis

Forms of data might include digital and analogue photographic or text-based artefacts that are generated through dialogue with the human and more than

Figure 29.3: Rita Irwin's hundred done otherwise

sssssllllllllllllllllllooooooooooooowwwwwwwhurryhurryhurryhurryhurryhurryhurryrhythmsarestillrhythms

Note: Posted to shared Padlet.

Source: Healy and Coleman (2022): 'InSEA Drawing in the Wild Gallery: regenerating through shared creative practice' [composition], University of Melbourne: https://doi.org/10.26188/19151450.v1.

human. This proposition invites you to explore deep listening and knowing on and through Country. In this setting, Country is understood through Indigenous relational ontology as a sentient phenomenon that presumes 'all living things to be related, destabilising imperial hierarchies of man versus nature in ways that demand holistic respect for nature or Country' (Harrison et al, 2017, as cited in Phillips et al, 2022, p 176; see also Black, 2011). Drawing and photographing work (Figure 29.4) in this way is initially an experimental data-making process (adapted from Coleman and Cook, 2021), and generally starts out through a place-making connection drawing and photographic data to show each other what Country you are working on:

1. Go outside and deeply listen with your ears and hands, hear the sounds that surround you and pay attention to those sounds and affects closest to you. Walk with the resonations you feel, do they make you tingle? Listen to the vibrating as the thinning swells and the movements linger. Can you hear differently? What's changed?
2. Stop, pause, listen, shift your movement. Now draw these sounds. Drag your feet. Point your toes and allow them to lead your movement. Listen deeply and carefully to the sounds that are furthest away from you as you trace walk with your body. What sounds are closest? What sounds overlap?
3. Draw these sounds surrounding you again over the top of your first drawing. Identify these sounds and places where you hear them and continue to draw and photograph the reverberations.
4. Share with your a/r/t collaborator and discuss the Country that you have been walking on.

Walking is adopted as a method by a range of sociocultural researchers. Rita Irwin's (2006) 'Walking to create an aesthetic and spiritual currere' is excellent introductory reading for the data working that underpins a/r/tographic walking methods. Rita is a leader in the field and has had a great influence on the metho-pedagogy and curriculum we teach and research within our sites of practice. This reading provides insight into how the ordinary act of walking in nature can become an active pedagogy of self – a way of re-forming the self in creative ways as research. The three themes she speculates on are: freedom, transformation, and flow. Speculation and analysis through trace walking (Coleman and Cook, 2024) allows for new opportunities to begin noticing and attuning to the world's educative intensities and dynamically shifting assemblages that enable the learning that Rita Irwin provokes. Trace walking acts as a blueprinting of place. It is a site-based collaborative and connected mark-making process (including digital marks) that makes it possible to creatively speculate through 'doing data'. In acknowledging that we strive for connection, despite significant distance, we choose to find alternate ways to co-locate and co-create. We identify a midway geographical site and plot, using map coordinates, the equidistance needing to be travelled. As we journey on our imaginary travels, we consider the creations along the way. We speculate and engage with our varying practices through imagineering. We document points of interest through visual representations and contribute our d(art)efacts to public repositories. We consider the speculative journey in terms of our artistry and our teaching practices while opening up the processes to others for engagement and contribution. The speculation becomes iterative and folds in new considerations when closing the gap digitally. The product is an endless resource of artistic and metho-pedagogical possibilities.

Mapping the process

We are drawn to and enjoy playful approaches to collaborative inquiry and scholarship; particularly those that help us make sense of a teaching/research problem by probing, poking, and provoking the tension between theoretical and practical aspects of art, research, and theory. The metonymic métissage is a multimodal representation that acknowledges the blurred arbitrary boundaries that occur in our theories and practices, as well as between our theories and practices. These boundaries both (re)define and promote inquiry into the multiple identities of being an artist/researcher/theorist, and those other identities that are yet to be known. The metaphoric boundaries – represented by the slants in a/r/t – are where the a/r/tographers inquire and creatively analyse the data that is created, collated, and curated. Put simply, this team converges around a shared interest in, and appreciation for, learning anew through metonymic qualities of word play, and this languaging plays an important role in our creation of data and analysis.

Our metonymic métissage emerges from co-conspiratorial acts of methodological and pedagogic interrogation and interchange. Through the entwining and integrative qualities of métissage, we inquire and conspire in ways that see us assemble, disassemble, and reassemble words to speculate and make meaning of our practice/s. In playing with the discursive slipperiness of metonymic métissage, we present a co-conspiratorial approach for co-inquiry that supports individual and collective movement between the becoming of things.

As a group of a/r/tographers who collaborate, our work continues to map and traverse through our ongoing practice *with* each other, *in* relation *to* data. In this

Figure 29.4: A/r/tographic blueprint for walking in four propositions

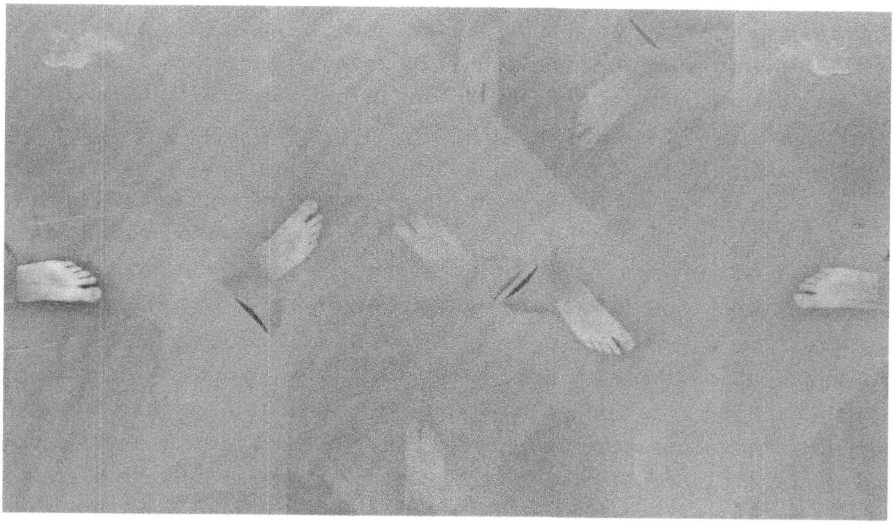

Note: Still from Coleman and Cook's (2021) video presentation of a/r/tographic blueprint for walking in four propositions. In this video they explore 'making kin as a/r/tographers that work in often contested spaces of conservative educational research and across disciplinary boundaries'.

chapter we map how the site of speculative a/r/tography and methodological and pedagogical possibilities mutually constitute each other, becoming what Gallagher et al (2011) located as metho-pedagogical. This chapter, as a site, creates opportunities to speculate, experiment, and inquire metho-pedagogically, and to inform how we conceive and practise creative inquiry in powerful and expansive ways for others. In its metonymic métissage of practices central to speculative a/r/tography, this chapter captures and maps a defining moment of becoming in, with, and through creative inquiry. The writing of this chapter into being constitutes a new métissage from which we draw, slip, and cut across examples from arts education, digital scholarships, activisms, world-centredness, and socially just practices.

Site

Speculative a/r/tography cracks open new possibilities and data sites, as our propositions explore. The metho-pedagogic data that we introduce in this chapter includes artefacts created in response to provocation through artistic, research-oriented creative inquiry in multiple sites.

As Australian educational researchers, we are inquirers, exploring the implications of a/r/tographic data to illustrate new onto-epistemic encounters as pedagogical acts *through* currere. Grumet and Pinar (1976) showed us that curriculum, perhaps most commonly understood as a course of study in formal education, comes from the Latin word currere, meaning to run the course. This action of movement or performance through space differs from how curriculum is often viewed: as a policy-driven text-based knowledge holder that must be taught as content and evidenced as assessment. Rather, currere is creative, personal, responsive, active, and always contextual. In an a/r/tographic currere, no site is ever the same and should never be treated as such.

Making, doing, performing, and co-generating visual and digital data *as* currere (Pinar, 1975) involves ethical and political decision making. It is a way of knowledge making and relational curriculum design in and through creative practice. Currere as a site is important for researchers interested in this practice because it is an underpinning for attunement. Being attuned to the world through the physical, virtual, material, and spiritual dimensions as artist–researcher–theorist allows us to wonder and to weave metho-pedagogic understandings of curriculum in place.

In slipping between playful, place-based curricula encounters in the academy, thresholds of precarity and possibility alight in powerful and transformative movements/moments of becoming ecological (Rousell, 2020). A central action of currere is the power to return to the past, to *re*-experience it, and to remain open to the forms a researcher might take through futurecasting. Currere then, allows us to move in between inner time, external time, and pedagogical time, and shifts how we do curriculum in a creative ecology (Harris, 2021).

Ways of doing the practice – determined by the practices

Given that a/r/tography inherently 'disrupts standardized criteria of research while evoking and provoking alternate possibilities for understanding' (LeBlanc and Irwin, 2019, para 1), we speculate on what we, a collective of artists/researchers/theorists immersed in teacher education, might offer others through our creative data practices. This chapter shows (rather than tells) how speculative a/r/tographic inquiry iteratively maps through the rituals of collaborative creative discourse, speculative maker exchanges, stories of teaching as practice, co-reflection, co-w/riting, and co-conspiratorial working. Our chapter conveys the foundations of speculative a/r/tography, and points to how this can advance creative research endeavours, practices, and products within webs of ever-shifting education relations (Braidotti, 2020).

We posit that sometimes methodology and pedagogy become indistinguishable in the process and practices of speculating. Other times, a pedagogy is borne from methodology, or a methodology from pedagogy; what is constant is they always exist in playful relation. This applies to speculating through an interplay of inquiry and creative analysis of data. As Biesta (2007) suggests, research can only provide an understanding of possibilities as we speculate on what the problem might be – it does not tell us what to do. The contribution of speculative metho-pedagogy opens opportunities in and for the humanities, arts, and sciences and beyond; be it overcoming institutional systemic obstacles or barriers – or the literal walls of the studio or gallery – or traversing the responsive novel terrain of ever-evolving digital platforms.

A return

The propositions presented are not intended as a step-by-step guide or scaffold; rather, they are sensitising concepts that orientate and compel speculative a/r/tographers towards a process that is risky, radical, iterative, and playful. They are propositions of speculative a/r/tographic work (process and practice) that intend to show how we *do* data as collaborators. Akin to an art process (exploration, experimentation, development, refinement, and resolution of ideas, materials, techniques, processes, visual language, and concepts), our ways of doing data work are rendered (Irwin and de Cosson, 2004). These renderings are processes or actions of practice that enable thinking with and doing through, so are not methods as such. They are iterative and involve a continuous cycle of rework, that is, remaking, redesigning, rethinking, retheorising, and re-analysing. Rework is a crucial part of the re-turn as you begin a creative inquiry. Where you began you must return as you play with data and speculate on the possibilities (McLeod et al, 2020). As you render your (re-)search through 'contiguity, living inquiry, openings, metaphor/metonymy, reverberation and resonance, excess and absence, and gaze' (Coleman, 2017) these practices are not yet known, and are messy, experiential, and playful, similar to the way some artists' work and how some artworks are created.

We believe there is much to be learned from our sharing of accounts for traversing change and complexity, as well as capturing the fears and hopes we lose and find ourselves along the way. We posit speculative a/r/tography as a highly evocative means for unfolding research across disciplines that slips playfully between acts of metonymy and métissage. Speculative a/r/tography's critical and slippery engagement with data provocations leads us to prompt theorists and practitioners to attune to and move between latent and temporal spaces of art, research, and theory. Speculative a/r/tography provides artists, researchers, and theorists with tools to play, imagine, question, activate, critique, and wonder about data with curiosity and imagination.

This is an important place to end a chapter concerned with creative data work because it positions where and how we do speculative a/r/tography in-between a/r/t and what the possibilities 'might' be as researchers begin to work *with* data in this space. Re-turning to the slant or in-between of a/r/t as metaphoric boundaries or interstitial spaces, researchers will discover a site to envisage, encounter, experiment, and elucidate. It is here in the messy place of not-yet-knowns where speculative a/r/tographic collaborators creatively inquire and analyse data.

References

Anderson, B., Aitken, S., Bacevic, J., Callard, F., Chung, K.D., Coleman, K.S. et al (2022) 'Encountering Berlant part 1: concepts otherwise', *The Geographical Journal*, 189(1): 117–42. Available from: https://doi.org/10.1111/geoj.12494.

Aoki, T. (2000) 'Locating living pedagogy in teacher "research": five metonymic moments', in E. Hasebe-Ludt and W. Hurren (eds), *Curriculum Intertext: Place/Language/Pedagogy*, New York: Peter Lang, pp 1–9.

Barad, K. (2007) Meeting the Universe Halfway: Quantum Physics and the Entanglement of Matter and Meaning, Durham: Duke University Press.

Bauwens, L., Miller, Q., and Świeżyński, K. (2021) *Speculative Facts*, Amsterdam: Onomatopee.

Berlant, L. and Stewart, K. (2019) *The Hundreds*, New York: Duke University Press.

Biesta, G. (2007) 'Why "What Works" Won't Work: Evidence-Based Practice and The Democratic Deficit In Educational Research', *Educational Theory*, 57: 1–22. Available from: https://doi.org/10.1111/j.1741-5446.2006.00241.x

Black, C. (2011) 'Maturing Australia through Australian Aboriginal narrative law', *South Atlantic Quarterly*, 110(2): 347–62.

Braidotti, R. (2020) '"We" are in this together, but we are not one and the same', *Journal of Bioethical Inquiry*, 17(4): 465–9.

Chambers, C., Hasebe-Ludt, E., Donald, D., Hurran, W., Leggo, C., and Oberg, A. (2007). 'Métissage: A research praxis', in J. G. Knowles, and A.L. Cole (eds), *Handbook of the Arts in Qualitative Research: Perspectives, Methodologies, Examples, and Issues*, Thousand Oaks: SAGE Publications, pp 141–53.

Coleman, K. (2017) 'An a/r/tist in wonderland: exploring identity, creativity, and digital portfolios as a/r/tographer', PhD dissertation, Melbourne Graduate School of Education, University of Melbourne. Available from: http://www.artographicexplorations.com

Coleman, K. and Cook, P. (2021) 'An a/r/tographic blueprint for walking in four propositions', Paper presented at University of Melbourne, 29 October. Available from: https://doi.org/10.26188/17089334

Coleman, K. and Cook, P.J. (2024) 'An a/r/tographic blueprint for walking in four propositions', in R.L. Irwin, N.Y.S. Lee, A.I. Baldus, D.T. Barney, J.M. Ursino, and Z. Valiary Eskandary (eds), *Pedagogical Propositions: Playfully Walking with A/r/tography. Book Three: Curriculum*, Viseu: International Society for Education Through Art (InSEA) Publications.

Gallagher, K. and Wessels, A. (2011) 'Emergent pedagogy and affect in collaborative research: a metho-pedagogical paradigm', *Pedagogy, Culture & Society*, 19(2): 239–58.

Harris, D. (2021) *Creative Agency*, Cham: Palgrave Macmillan. Available from: https://doi.org/10.1007/978-3-030-77434-9

Harrison, N., Bodkin, F., Bodkin-Andrews, G. and Mackinlay, E. (2017) 'Sensational pedagogies: learning to be affected by country', *Curriculum Inquiry*, 47(5): 504–19.

Healy, S. and Coleman, K. (2022) 'InSEA Drawing in the Wild Gallery: regenerating through shared creative practice', composition, University of Melbourne. Available from: https://doi.org/10.26188/19151450.v1.

Healy, S. and Edwards, A. (2020) 'The 100s: Writing as method workshop', media, University of Melbourne. Available from: https://doi.org/10.26188/13121837.v1

Healy, S., Coleman, K., Johnson Sallis, R. and Belton, A. (2022) 'Encountering a pedagogy of the world in a university setting', in S. Riddle, A. Heffernan, and D. Bright (eds), *New Perspectives on Education for Democracy: Creative Responses to Local and Global Challenges* (1st edn), Abingdon: Routledge, pp 77–93. Available from: https://doi.org/10.4324/9781003145806

Irwin, R.L. (2006) 'Walking to create an aesthetic and spiritual currere', *Visual Arts Research*, 32(1): 75–82.

Irwin, R.L. (2017). 'A/r/tography A Metonymic Métissage', in M.R. Carter and V. Triggs (eds), *Arts Education and Curriculum Studies: The Contributions of Rita L. Irwin* (1st edn), London: Routledge, pp 139–49

Irwin, R.L. and de Cosson, A. (eds) (2004) *A/r/tography: Rendering Self through Arts-based Living Inquiry*, Vancouver: Pacific Educational Press.

Kalin, N.M. (2014) 'A/r/tography as indisciplinary turning', *Visual Inquiry*, 3(2): 131–46.

LeBlanc, N. and Irwin, R. (2019) 'A/r/tography', in G. Noblit (ed), *Oxford Encyclopedia of Qualitative Research Methods in Education*, New York: Oxford University Press, pp 1–21. Available from: https://doi.org/10.1093/acrefore/9780190264093.013.393

MacDonald, A., Coleman, K., Healy, S., and Diener, M. (2022) 'How does pedagogical slipperiness enable speculation in/for teacher professional learning?' in E. Baumgartner (ed), *A Retrospective of Teaching, Technology, and Teacher Education during the COVID-19 Pandemic* [online book], Association for the Advancement of Computing in Education (AACE), pp 45–9. Available from: https://www.learntechlib.org/p/216903

McLeod, K., Thakchoe, S., Hunter, M.A., Vincent, K., Baltra-Ulloa, A.J., and MacDonald, A. (2020) 'Principles for a pedagogy of unlearning', *Reflective Practice*, 21(2): 183–97. Available from: https://www.insea.org/wp-content/uploads/2022/05/IMAG_issue_13.pdf

Muff, K. (2014) 'The collaboratory: a common transformative space for individual, organizational and societal transformation', *Journal of Corporate Citizenship*, 62(1): 91–108.

Phillips, S., McLean Davies, L., and Truman, S.E. (2022) 'Power of country: Indigenous relationality and reading Indigenous climate fiction in Australia', *Curriculum Inquiry*, 52(2), 171–86.

Pinar, W. (1975) 'The method of "currere"', Paper presented at the annual meeting of the American Research Association, Washington DC, April. Available from: https://files.eric.ed.gov/fulltext/ED104766.pdf

Pinar, W.F., and Grumet, M.R. (1976). *Toward a Poor Curriculum.* Kendall Hunt,

PlayTank Collective, Healy, S., Mayes, E., Flynn, E., and Edwards, A. (2022) 'Entering into sympogogies', *Cultural and Pedagogical Inquiry*, 14(1): 166–88. Available from: https://doi.org/10.18733/cpi29658

Rae, J. and Coleman, C.G. (2021) *Refugium*, Melbourne: Centre for Reworlding.

Reason, M. (2018) 'Drawing', in C. Lury, R. Fensham, A. Heller-Nicholas, S. Lammes, A. Last et al (eds), *Routledge Handbook of Interdisciplinary Research Methods* (1st edn), London: Routledge. Available from: https://doi.org/10.4324/9781315714523

Ross, J. and Collier, A. (2016) 'Complexity, mess and not-yetness: teaching online withemerging technologies', in G. Veletsianos (ed), *Emergence and Innovation in Digital Learning: Foundations and Applications*, Athabasca: Athabasca University Press, pp 17–33.

Rousell, D. (2020) 'Cosmopolitical encounters in environmental education: becoming-ecological in the intertidal zones of Bundjalung National Park', *Journal of Environmental Education*, 52(2): 133–148.

30

Between two worlds: a shared (shamanic) journey of creative qualitative data analysis and researcher identity

Karen Hammond and Nick Fuller

Introduction

Building on a developing practice of first-person writing as critical method and a theory of knowledge as partial and situated, this chapter will take the form of a conversational autoethnography between two early career researchers, Dr Karen Hammond and Dr Nick Fuller. Inspired by the work of Ellis and Bochner (2006), this chapter highlights the potential of autoethnography to aid in the development and dissemination of creative forms of qualitative data analysis. It documents our exchanges and experiences in the context of two recent doctoral projects carried out in the University of the West of Scotland and the University of Stirling.

As we both occupy a marginal position between two different and apparently contradictory worlds, we think about how our other identities and, specifically, our spiritual and alternative healing practices outside of the academy have shaped our approaches to data analysis, with a focus on Nick's unique use of the shamanic journeying method. Together, we explore the challenge of developing creative approaches to qualitative data analysis in the context of mainstream research spaces where conforming to institutional and disciplinary norms is rewarded with research capital. Our dialogue seeks to address important ethical and practical considerations; as we render visible some of the difficulties in describing and engaging with highly individual experiences, we reflect on if and how such creative methods that require specific skillsets and qualities can be taught or passed onto other researchers. It is hoped that the conversational style will make learning data analysis methods more accessible and engaging for students and experienced researchers alike.

The chapter concludes by discussing ethical issues and implications arising from the shamanic approach, including concerns about cultural appropriation, and positioning the shamanic approach within wider spiritual epistemologies as well as identifying future developments and applications. In closing, we suggest our creative forms of analysis might address the inadequacy of ordinary language that is so often a problem in the fast-developing field of psychedelic research, and it is our hope that our contribution to this volume on creative data analysis will invite further and much needed discussion on the role of researcher identity and positionality in the analytical and research process.

Part one: setting the scene

Karen: Agreeing with Brown (2019) that creative research methods are not developed in a vacuum, I joined the Scottish Psychedelic research group as I was interested in collaborating with like-minded others to discover creative ways to represent psychedelic experiences. My recent narrative ethnography on a hidden population of women who use drugs had highlighted the limitations of conventional research methods and the apparent paucity of our everyday language to do justice to non-ordinary states of consciousness.[1]

That said, my own approach to qualitative data analysis had been somewhat creative as I felt that conventional and objectivist methods of data processing and analysis failed to account satisfactorily for researcher subjectivity and how this shaped the research process. Dialogical narrative analysis (DNA) views method as guidelines, a heuristic guide to interpretation instead of the codified and easy-to-follow steps that one may find for other more mainstream types of qualitative research, such as thematic analysis or grounded theory (Smith, 2016). Rather than following precise steps to arrive at the singular truth of a phenomenon, narrative researchers agree there are multiple valid interpretations and, as such, the task becomes one of building a persuasive case for a particular interpretation (Squire, 2013). In an attempt to resolve my crisis of positioning, I complemented DNA with reflexive autoethnography, which allowed me to include some of the relational, sensory, and embodied aspects of my knowing.

During the period when I was using DNA to interpret the women's stories, I recall waking up and experiencing a flash of insight into the data, as though my mind was processing the complexity as I slept. After debating whether to share this with my supervisory team, eventually I decided against it, instead packaging up neatly what had been a messy and at times unconscious process. I also rejected the view that analysis was a stage that could be viewed as separate from data collection. In seeking to rehumanise the practice of research, Absolon and Willet (cited in Brown and Strega, 2005) talk about fostering a knowledge creation process which takes into account the hidden and underlying factors. The processing of data during sleep seems to me to be a particularly interesting and overlooked aspect of what is not openly discussed, beyond the positionality and location of the researcher. Indeed, for Gilliat-Ray (2021, p 147) 'the omission of any extensive discussion about sleep (or dreams) as sources of insight' in qualitative textbooks is somewhat 'perplexing'; but as Bochner (2012) points out, we are groomed to engage in particular modes of research and to create very specific types of research texts in the interests of conformity. For me, there was a definite tension between really wanting to explore these deeper, darker pathways to understanding, yet at the same time the competing desire to conform to institutional and disciplinary norms in order to be accepted within the academy.

As a reflexive researcher, I argue that positionality is an important way of being accountable in the knowledge creation process, which requires an exploration of relationships to 'land, language, spiritual, cosmological, political, economic, environmental, and social elements in one's life' (Brown and Strega, 2005, p 98). Adopting a reflexive approach requires being honest and open about your 'own views and positions and how these might have influenced the design, execution and interpretation of the research data findings' (Greenbank, 2003, p 9). Therefore, I attempted to join the dots between my experiences of multiple marginality and how it [marginality] might have come to bear on my narrative ethnography of women who use drugs. Yet, in looking back it is clear that I neglected to share how my spiritual practice had informed my approach to data analysis. As a working-class White woman whose autoethnography of drug use reveals a degree of insider status, I already have to navigate my professional and social world as *that* woman who has embodied a range of mad, bad and sad dark feminine figures. Occupying that 'treacherousspace' (Vickers, 2002, p 608) it was here that I appear to have drawn the boundary along my pathway down the 'vulnerability route in creating the autobiographical I' (Allen-Collinson, 2013, p 5).

So, when I first heard that you had incorporated your shamanic practice into your research I was fascinated, and I thought you were brave. How did it all come about Nick, and can you explain what it entails?

Nick: The origins of the approach arose from my dual identity as an academic and a shamanic practitioner and from my intention, whilst undertaking my PhD research,[2] to create alignment between my research topic of spirituality and my research method. My professional identity is an academic nurse, with a clinical background in addictions and research interests in spirituality, non-ordinary states of consciousness, psychedelic substances and shamanism. I have a second, parallel, identity as a shamanic practitioner, having undertaken training with shamanic teachers in the UK and overseas to perform shamanic practices and healing rituals. The core practice of shamanism is the shamanic journey, which forms a central element of my spiritual practice and has become a source of guidance and inspiration.

Shamanism is identified as a set of ethnically contextualised practices employing altered states of consciousness for healing and tribal group cohesion (Wright, 1989). Shamanic journeying is typically characterised by a light trance state, typically attained through rhythmical drumming (Eliade, 1964; Kjellgren and Eriksson, 2008) with the aim of experiencing what Micheal Harner describes as non-ordinary reality or the 'shamanic state of consciousness' (Harner, 1980, p 29). Although the basis of shamanic journeying is contested, many aspects of the experience can be understood in neuro-psychological terms. For example, the characteristic use of a rhythmical drum beat at around 200 beats per minute produces a greater proportion and amplitude of alpha band (8–13 Hz) activity, which is associated with a trance-like state and transcendent experiences (Yaden et al, 2017). Nevertheless, many shamanic practitioners ascribe transcendent characteristics to the experience and attest to the ability of shamanism to access unseen realms of non-ordinary reality. Flor-Henry et al (2017) highlight the relevance of cultural context in ascribing meaning to shamanic experiences. In recent decades, it has emerged as a contemporary spiritual practice, informed by the 'core shamanism' or 'neo-shamanic' approach developed by Harner (1980).

Whilst undertaking Interpretive Phenomenological Analysis research into the relevance of spirituality to processes of addiction recovery, I was eager to find a way of aligning methodology with research topic. The centrality of spirituality in my study invited exploration of the relevance and legitimacy of including spiritually derived knowledge within the research process, as argued for by Carrington (2010a, 2010b, 2014). With that in mind, I developed a process of shamanic journeying, aimed at deepening and enhancing insight into qualitative data, as an adjunct to conventional qualitative data analysis procedures, and as a tool for researcher reflexivity.

The relevance of shamanic journeying to researcher reflexivity lies with its potential to access aspects of the deep unconscious and, thereby, to reveal concealed researcher bias; while its potential contribution as an adjunct to data analysis relates to revealing patterns, connections, and interpretations within the data. To better understand the potential of the shamanic process within qualitative research, I developed a process for recording shamanic journeying, adapted from methods developed by Rock and colleagues (Rock et al, 2006; Rock and Krippner, 2008). This was deployed with the dual aims of bracketing researcher bias through awareness-raising and, following the methods of Tufford and Newman (2012) to enhance analytic insight consistent with Carrington's integrated spiritual theory concept (Carrington, 2014). This process provided a means of incorporating a spirituality epistemology within the analysis process, thereby reflecting consistency between methodology and research topic.

I undertook the shamanic process prior to starting initial free coding for each transcript with the aim of informing the subsequent analysis process. I used a monotonous drum beat to journey for between 10 and 20 minutes in accordance with the 'core shamanism' method initially described by Harner (1980) and developed by Ingerman (1991). Holding a clear intention is understood as a necessary condition for successful shamanic practice (Ingerman, 1991). Accordingly, the journey included two elements, first a general shamanic reflection on the transcript in question, aimed at highlighting material for consideration during data analysis. Secondly, a reflexive element, aimed at highlighting possible sources of researcher bias, through shamanic responses to questions, adapted from Roller (2015). Outputs from the shamanic process include reflexive summaries of the shamanic journey and hand-drawn artistic depiction of the shamanic journey made immediately following the process (see Figure 30.1). These outputs were positioned alongside research journal activity and the 'standard' Interpretive Phenomenological Analysis process (Smith et al, 2009).

After you and I had met via video call several times to discuss collaborating around concepts of creative research and researcher identity, I was excited to introduce you to this creative analytical process and curious about what insights you might gain from the experience. I also envisioned the experience as a prototype workshop to introduce the shamanic process to a wider academic audience in the future, and so was anxious to know how accessible you would find the experience.

Part Two: Karen's shamanic journey *A-Ho! A-Ha!*

As I disembark the train and made my way through unfamiliar West of Glasgow streets, carefully following the point in the map on my phone, I become aware that the time for me to make a decision is running out. In our last conversation, Nick asked me to think of a question related

Figure 30.1: After a shamanic journey

Note: Example of art piece produced following a shamanic journey to explore a qualitative research interview.

to my data that I would like to find an answer to, to give focus to our journeying session. In my mind I throw around a few possibilities, but somehow struggle to settle on any particular aspect. This is surprising, for my experience of accessing and engaging with a hidden population of women who use drugs raised many interesting and important questions. I keep going, resisting the feeling that I have come ill-prepared.

The space Nick has provided for the session is small, but very comfortable and safe. Nick has a very grounding and calm energy, and it does not take long for me to feel at ease in his company.

We talk about our personal lives for a while, perhaps to build a sense of familiarity and trust.

'So how do you imagine today will go?' I ask Nick when I feel we have arrived at such a point.

Nick responds by explaining that his intention is to let me experience a shamanic journey first of all, with guidance and support from him.

He enquires about the extent of my existing knowledge about the shamanic world, briefly outlining the shamanic worldview and the concept of the three worlds. I tell him that I have a little experience and the fact that it is not entirely new to me seems somewhat pleasing to him, given that the purpose of this session is to get a sense of the possibilities for teaching this method to other researchers. A good place to start, perhaps, with limited scepticism to overcome.

Nonetheless I still have very real and specific anxieties; will I be able to do it correctly? What if nothing happens? Nick listens carefully and then reassures me. He tells me that the imagination is the gateway to the spiritual dimension and that I should simply do my best to 'picture' the scene in my mind and allow myself to go with it.

He emphasises the importance of every journey having a clear intention to guide it and suggests that for this initial journey that could be to journey to the lower world and ask to meet an animal that will serve as a guide for our session today. Nick talks me through how to go about this, by picturing an opening in the earth somewhere in nature that feels good to me, like a forest or beach. Instantly I know where my entry point will be. It is a nearby cemetery close to where I live that has a giant pine tree with a rabbit hole at the bottom of the trunk. Having walked there a few nights prior it is very fresh in my mind, the sights and smells, the textures. It seems perfect.

To open sacred space, Nick uses sage and a beautifully scented oil which is first rubbed in the hands and then passed over the head, to cleanse and prepare us to enter sacred space. He calls in the spirit of the four directions and states loudly my intention for the journey.

I lower my back to the ground and find a position that is comfortable before closing my eyes. A soothing rattling sound begins and soon gives way to a rhythmic drumming. Despite my comfort I am aware there is a strangeness to what we are doing and for a moment I experience an acute sense of self-consciousness as I lay here flat on the floor under a blanket in this outhouse while Nick watches over me and steadfastly strikes the drum.

The sound being so close and visceral thuds through my chest, and I try hard to relax and be completely present with the process, actively returning my focus to my breath. In my mind's eye I see the entrance at the base of the pine tree, and squeezing through it, I leave behind the light of day. I imagine the smell of earth and the feeling of dirt between my fingers as I descend, grabbing onto loose hanging roots. After a few breaths, I stop holding on and allow myself to fall through this black hole in space. For a second I am Alice in Wonderland and Nick, the white rabbit.

There is something about the drumming that facilitates this feeling of movement, despite being perfectly stationary. My sense of direction rapidly shifts as though tumbling and for a bit I feel like I am falling head-first. From what Nick had explained I expected to exit the tunnel at some point and arrive at the lower world, where the animal with its gift of wisdom would be waiting. But the sense of travelling was seemingly endless.

The drumming continues – boom boom boom-boom, boom boom boom-boom.

When will I get there? I project this thought outwards and suddenly have a vision of Native Americans dancing at the bottom of a huge tree and they seem to be thoroughly amused. A-Ho!

Then, for a fleeting second, I see a bull flash past and my attention seems pulled towards its horns. No sooner does the bull disappear when I am greeted with another clear image – this time of beavers building a dam in the middle of a forest. But this too is fleeting and before I know it the beat quickens, and my heartbeat responds accordingly – the call from Nick to come back. With this new sense of urgency, I abandon my endless descent and imagine climbing back up through the giant hollowed out tree. The way back seems so much quicker.

As the hastened drumbeat comes to a stop, I bring my awareness completely back into the room. The world in my mind I was leaving behind was not entirely without colour, but the stark contrast of the daylight causes me to screw up my eyes upon return. Nick welcomes me back and makes sure that I am okay. I can see he is as eager to hear how it was just as much as I am eager to tell him.

Post-journey reflection

Karen: I felt relief that I had experienced something and that my biggest fears over experiencing nothing were unwarranted. However, I also experienced a mild sense of disappointment. Why did I not arrive at my destination in the lower world? Why did it feel like I had failed to get 'there'? I had wanted to see the Native Americans and the bull and the beavers for longer and find out what message they had for me. I decided to be entirely honest and share these feelings with you, and as I did something in me clicked and I suddenly understood why the Native Americans were amused at my question.

In that dialogical space of knowing we had created I realised that my endless shamanic journey was a symbol for how I experienced my process of data analysis in my PhD. Often struggling to arrive at a definitive interpretation in the women's stories, grappling with a sense of meaning that seemed to constantly shift and escape what Arthur Frank calls finalisation (Frank, 2012). Frank's approach to research methodology and the politics of interpretation, which he describes as being 'haunted and provoked' by Russian philosopher Mikhail Bahktin (Frank, 2005, p 964), problematises the manner in which research subjects can be overly determined by claims to have 'captured' the essence of their character. Resisting finalisation is thus a way of preserving complexity and ambiguity and in my thesis I applied this concept to make space for the contradictory meanings and subjectivities which ' refuse to stand still. Consequently, my analysis of my storied data, much like my visit to the lower world, felt frustratingly incomplete. But it served very well to illuminate my dialogical sensibility and reflexive proclivity for process over outcome. It is hard to put into words the sense of knowing that I experienced in that moment. It was pleasurable like a well-executed plot twist. An aha-moment if you like.

Nick: I knew that you had had some previous shamanic experience, but I found that you were naturally adept at journeying. It seems important to point out that while there are common elements to the shamanic journeying process, there is also significant variation between practitioners. My initial intention was to introduce you to shamanism as an adjunct to data analysis, like the process I had used in my own research; however, your experience served to widen my understanding of the potential value of journeying in qualitative research. It became clear that, for you, the focus lay with your positioning as a researcher and the tools, approaches and personal qualities which make you an effective researcher.

Your journey also featured beaver, which seemed to represent an important discovery for you, but is also a great illustration of the relevance of *power animals* in shamanic practice. In the shamanic understanding, power animals, or animal totems, guide and support us through our lives. It was clear that beaver was appearing to work with you in your research. This reminded me that in shamanism, power animals represent the essence of the animal, rather than a specific creature. It was not, therefore 'a beaver' that presented itself, but rather a representative of beavers in general. So, I suggested we do a further shamanic journey to clarify the specific qualities that beaver was bringing to support you in your research, which we decided revealed the ability of beavers to 'gnaw through' hard, fibrous material (data) from which they create elaborate structures (themes and theories). In reflecting on your experience, I suggest that shamanic journeying has potential relevance at different levels of abstraction in the qualitative research process. At the highest level of abstraction, journeying facilitates insight into researcher positioning, approach and skills – and your journey provides an excellent example of this insight. Closer in, journeying has potential to provide an overview of a research project – to map the 'landscape' of a project and provide guidance on the direction of travel. Finally, journeying can contribute to data analysis by providing deep insight into data, in the way that I did in my PhD research.

Shamanic journeying is an unusual adjunct to data analysis and is therefore potentially lacking in acceptability within the wider qualitative research world. While it is accepted that qualitative analysis involves a degree of subjectivity, shamanic journeying is an entirely individual and personal experience. It is not possible to verify the 'results' or even to observe the process, as it is a largely inner experience. Furthermore, the interpretation of the shamanic 'vision' is often the subject of a felt sense, rather than anything tangible, meaning that it is challenging to communicate the findings and relevance to anyone other than the researcher. The inclusion of shamanic journeying might, therefore, undermine the acceptability of the process and thus the (perceived) validity of the results. For some researchers, the inclusion of this untested method may undermine the credibility of the findings. From this critical perspective the subjectivity of the shamanic process carries an unjustifiable risk of bias. Certainly, further testing is required before this approach is afforded wider credibility.

That said, in my PhD project, I suggest that spiritual epistemologies have potential to enhance qualitative research. This touches on a wider discourse concerning the division between academia and spirituality. Spirituality is variously conceptualised as encompassing transcendence, meaning and purpose in life, connection and religious or ritual observance. Central to many interpretations of spirituality is the idea that spirituality is fundamentally an individual experience, which is highly subjective. This runs counter to the aspiration of most mainstream academic practice, which is to adopt an objective stance to the topic under investigation. Legitimising spiritual epistemologies within qualitative research invites innovative and creative approaches to qualitative research, one of which is accessed through the practice of shamanic journeying.

Karen: I think it is a very interesting and necessary discussion which contributes something unique to the critique of conventional objectivist modes of inquiry. Indeed, I have become increasingly aware just how much the legacy of positivism has at times, held me back from pursuing alternative pathways. In line with a developing clarity and confidence that I have gained only in relation to the wider qualitative research community, like Gergen and Gergen (2003) I would argue that true bias lies, not in the inescapable subjectivity of the researcher, but in the failure to be explicit about social and political backgrounds. But

I accept that in the context of increasing institutional demands to produce research which translates easily into public policy, this approach carries little in the way of research capital. In seeking to determine the value of the shamanic method, it is perhaps more fruitful to utilise a more expansive lens of what constitutes quality, such as the eight-point conceptualization offered up by Sarah Tracy which 'leaves space for dialogue, imagination, growth, and improvisation' when asking what makes for good qualitative research (Tracy, 2010, p 837).

For me, the greatest value has been the sense of clarity I now have about my own analytic processes which I feel will only improve my research practice going forward. I am excited to know how other researchers find it and indeed, one of the things we set out to discover through our shared endeavour was whether this method could be taught to other researchers. What are your thoughts about this now?

Nick: As I am someone who is trained both in qualitative research and in shamanic journeying, I think this process was relatively straightforward. However, I recognise this is probably a rare combination and for existing qualitative researchers to become familiar with the process would entail learning and practising the shamanic process before subsequently developing the skill to assist data analysis. Thought must be given to how to make the shamanic approach accessible to researchers with no prior experience of shamanism.

Part three: resonance, identity, ethics

Karen: In one of our earliest Zoom conversations, I asked if you were worried about revealing your other identity as a spiritual healer to others in the academy. Having spent much time engaged with hidden populations I have become finely attuned to the exercising of what Shotter (2014, p 318) calls an 'acute discriminative awareness' which acts as a guide to what we express to others, suggesting an implicit understanding of what is appropriate and what effect one wishes to have on those being addressed. It is evident that we have both exercised such caution in giving away information about our simultaneous participation in that 'other world' of alternative healing modalities and spirituality and certainly for me this has been out of fear of negative judgement or even derision. When we discovered a similar positionality, between two worlds, it was thus experienced as both validation and relief. This parallels with the notion of coming out that has been discussed in other marginal research spaces, for example in the network of drug researchers who use drugs (Ross et al, 2020).

In my thesis I coined the term the 'resonance effect' to describe the process whereby researchers and research participants in ethnography are drawn together on the basis of a shared narrative habitus which describes 'the embedding of stories in bodies' an 'embodied sense of attraction, repulsion or indifference, and it represents a world of which we are a part or in which we have no stake' (Frank, 2010, p 53). From this perspective stories seem to find their way to the right people, implying an almost magnetic quality that guides one to those who share a similar 'inner library' (Frank, 2010, p 55). It strikes me that this concept might also explain the process of our meeting and collaboration, and so may also work to identify suitable participants for the workshop. That said, the qualitative research community is characteristically diverse (Tracy, 2010) and I do not think this method will be for everyone.

Ethical considerations

Karen: Did you experience any ethical challenges or forms of resistance when using this method in your research?

Nick: In my experience the shamanic research process raises interesting, and perhaps unexpected, ethical questions. The first of which concerns participant consent. While it is not common practice for qualitative researchers to provide potential participants with a detailed description of the analysis process planned for their data, it is arguable that the shamanic process may be different. Although shamanism is best categorised as a spiritual, rather than religious, practice, some potential participants, for instance those with strong fundamentalist religious convictions, might not be happy with their data being the subject of a shamanic process. This question undoubtedly requires further debate, however at this early stage of development it is probably right to argue that the ethical principle of autonomy requires that where a study involves the use of a shamanically, or spiritually, informed data analysis method, potential participants should be fully informed about the nature of the process.

Karen: When it comes to teaching it to other researchers, I think it is also important to acknowledge that perhaps the success of our experiment owed much to the sense of trust that we had cultivated prior to the event. This was instrumental in fostering the degree of openness that was so necessary to letting go and immersing myself in the process. Even so, the vulnerability that can be read from my reflective account should not be overlooked. For me, this is definitely something to consider if introducing it to the wider qualitative research community and one of the necessary requirements to keep the ethical standards high would be to be really clear about what the process involves so that researchers with trauma histories can make an informed decision.

For those who may find the spiritual connotations uncomfortable I think there is also a degree of flexibility in terms of how it is conceptualised and understood. As you say, the journeying could be approached as a psychological phenomenon akin to a meditation or hypnosis where thoughts and impressions on the data could be processed at a deeper level.

Epistemologically, it could also be situated within the bracket of marginalised research methods which incorporate queer and feminist approaches alongside Indigenous research which legitimise alternative ways of knowing such as storytelling, singing and dreaming (Brown and Strega, 2005; Rowe, 2014). According to Kuokannen (2000, p 417), one of the main characteristics of Indigenous research is a holistic approach which 'strives towards a balance between different areas of life and which does not separate intellectual, social, political, economic, psychological and spiritual forms of human life from each other'. It seems that we, as researchers who at times struggle to reconcile or even identify our own Eurocentric worldview replete with all its dualisms could thus learn much from Indigenous scholars in particular in order to resolve this crisis of positioning and identity that we have both experienced within the academy. It is of course important to note that while Western researchers can be guided by similar principles and objectives, particularly those who draw upon postmodern and poststructural approaches in which there are clear areas of overlap, the title of Indigenous scholar should be reserved for those with Indigenous heritage.

Whichever way the shamanic approach to data analysis is conceptualised there are interesting conversations around epistemology and ethics in particular. For example, what do you think the Indigenous community would have to say about our use of shamanic journeying to aid our work within the academy? Do you think we have to answer to potential accusations of cultural appropriation?

Nick: It is true that neo-shamanism, at least as practised in the English-speaking Western world, derives from various Indigenous traditions, especially Native American culture, raising questions about the appropriateness of applying these spiritual techniques in academic research. For example, as a White, middle-class academic, is it acceptable for me to deploy Indigenous traditions and techniques as part of my research?

I can only respond from my own experience, which has included the privilege of receiving teachings from several Native American elders: men and women who have generously shared aspects of their tradition and beliefs. Through these experiences, I have learnt much about the reverence with which Indigenous people view ritual and ceremony. Understandably, Indigenous peoples are selective about what they share of their traditions and experiences, and many aspects remain secret. Indeed, some elders disagree with the sharing of their traditional cultural and spiritual knowledge outside the Indigenous community at all. Nevertheless, my experience has been one of generosity of spirit, and a belief that traditions should be shared to facilitate mutual understanding and cultural and personal enrichment.

I believe choosing to draw from Indigenous traditions requires acceptance of a responsibility to honour and respect the integrity of those traditions. Here we can indeed learn from Kuokkanen (2000), who advances the concept of an 'Indigenous Paradigm' which rejects Western dualism in favour of Indigenous holistic philosophies in research. Such an approach requires acceptance of our own Indigenous identity to foster an egalitarian, rather than colonial, exchange of knowledge (Kremer, 1999). Such an ambition resonates with the current revival of northern European nature spirituality which seeks to reclaim Celtic shamanism, Druidry, Wicca and Paganism, and which chimes with renewed reverence for nature in Euro-Western society more widely. This is reflecting in the results of the 2021 UK Census, which reported a 13.1 percentage point decrease in individuals describing themselves as Christian and increases in individuals describing their religion as Pagan (74,000), Wicca (13,000) and Shamanism (8,000) (ONS, 2022). Undoubtedly, applying shamanism in qualitative research raises ethical questions around cultural appropriation; however, if approached with sensitivity towards the Indigenous traditions that inform the approach, it has potential to offer an enriching addition to the field.

Conclusion

We believe the shamanic process shows great potential for enhancing researcher insight into data at depth as an adjunct to conventional data analysis tools. The experience of using the process was characterised by an immediate and intuitive understanding of the data that readily revealed meaning and linkages between themes. This application of the shamanic journeying method holds the potential to make a novel contribution to methodology in the qualitative research field and provide a framework within which to test the legitimacy of spiritually derived epistemologies within academic research (Carrington, 2010a, 2010b, 2014). The potential benefits include enhancing researcher depth of understanding and insight into data and proving structure, especially in identifying themes and linkages between themes.

The next stage is to develop a training approach for qualitative researchers and a method for evaluating the impact of the shamanic approach in the research context. It may not be an appropriate or desired method in some situations; we invite those of you in particular who feel an embodied resonance and a calling to explore those deep and darker pathways to knowing in qualitative data analysis to join us and take the first step in your own journey between those two seemingly disparate worlds of academia and spirituality.

In a broader sense it is our sincere hope that our contribution to this exciting volume will invite further and much needed discussion on the role of researcher identity and positionality in the analytical processes of creative qualitative research.

Notes

1 *Stories Told and Untold*: a narrative ethnography of a hidden population of women who use drugs in Scotland. Using a blend of autoethnographic, ethnographic, and narrative methods, this study analyses the experiences and storytelling of 20 socially included women who use illicit drugs, with a focus on how they narratively construct and perform socially acceptable identities and manage potential stigma and discrimination. The study was given ethical approval by the University of the West of Scotland.

2 *Following Ariadne's Thread*: a qualitative exploration of the relevance of spirituality in experiences of problem substance use and of spiritual engagement in processes of recovery, with particular reference to the experience of men living in Scotland. The study explored the experiences of men, living in Scotland, with personal experience of drug or alcohol addiction, who described using spiritual processes as a central part of their recovery journey. The study was given ethical approval by the University of Stirling.

References

Allen-Collinson, J. (2013) 'Autoethnography as the engagement of self/other, self/culture, self/politics, selves/futures', in S. Holman Jones, C. Ellis, and T. E. Adams (eds), *Handbook of Autoethnography*, New York: Routledge, pp 281–99.

Bochner, A.P. (2012) 'On first-person narrative scholarship: autoethnography as acts of meaning', *Narrative Inquiry*, 22(1):155–64.

Brown, N. (2019) 'Emerging researcher perspectives: finding your people: my challenge of developing a creative research methods network', *International Journal of Qualitative Methods*, 18: 1–3.

Brown, L.A. and Strega, S. (2005) *Research as Resistance: Revisiting Critical, Indigenous, and Anti-Oppressive Approaches* (1st edn), Toronto: Canadian Scholars' Press.

Carrington, A.M. (2010a) 'A researcher in wonderland: a spiritual approach from paradigm to practice', PhD dissertation, James Cook University. Available from: https://researchonline.jcu.edu.au/18995/

Carrington, A.M. (2010b) 'Spiritual paradigms: a response to concerns within social work in relation to the inclusion of spirituality', *Journal of Religion & Spirituality in Social Work: Social Thought*, 29(4): 300–20. Available from: https://doi.org/10.1080/15426432.2010.518869

Carrington, A.M. (2014) 'Expanding the debate: a comprehensive exploration of the physical and spiritual ways of knowing, methods and measures', *Qualitative Research Journal*, 14(2): 179–96. Available from: https://doi.org/10.1108/QRJ-08-2013-0049

Eliade, M. (1964) *Shamanism: Archaic Techniques of Ecstasy* (2nd edn), Princeton: Princeton University Press.

Ellis, C.S. and Bochner, A.P. (2006) 'Analyzing analytic autoethnography: an autopsy', *Journal of Contemporary Ethnography*, 35(4): 429–49.

Flor-Henry, P., Shapiro, Y., and Sombrun, C. (2017) 'Brain changes during a shamanic trance: altered modes of consciousness, hemispheric laterality, and systemic psychobiology', *Cogent Psychology*, 4(1): 1–25. Available from: https://doi.org/10.1080/23311908.2017.1313522

Frank, A.W., (2005) 'What is dialogical research, and why should we do it?', *Qualitative Health Research*, 15(7): 964–74.

Frank, A.W. (2010) *Letting Stories Breathe: A Socio-Narratology*, London: University of Chicago Press.

Frank, A.W. (2012) 'Practicing dialogical narrative analysis', *Varieties of Narrative Analysis*, 2010: 33–52.

Gergen, K.J. and Gergen, M. (eds) (2003) *Social Construction: A Reader*, London: Sage.

Gilliat-Ray, S. (2021) '"Sleeping on the job": where qualitative fieldwork meets the sociology of sleep', *Qualitative Research*, 21(2): 45–160.

Greenbank, P. (2003) 'The role of values in educational research: the case for reflexivity', *British Educational Research Journal*, 29(6): 791–8.

Harner, M. (1980) *The Way of the Shaman*, New York: Harper & Row.

Ingerman, S. (1991) *Soul Retrieval: Mending the Fragmented Self*, New York: HarperCollins.

Kjellgren, A. and Eriksson, A. (2008) 'Altered states during shamanic drumming: a phenomenological study', *International Journal of Transpersonal Studies*, 29(2): 1–10.

Kremer, J. (1999) 'Shamanic inquiry as recovery of indigenous mind: toward an egalitarian exchange of knowledge', *Journal for Ethnomedicine*, special volume, 125–40.

Kuokkanen, R. (2000) 'Towards an "Indigenous paradigm" from a Sami perspective', *Canadian Journal of Native Studies*, 20(2): 411–36.

ONS (Office for National Statistics) (2022) 'Religion, England and Wales: Census 2021'. Available from: https://www.ons.gov.uk/peoplepopulationandcommunity/culturalidentity/religion/bulletins/religionenglandandwales/census2021

Rock, A.J. and Krippner, S. (2008) 'Proposed criteria for the necessary conditions for shamanic journeying imagery', *Journal of Scientific Exploration*, 22(2): 215–26.

Rock, A.J., Baynes, P.B., and Casey, P.J. (2006) 'Experimental study of ostensibly shamanic journeying imagery in naive participants I: antecedents', *Anthropology of Consciousness*, 17(2): 65–83. Available from: https://doi.org/10.1525/ac.2005.16.2.72

Roller, M.R. (2015) *Qualitative Research Design*, Gloucester, VA: Roller Marketing, p 104.

Ross, A., Potter, G.R., Barratt, M.J., and Aldridge, J.A. (2020) '"Coming out": stigma, reflexivity and the drug researcher's drug use', *Contemporary Drug Problems*, 47(4): 268–85.

Rowe, G. (2014) 'Implementing Indigenous ways of knowing into research: insights into the critical role of dreams as catalysts for knowledge development', *Journal of Indigenous Social Development*, 3(2): 1–17.

Shotter, J. (2014) 'Agential realism, social constructionism, and our living relations to our surroundings: sensing similarities rather than seeing patterns', *Theory & Psychology*, 24(3): 305–25.

Smith, B. (2016) 'Narrative analysis', *Analysing Qualitative Data in Psychology*, 2: 202–21.

Smith, J.A., Flowers, P., and Larkin, M. (2009) *Interpretative Phenomenological Analysis*, London: SAGE.

Squire, C. (2013) 'From experience-centred to socioculturally-oriented approaches to narrative', *Doing Narrative Research*, 2: 47–71.

Tracy, S.J. (2010) 'Qualitative quality: eight "big-tent" criteria for excellent qualitative research', *Qualitative Inquiry*, 16(10): 837–51.

Tufford, L. and Newman, P. (2012) 'Bracketing in qualitative research', *Qualitative Social Work*, 11(1): 80–96. Available from: https://doi.org/10.1177/1473325010368316

Vickers, M.H. (2002) 'Researchers as storytellers: writing on the edge – and without a safety net', *Qualitative Inquiry*, 8(5): 608–21.

Wright, P.A. (1989) 'The nature of the shamanic state of consciousness: a review. *Journal of Psychoactive Drugs*, 21(1): 25–33. Available from: https://doi.org/10.1080/02791072.1989.10472140.

Yaden, D.B., Le Nguyen, K.D., Kern, M.L., Belser, A.B., Eichstaedt, J.C., Iwry, J. et al (2017) 'Of roots and fruits: a comparison of psychedelic and nonpsychedelic mystical experiences', *Journal of Humanistic Psychology*, 57(4): 338–53. Available from: https://doi.org/10.1177/0022167816674625

31

Conclusion

Alastair Roy, Dawn Mannay, and Helen Kara

Searching for accessible examples of creative analysis

In the 'Creativity in Research' webinar we delivered in 2021, we presented examples of our own creative research methods. We spoke about the ways in which we had found, unearthed, and stumbled towards or across approaches to doing data analysis creatively within the context of particular projects. Although we knew that many researchers were already doing data analysis creatively, we saw that some people were struggling for permission to do things differently as well as seeking to find formal writing on the subject. We were struck by a realisation that many were looking for usable examples of doing data analysis differently; for practical resources that they could pick up and apply; and for texts that could describe, simply, how creative approaches have been undertaken and to what effect epistemologically, methodologically, and substantively.

This search for formal writing on the subject reflects an appetite for learning from others and we are confident that creative methods of analysis will become central to new developments in research in the coming years, not least because the move towards transdisciplinary research requires the kinds of development, adaptation, and mixing of methods found in this book. The idea of being engaged in a search also captures something of our intent as editors of this collection. We recognise that we need this handbook as surely as those who attended the webinar and many of those who have contributed chapters have told us that they need this handbook too.

One directive we gave to all authors in preparing their chapters was to outline a case example of a method of analysis in enough detail for another researcher to be able to use their approach. The authors all made serious attempts to deliver on this request and each chapter provides some form of 'map' of the journey taken. Ingold (2011, p 204) argues that maps are often more useful in 'revealing the inner reality of the world' than providing some coherent view of its 'outer surface'. We believe that that the chapters in this collection do both and we see how researchers have had to struggle to find an approach that they can live with themselves, but which will also be seen as acceptable in their institutional and disciplinary context. Each chapter provides a valuable example of how knowledge about creative data analysis is being forged within the practice(s) of doing research.

In writing about imaginative criminology, Seale and O'Neill (2019, p 15) seek to raise 'awareness of the researcher as a moving, interacting, relational being, in and through space and place'. They further point out that these characteristics of the researcher are also characteristics of the knowledge base, which is also moving and moveable. Many of the chapters in this handbook make similar observations, often using the language of physical movement to describe the ways in which creative approaches have worked for them. For example, Louise Couceiro (Chapter 21) described finding a new 'vantage point'; Kathryn Coleman and colleagues (Chapter 29) describe how their approach offered 'an in-between place'; Louise Gascoine, Kate Wall, and Steve Higgins (Chapter 5) talked in terms of 'unpicking and digging deeper'; and Lauren White, Adam Carter, and Katherine Davies (Chapter 16) reflected on the capacity of their approach to spark the team's 'analytical imaginations'.

It is our understanding that this is the first English-language publication to provide a detailed exploration of creative data analysis. It is encouraging that many of the authors in this collection have found a receptive audience for their work, even among scholars and publics who have expressed reservations about the approaches they have used. The examples offer a rich starting point for researchers interested in creative methods, whether they are just getting started in research, undertaking doctoral studies, or experienced and looking for new ideas.

From anxiety to imagination

In recent years there has been significant growth and interest in creative and imaginative research. This is reflected in books, conferences, blogs, and academic journals that document the development and application of a wide range of creative activities as part of the research process (see Roy et al, 2015b, 2020; Seale and O'Neill, 2019; Mannay, 2016; Kara, 2020; Kara et al, 2021; von Benzon et al, 2021). Creative and imaginative research is not a guaranteed route to improved understandings, democratising processes, power-free relations, or change/impact (Roy, 2012; Roy et al, 2020; Kennelly, Larkins, and Roy, 2023). However, creativity and imagination can be part of a reflexive response to the context in which research takes place, to the humans and other content of that context, as well as to the questions or issues the research seeks to better understand. As the previous chapters have demonstrated, creative approaches to analysis can also encourage researchers to examine their positionality and subjectivity; and changing analysis practices can help researchers to see new things and make new connections.

We have marvelled at the different ways in which people have worked to produce, or craft, approaches to making interpretations and meanings with their data. The chapters provide useful insights into the authors' experiences of using creative methods to facilitate new ways of approaching their data, establishing novel connections between different elements, generating different forms of dialogue, as well as extending a commitment to dialogue throughout all phases of the research process. As illustrated across the collection, these approaches included adaptations of existing methods, merging different approaches and explicit use of arts-based or other creative processes. Reading about the ways in which the authors have struggled, collaborated, and adapted in order to develop, define, and defend their approaches has reminded us of two ideas developed by C. Wright Mills in honing the idea of the sociological imagination. First, about the relationship between 'private troubles' and 'public issues', which, as the contributions make clear, is a live issue for many social researchers (Wright Mills, 1959, p 8) and second, that it is imagination that distinguishes the researcher from a mere technician (Shilling, 2012).

Being brave and acquiring confidence

All research involves learning practices, which can include how to undertake different forms of statistical analysis, coding qualitative data, group panel analysis, narrative interviewing, learning the principles and practices of participatory action research, and so on. Deciding to develop or use creative approaches to analysis can require a level of confidence, not least because such approaches involve new ways of doing things, new combinations, or mixing established modes of analysis with creative approaches. In this collection these new approaches include textile arts, illustration, stitch work, graphic novels, shamanism, collaborative analysis, and many others; and the combinations include emoji being used with coding (Chapter 14) and using word clouds with statistical analysis (Chapter 5).

In Chapter 1, we used the language of mystery and magic to describe the sense of struggle that many researchers are wrestling with in seeking, locating, defining, and defending creative approaches to analysing their data. In Chapter 24, Kyla Tully articulated this struggle with the idea that there may be a 'secret research handbook' that, if one could only locate it, might help to solve these mysteries and reveal the magic. This conveys something important about the sense of anxiety that can accompany decisions about defining our approach to analysing data. Mike Crang (2003, p 127) describes this moment when we start to wrestle with how to work with the materials we have generated, or are generating, in our research:

> There is a certain moment of pleasure that often occurs in projects when we complete fieldwork and with satisfaction look at the mass of accumulated materials – be they questionnaires or field notes, tape or transcripts, copied documents, pictures or whatever – and think of what we have achieved. This is the lull before the storm, the moment before the rising anxiety starts tapping on our shoulders (well, it does on mine anyway) and asks what are we now to do with all this stuff? How are we to turn this mass of material into some cogent, hopefully illuminating, maybe even impressive, 'findings'? And, of course, we realise the one thing they are not is findings – findings, like questions, require work. It is better to think that through analysis we make interpretations, not find answers.

> The process … is one of producing order out of our materials, of making sense. And this making sense is a creative process.

Like many of our contributors, we want to trouble the idea that data generation is always a distinctive phase of activity that is completed prior to analysis. Analysis is an integral aspect of the 'knowledge production process', which 'emerges at moments' throughout research and which interacts with the past and the future through the being(s) of the researcher(s) (Pink, 2015, pp 141–2). However, we do identify strongly with the pleasures and anxieties that accompany the process of doing research, as well as Crang's idea that 'findings require work'.

We have deliberately avoided the use of the term 'skill' in our discussion of creative approaches because it evokes the *technician* rather than the *imaginative researcher*. It is important to recognise that in most of the cases in this book, the value of the creative process lies in its capacity to help people see or think differently and/or to have new forms of dialogue with people, objects, and ideas, and hence make new connections. In most cases the value of these creative activities could not be foreseen in advance and emerged through taking a chance and trying things out. A few years ago, Alastair interviewed socially engaged artist Heather Morison from Studio Morison as part of a collaborative research project (Roy et al, 2015b). Studio Morison's artistic practice involves both physical making in the studio and relational work in communities. In the interview, Heather conveyed that there is a clear connection, in process terms, between the physical making she does in the studio and the relational work she does in communities, noting that both are central to the thinking and the outcome of the project. Heather said,

> When you're in the studio what you do is you begin. So, you get a material and a form and … you begin it. But you make something, you look at it, you reflect on it, you make another one. You go back. You add that one to this one and then you bring in that piece of work that you made before and then you add those two things together. You start to build something, and it makes sense. You go backwards and forwards and it starts to become this.

One thing that is particularly evocative about this quote is the link between the physical materiality of making in the studio, the relational work with communities, and the thinking that emerges through both. In this book we saw how Naomi Clarke (Chapter 18) used the material practice of hand making and stitching hexagons to represent the data generated by each participant, an approach that slows down time, and links making and thinking. In a similar way Charlotte Barratt (Chapter 13) described her approach in terms of 'thinking through drawing'. What these creative practices appear to facilitate is a form of syncretistic thinking, which allow researchers to move into a different mode of thinking and relating (Ehrenzweig, 1970). In the book *Playing and Reality*, Winnicott (1971) links creativity to play. He suggests that creativity is situated 'in the *potential space* between the individual and the environment (originally the object)' and he opposes it to compliance, which he sees as 'a sick basis for life' (Winnicott, 1971, p 65). The point is that although many accounts of play focus merely on fun, play is always meaningful to the players.

The chapters provide many other examples in which practices such as making and moving link doing with thinking. We feel that the contributions to this handbook should encourage people to be brave and playful and to take a chance. This said, all decisions take place in a context and creative forms of analysis can sit outside of accepted disciplinary conventions, and it is to this that we now turn.

Transgressing institutional and disciplinary norms

Many contributors have described in some detail how their own decisions, about whether and how to do data analysis creatively, have inevitably been framed by the context in which the research takes place and in which the researcher(s) must live, relate, and hope to thrive. These contexts have features defined and framed by sociopolitical geography and disciplinary histories/herstories, and are also influenced by gender, ethnicity, spirituality, and sexuality, among other things.

Researchers must all learn to craft and present stories about their work. All stories seek to persuade, and many are written with a particular audience in mind. In the landscape of research, the plausibility of stories is often judged by funders, commissioners, supervisors, reviewers, and other academic colleagues (Roy, 2009). And, as Bourdieu (1993, p 55; original emphasis) suggests, *who* tells a story, often matters as much as *what* the story is: 'the competence adequate to produce sentences that are likely to be understood may be quite inadequate to produce sentences that are likely to be *listened to*.'

Researchers who, by virtue of practice and/or personal characteristics, find themselves symbolically, physically, or relationally located at the margins of their disciplines and/or institutions must sometimes wrestle with unwelcome forms of visibility and invisibility, and sometimes suspicion and censure. Visibility and suspicion, if their work attracts unwelcome forms of attention; and invisibility and censure if, for example, the work is presented in a form that is not accepted by senior colleagues and/or by major journals in the discipline. Karen Hammond and Nick Fuller (Chapter 30) explicitly reference their methodological decisions in terms of acts of suppression, describing the need to conform in some research situations in the service of convention and the pursuit of acceptance. The most difficult conflicts that researchers have described in this collection have emerged when disciplinary narratives have closed down the possibility of other stories and have prevented or foreclosed dialogue (Frank, 2010). For example, Shehr Bano Zaidi (Chapter 22) described how the pressure to conform is reinforced in a context that validates some forms of research over others. In her chapter, which is titled 'How to be creative when creativity is policed', Shehr talks about being 'faced with tight-rope walking' as she had to negotiate the discomforts of disciplinary norms, while also 'straddling ... Western and the Eastern sensibilities and approaches'.

Many other examples in the book show how submission is not always a conscious decision but rather the result of innumerable micro-moments, which tend to reinforce the idea that there are shared rules of the game. It can be difficult to question the legitimacy of these rules or the institutional and contextual conditions that generate them (Roy et al, 2020). Bingham (2003, p 149) suggests that many disciplines have clear and sometimes inflexible expectations about the characteristics of acceptable methodological storylines, and often what might be expected is 'A clean and tidy, post-hoc, rational version ... [and] What usually won't be welcomed is a messy, changing with events, pragmatic version that better describes the process as it inevitably actually happened'.

What Bingham (2003) suggests, is that there can be pressure to exclude from view the messy and uncertain realities of the process of undertaking research. The fear that accompanies a decision to work in ways that sit outside of disciplinary and institutional norms can also lead people to question and police their own motives and practices. For example, Louise Couceiro (Chapter 21) described entertaining the possibility that her approach to data analysis might be a 'frivolous procrastination strategy' and Kyla Tully (Chapter 24) noted 'explicit discouragement regarding creative practice from institutional leaders'. Understanding how a story fits (or does not fit) within an institutional or disciplinary narrative means positioning it within its relevant context, and, for Frank (2010, p 122), '[A] story outside any narrative is a fish out of water: it can't breathe and usually will have a quick end.'

The point is that when (research analysis) stories sit outside of existing disciplinary narrative(s), it can make these stories harder for people to grasp, accept, and understand. This may be why several authors in this collection have used terms such as transgression to describe the development, implementation, and dissemination of their approaches. These experiences encapsulate the bravery that can be needed by some researchers to choose to do things differently and to defend these decisions, both to themselves and to others. For example, in a chapter written by a group of female academics working creatively within a science, technology, engineering, or mathematics (STEM) discipline, Jennifer Leigh and colleagues (Chapter 6) draw on the work of Popper (1959), arguing that:

> One of the main challenges for us has been documenting, disseminating, and justifying the analytic processes in a way that is understandable and deemed rigorous enough for a scientific audience. Scientific ideas of objectivity and meritocracy dominate within STEM disciplines. (p 88)

Nonetheless the chapters provide fascinating accounts of how people have been able to live with and negotiate these issues and tensions, and many of the stories are hopeful.

Edging towards an ending

While preparing this conclusion we came across a piece written about a new exhibition that explores Toni Morrison's complex creative process ('Toni Morrison: Sites of Memory' – Naquian Wheeler, 2023). Autumn Womack, Assistant Professor of English and African American Studies at Princeton, describes how the exhibition presents what normally remains out of view in fiction, 'the labor, the study, and the rigor that is at the heart of the work itself'. The exhibition shows how rigour and systematic approaches can be at the heart of the creative process of writing literature. The examples in this book demonstrate that creative analysis can involve data gathered using creative or conventional methods (Kara et al, 2021) and that some modes of creative analysis also rely on structured and systematic approaches. Hence, for example, we see creative adaptations of existing methods (see Jacqueline Dodding and Hazel Partington – Chapter 15), as well as of creative analysis being combined – or added onto – quantitative or other systematic approaches (see Louise Gascoine, Kate Wall, and Steve Higgins – Chapter 5; Karen Gray and Emma Lazenby – Chapter 19).

The chapters have emerged from a wide variety of disciplinary, institutional, and international contexts, and they describe projects that have engaged diverse subjects using a range of theories and methodologies. Hence, they reflect the broader growth in trans- and interdisciplinary research, as well as the continued development of different modes of participatory research since the late 1990s. The objectives of different researchers have included the desire to provide access to different researchable entities, to change the objects of research inquiries, as well to change the relationships between people and objects within the process of conducting research (Bücher and Urry, 2009). Working across and between disciplines, or forging new disciplines, requires an attitude in which borders are seen as sites of exchange rather than as boundaries. In this respect the adaptability, mixability, and moveability of modes of analysis will be strengths of future research. This handbook acts as a source of information, inspiration, and imagination for those interested in exploring the affordances of creative data analysis.

References

Bingham, N. (2003) 'Writing Reflexively', in M. Pryke, G. Rose, and S. Whatmore (eds), *Using Social Theory: Thinking through Research*, London: SAGE, pp 163–80.

Bourdieu, P. (1993) *Language and Symbolic Power*, ed J. Thompson, trans G. Raymond and M. Adamson. Cambridge, MA: Harvard University Press.

Büscher, M. and Urry, M. (2009) 'Mobile methods and the empirical', *European Journal of Social Theory*, 12(1): 99–116.

Crang, M. (2003) 'Telling materials', in M. Pryke, G. Rose, and S. Whatmore (eds), *Using Social Theory: Thinking through Research*, London: SAGE, pp 127–44.

Ehrenzweig, A. (1970 [1967]) *The Hidden Order of Art*, London: Paladin.

Frank, A. (2010) *Letting Stories Breathe: A Socio-Narratology*, Chicago: University of Chicago Press.

Ingold, T. (2011) *Being Alive: Essays on Movement, Knowledge and Description*, London: Routledge.

Kara, H. (2020) *Creative Research Methods: A Practical Guide* (2nd edn), Bristol: Policy Press.

Kara, H., Lemon, N., Mannay, D., and McPherson, M. (2021) *Creative Research Methods in Education: Principles and Practices*, Bristol: Policy Press.

Kennelly, J., Larkins, C., and Roy, A. (2023) 'Doing youth participatory action research (YPAR) with Bourdieu: an invitation to reflexive (participatory) sociology', *Sociological Review*, 0(0). Available from: https://doi.org/10.1177/00380261231177585

Mannay, D. (2016) *Visual, Narrative and Creative Research Methods: Application, Reflection and Ethics*, Abingdon: Routledge.

Pink, S. (2015) *Doing Sensory Ethnography* (2nd edn), London: SAGE.

Popper, K. (1959) *The Logic of Scientific Discovery*. London: Routledge & Kegan Paul.

Roy, A. (2009) 'New methods – old problems: a practical and philosophical analysis of participatory approaches to qualitative drugs research', in Z. Demotrovics, J. Fountain, and L. Kraus (eds), *Old and New Policies, Theories, Research Methods and Drug Users across Europe*, Lengerich: PABST Science Publishers, pp 112–27.

Roy, A. (2012) 'Avoiding the involvement overdose: drugs, race, ethnicity and participatory research practice', *Critical Social Policy*, 32(4): 636–54.

Roy, A., Froggett, L., Manley, J., and Wainwright, J. (2015a) *Developing Collaborative and Social Arts Practice: The Heart of Glass Research Partnership 2014–2017*, Preston: Heart of Glass.

Roy, A., Hughes, J., Froggett, L., and Christensen, J. (2015b) 'Using mobile methods to explore the lives of marginalised young men in Manchester', in L. Hardwick, R. Smith, and A. Worsley (eds), *Innovations in Social Work Research*, London: Jessica Kingsley, pp 155–72.

Roy, A., Kennelly, J., Rowley, H., and Larkins, C. (2020) 'A critical discussion of the use of film in participatory research projects with homeless young people: an analysis based on case examples from England and Canada', *Qualitative Research*, 26(6): 957–74.

Seale, E. and O'Neill, M. (2019) *Imaginative Criminology of Spaces Past, Present and Future*, Bristol: Bristol University Press.

Shilling, C. (2012) 'Physical capital and situated action: a new direction for corporeal sociology', *British Journal of Sociology in Education*, 25(4): 473–87.

von Benzon, N., Holton, M., Wilkinson, C., and Wilkinson, M. (eds) (2021) *Creative Methods for Human Geographers*, London: SAGE.

Wheeler, A.N. (2023) 'At Princeton, a new exhibition unpacks Toni Morrison's complex creative process', *Vogue*, 23 February.

Winnicott, D. (1971) *Playing and Reality*, London: Tavistock Publications.

Wright Mills, C. (1959) *The Sociological Imagination*, Oxford: Oxford University Press.

Index

References to figures and photographs appear in *italic* type; those in **bold** type refer to tables. References to endnotes show both the page number and the note number (267n3).

J

K

S

T

U

Printed and bound by CPI Group (UK) Ltd, Croydon, CR0 4YY

15/09/2026

14962389-0002